INSTRUMENTATION FOR FLUID-PARTICLE FLOW

INSTRUMENTATION FOR FLUID-PARTICLE FLOW

Edited by

Shao Lee Soo

University of Illinois at Urbana-Champaign
Urbana, Illinois

NOYES PUBLICATIONS
Park Ridge, New Jersey, U.S.A.

WILLIAM ANDREW PUBLISHING, LLC
Norwich, New York, U.S.A.

Library of Congress Catalog Card Number: 99-26198
ISBN: 0-8155-1433-6
Printed in the United States

Published in the United States of America by
Noyes Publications
169 Kinderkamack Rd., Park Ridge, NJ 07656

Transferred to Digital Printing, 2010

Printed and bound in the United Kingdom

Library of Congress Cataloging-in-Publication Data

Instrumentation for fluid-particle flow / edited by Shao L. Soo.
p. cm.
Includes bibliographical references and index.
ISBN 0-8155-1433-6
1. Fluid dynamic measurements. 2. Flow meters. 3. Particle--Measurement. 4. Measuring instruments. I. Soo, S. L. (Shao-lee), 1922-
TA357.5.M43I57 1999
681'.28--DC21 99-26198
CIP

PARTICLE TECHNOLOGY SERIES

Series Editor: Liang-Shih Fan, Ohio State University

FLUIDIZATION, SOLIDS HANDLING, AND PROCESSING: Edited by Wen-Ching Yang

INSTRUMENTATION FOR FLUID-PARTICLE FLOW: Edited by Shao Lee Soo

Dedication

In memory of Professor Shao L. Soo (1922-1998) for his significant contributions to the field of multiphase flow.

G. M. Colver

Preface

An essential element in the progress of research and engineering of multiphase flow systems and specifically particle-fluid flow systems is improved instrumentation for measurements. They make possible validation of basic concepts in the formation, determination of design parameters, and design of systems.

This volume brings together the most original and productive specialists who have conducted research on various aspects of instrumentation for particle-fluid flow systems. They might be associated with universities or industries, in the disciplines of chemical, mechanical, civil, aerospace engineering, and environmental and material science, as well as pharmaceutical processing.

Particle-fluid flow has been in existence in industrial processes since the nineteenth century. Applications include pneumatic conveying, which deals with pipe flow of solid material transported by a gas, slurry transport and processing of solids in a fluid. The necessity of predicting blower or pumping power for a given amount of material to be conveyed led to measurements of pressure drops and attempts in the correlation of physical parameters. That anomaly exists in the correlation in terms of simple parameter is one of the motivations for the exploration into the details of distributions in density and velocity and the present state of development of instrumentation.

A trend as seen in this book is the increased usage of electronic computers and the availability of high sampling frequency for data coding in their fluctuations. The trend is also toward non-invasive measurements using acoustic, laser, nuclear and electromagnetic devices. Results from these advanced techniques have given a new perspective on the primary standard of isokinetic sampling. At the same time, measurements on particle cloud properties has extended from observation of average dynamic properties of a cloud of particles to local instantaneous properties. This has made possible determination of process parameter such as transport properties, stress systems in particle clouds, and other parameters of phase interactions.

The text has been arranged in the following sequence from basic to futuristic:

- Introduction
- Isokinetic sampling
- Electrostatic measurements
- Fiber optics
- Sonics and ultrasonics
- Electromagnetics and pulsed neutron
- Laser measurements
- Particle imaging velocimetry
- Radioactive tracer

These techniques cover a wide range of particle sizes and concentrations, from tracers of fluid motion to packed beds.

The high scientific level of these contributions does not preclude considerations of applying various techniques by engineers for applications in field measurements and process evaluation. Thanks are due to all participants in the preparation of this volume of stimulating ideas that should lead to many future innovations. The extra help of Professor Robert S. Brodkey of the Ohio State University is greatly appreciated.

Urbana, Illinois
September 8, 1998

Shao L. Soo

Contributors

Robert S. Brodkey
Department of Chemical Engineering
Ohio State University
Columbus, OH (Chapter 8)

Michael Ming Chen
Department of Mechanical Engineering
University of Michigan
Ann Arbor, MI (Chapter 9)

Hual-Te Chien
Energy Technology Division
Argonne National Laboratory
Argonne, IL (Chapter 5)

Gerald M. Colver
Department of Mechanical Engineering
Iowa State University
Ames, IA (Chapter 3)

Mooson Kwauk
Institute of Chemical Metallurgy
Chinese Academy of Science
Beijing, P.R. China (Chapter 4)

Shaozhong Qin
Institute of Chemical Metallurgy
Chinese Academy of Sciences
Beijing, P.R. China (Chapter 4)

Apostolos C. Paul Raptis
Energy Technology Division
Argonne National Laboratory
Argonne, IL (Chapter 5)

Shu-Haw Sheen
Energy Technology Division
Argonne National Laboratory
Argonne, IL (Chapters 5, 6)

Martin Sommerfeld
Institute für Mechanische Verfahrenstechnik und Umweitschutztechnik
Martin-Luther Universität Halle Wittenberg
Halle, Germany (Chapter 7)

Shao L. Soo
University of Illinois at Urbana-Champaign
Urbana, Illinois (Chapter 1)

Jian Gang Sun
Energy Technology Division
Argonne National Laboratory
Argoone, IL (Chapter 9)

Cameron Tropea
Institute of Fluid Mechanics
Technische Universität Darmstadt
Darmstadt, Germany (Chapter 7)

Yang Zhao
Department of Chemical Engineering
Ohio State University
Columbus, OH (Chapter 8)

Chao Zhu
Department of Mechanical Engineering
The Hong Kong Polytechnic University
Kowloon, Hong Kong, P.R. China
(Chapter 2)

Contents

1

Introduction

Shao L. Soo

Nearly all manufacturing processes and energy systems include at some stages, processing of particles ranging from nanometer particles to bulk solids. A recent monograph by Roco (1993) covers the whole spectrum of particle density in a flowing mixture of solid particles and fluids:

(1) Particles as tracer of fluid motion - Particles are less than 4 μm, their motion is representative of fluid motion. Aerosol dynamics may include Brownian motion and agglomeration.

(2) Particle dispersion - Single particle motion as influenced by turbulence or vortices.

(3) Dilute suspension - Particle-wall interaction predominates, interparticle spacing greater than 10 particle diameters. This is the normal range of laser Doppler velocimetry (LDV). Application is in conservative pneumatic conveying.

(4) Dense suspension (Range I) - Particle-particle interaction is significant with interparticle spacing less than 10 diameters. Grouped particle motion occurs leading to fluctuating motion. Application is in optimum pneumatic conveying designs and circulating fluidized beds.

(5) Dense suspension (Range II) - Volume fraction of particle is greater than 10%. Cases include fluidized beds and dense slurries.

(6) Granular flow - Particles in direct contact, presence of fluid gives lubrication. A simple case is granular flow in a vacuum.

Ranges (4) to (6) are covered by fiber optics, nuclear magnetic resonance imaging (NMRI) and radioactive tracer technique; ranges (1) to (4) are covered by the rest of the instrumentation in this volume.

Instrumentation serves various aspects of research, design, and evaluation of a process or system. Measurement and correlation of experimental data give parameters to facilitate and validate theoretical formulation for design calculations and computer modeling.

The use of particles as fluid tracers calls for a knowledge of single particle behaviors only. Theoretical formulation of multiphase flow in general and fluid-particle flow in particular has progressed along two paths. The one based on analogy to kinetic theory calls for experimental determination of details of particle-fluid and particle-particle interactions. The one through an extension of continuum mechanics needs an input of transport parameter which have to be determined by data of local instantaneous properties of particle clouds. Continuum mechanics based on averages of flow properties cover the whole range of dense and dilute suspensions because intractions are inclusive in the transport properties. The kinetic approach tends to be successful in treating dilute suspensions.

1.1 Averages and averaging

Measurements of local instantaneous velocity, density, and mass flow of phases of a gas-solid suspension are needed in determining transport properties, validating theoretical predictions, and formulating design procedures.

Much has been discussed on the basic concepts of multiphase flow and the interrelations of time- and volume-averaged formulations (Soo, 1991). Conceptually, volume-averaging is more direct than time-averaging, but nearly all measurements are based on time-averaging of passage of a phase through a given area (mass flow) to deduce events in an elemental volume (such as density). The emphasis will be on instantaneous measurements of local flow properties and the bases of time-averaging of events in an averaging volume.

Measurements by probes have relied on the triangular relation of mass flow, $\rho_p U_p$, density, ρ_p, and velocity, U_p, of particles, i. e., determine the third quantity by measurement of the other two (Soo, 1982). Fully developed duct flow has the advantage of knowing the principal direction of U_p. There are limitations in using two individual measurements on a dense suspension for accurate determination of the local averages and intensities of these quantities. One of them is in the determination of average velocity or density from measured average mass flow; the other is in the determination of intensity of particle motion. This is because of fluctuations in these quantities which can be expressed in terms of local (time) averages ('angular bracket' quantities) and fluctuations ('prime' quantities):

$$\rho_p = \langle\rho_p\rangle + \rho_p', \tag{1}$$

$$U_p = \langle U_p\rangle + U_p'. \tag{2}$$

Averaging theorems give, for mass flow of the particle phase:

$$\langle\rho_p U_p\rangle = \langle\rho_p\rangle\langle U_p\rangle + \langle\rho_p' U_p'\rangle, \tag{3}$$

in terms of product of averages and averages of fluctuation. The last term in Eq. (3) is the covariance of mass flow and $\langle\rho_p' U_p'\rangle/\langle\rho_p\rangle\langle U_p\rangle$ is called normalized covariance (Zhu et al., 1991). Available instrumentation such as the electrostatic ball probe (Cheng and Soo, 1970) permits the determination of mass flow in terms of averages and fluctuations as:

$$\rho_p U_p = \langle\rho_p U_p\rangle + (\rho_p' U_p'). \tag{4}$$

The averaged mass flow was calibrated by isokinetic sampling (Soo et al., 1969). With an independent measurement of local density and its average $<\rho_p>$ by laser beam attenuation (Lucht and Soo, 1991), the covariance is a measure of the accuracy of determination of $<U_p>$ by dividing $<\rho_p U_p>$ by $<\rho_p>$.

The corresponding scale relations of both volume and time averaging were identified in Soo (1991). It was noted that almost all measuring techniques in particulate flows are based on time averaging. The latter requires that the criterion of time scale relation based on passage of interfaces at an observation point is satisfied according to (Soo, 1991):

$$U_s/(\mathrm{d}U_s/\mathrm{d}t) > T > \Delta t_i \tag{5}$$

where U_s is the velocity of an interface, t is the time, such that the acceleration time be longer than the averaging time T, and Δt_i is the passage time of a phase i, such that the acceleration time be longer than the averaging time, and the latter greater than the passage time of a phase. There is no question that Eq. (5) is sufficient when dealing with one-dimensional motion, although there is no conceptual difficulty in extending to three-dimensions. It is also sufficient for determining long time averages in the way of isokinetic sampling of the average mass flux of particles of a suspension. However, provision for determining the local instantaneous density and velocity will have to be made to determine the correlations and transport properties. Removal of this limitation has been one of the principal aims of research in fluid-particle flow which also lead to the accurate determination of local instantaneous velocity of a particle cloud with a distribution of particle sizes. Fundamental concepts and experimental procedures are illustrated in the sections to follow according to these topics:

Basic relations concerning probe dimensions in relation to time and volume averaging;

Small measuring volume in some of the laser devices and its relation to the concept and formulation for the continuum counter part of time averaging to give instantaneous local density and velocity.

1.2 Effect of probe dimension

A previous survey (Soo et al., 1994) showed that most of the characteristic dimensions of probes for particle mass flow such as isokinetic sampling (Soo et al., 1969) and electrostatic ball probe (Zhu and Soo, 1992) were much larger than the size of particles. Therefore, the restrictions of scales of volume (or area) averaging given in (Soo, 1989) were satisfied first of all, or

$$L_s > l > v_i^{1/3} \tag{6}$$

for volume-averaged local instantaneous flow properties such as mass flux and density, where L_S is the characteristic dimension of the physical system, l is that of the control volume, and v_i is the characteristic volume of phase i or a multiple of its interparticle spacing in a particulate suspension for volume-averaged local instantaneous flow properties such as mass flow and density. The characteristic dimension of the physical system should be greater than that of the control volume, and in turn the characteristic volume of a phase a multiple of its interparticle spacing in a particulate suspension. It was taken for granted that the inlet opening (diameter D) of an isokinetic sampling (Soo et al., 1969) probe be larger than each of the particles (diameter d). This condition is conceptually significant in measurements based on time-averaging, with mass flow of particles of various sizes determined by the area $\pi D^2/4$. However, when d is large compared to D, the effect of collision with the rim of the sampling tube becomes significant. The accuracy of measured mass flux and the deduced density from average velocity of particles become questionable when the former is based on the probe diameter alone. This is because many large particles whose centers are within the projected area or the area of exclusion of the probe would be excluded, thus giving a mass flow lower than the actual value.

Calibration of the electrostatic ball probe by the isokinetic sampling device does not provide for the correction of the above effect of large (compared to probe diameter) particles from the results of one to the other through long time averages of density or mass flow. However, by using a larger probe diameter than the particle diameter, one effectively satisfies the scale relation in Eq. (6) to average the electrostatic charge transfer of a number of particles over the projected area of the probe. The electrostatic probe thus calibrated can be used to determine the instantaneous mass flux of particles based on the projected area $\eta\pi D^2/4$, η being the fraction impacted based on the projected area of exclusion of particles (see for instance, Soo, 1989). For $D>>d$, instantaneous mass flow is given by the probe current due to simultaneous collision of particles of average number N_c :

$$N_c = n^{-1/3}/\lambda_B \qquad (7)$$

where λ_B is the collision free path of the probe by particles ($= 4/n\pi D^2$), and $n^{-1/3}$ is the mean interparticle spacing (Zhu and Soo, 1992). For a local average density of $<\rho_p>$ of particles of material density $\bar{\rho}_p$, the interparticle spacing is given by $n^{-1/3} = [(\pi/6)(\bar{\rho}_p/<\rho_p>)]^{1/3}d$ and $\lambda_B = (2/3)(\bar{\rho}_p/<\rho_p>)(d/D)^2 d$. N_c is of order 10 for a 2.4 mm diameter probe and 44-62 μm glass particles in the present study at a mass flux to give a local mass flow ratio of nearly 1 kg particles/kg of air or a volume fraction of particles of 10^{-3}, with an interparticle spacing of nearly 10 particle diameters.

A different situation arises when the particle size is large when compared to the diameter of an electrostatic ball probe; the effective projected probe area would be $\eta\pi(D+d)^2/4$; unless this is accounted for, the probe tends to read a higher mass flux of the larger particles if the probe current is based on the probe diameter only. This illustrates the significance of probe dimension in comparison to the particle size in the above calibration procedure.

Over an averaging time Δt, the number of collisions k for $D>>d$ is given by k = $(\pi/4)D^2 n<U_p>\Delta t$; $<U_p>$ being the local mean velocity of the particle cloud. For the above numerical example when Δt = 1 ms, k ~ 300, a large enough number to give a statistically meaningful local instantaneous mass flow. The length of path was nearly 16 mm, and the averaging volume was then nearly $7x10^{-5}$ m^3. Therefore, the local averaging time is not crucial unless the suspension is highly dilute (say, below 10^{-5} volume fraction of particles) such that less than one particle (N_c ~ 1/4) collide with the probe at any given time, a state corresponding to that of molecules in a rarefied gas.

Instantaneous local data are readily determined. Sampling frequency of the order of 10^4 cycles/s or higher are available from current data system.

1.3 Effect of measuring volume

The above examples raised the question on averaging procedure when the probe dimension or probing volume has a similar size as that of the particles. An earlier exception was the case of an electrical conductance probe of diameter smaller than that of bubbles in liquid (Barnea et al., 1980). In recent time, an important technique is the measurement of particle velocity and mass flow by laser devices. For tracer particles representing the fluid velocity, the use of laser Doppler velocimetry (LDV) for the determination of local instantaneous fluid velocity is straight forward (McLaughlin and Tiederman, 1973). However, studies using phase Doppler particle analyzer (PDPA) to make measurements on a gas-particle suspension gave rise to the situation of the characteristic dimension of probe probing volume of similar order of magniture of the particle size. This is because the measuring volume of laser beams of a characteristic dimension (Gaussian diameter of the beams) of 100 μm was used to determine the density of a particle cloud from measurment of size, number and velocity of particles of around 50 μm to give the density of a particle cloud. The long time average density thus determined, with some careful consideration of the effective flow area (Saffman, 1987), is accurate when the flow satisfied the time averaging criterion in Eq. (5). In this case, the continuum counterpart of time averaging (van de Wall and Soo, 1994) had to be applied in treating its data.

The procedure of continuum counterpart of time averaging for successive arrival time of particles Δt_i and averaging window of time Δt covering $O[10^2]$ particles to control the statistical error.

$$\Delta t = \sum_{1}^{k} \Delta t_i \qquad (8)$$

for the procession of particles through the measuring volume. The effective maximum frequency of fluctuation up to $1/\Delta t_j$ was further preserved with a window shift time Δt_j (van de Wall, 1996). We have the following relations for particle cloud properties at the n-th time (n_j = 0 to n) shift, starting with the first interval at $t = (1/2)\,\Delta t_j$, the instantaneous value at n_j intervals with $t = (1/2)\,\Delta t_j + n_j\,\Delta t_j$:

$$\rho_p(t) = (1/\Delta t)\left[\sum_{i=1}^{i=k} (m_{pi}/W_{pi}A_{zi})\right]_{nj} = <\rho_p> + \rho_p' \qquad (9)$$

A_{zi} is the particle flow area allocated to particle i in the principal flow direction (axial). m_{pi} or d_i for spherical particles, A_{zi}, and W_{pi} (axial component of particle velocity U_p)were given by the PDPA using, say, the green beams of the LDV. The details constituting A_{zi} (Saffman, 1987) as applied to our study were discussed in van de Wall and Soo, 1994, along with the relation between particle concentration and probe volume. The local axial component of the particle cloud velocity is now:

$$W_p(t) = \left[\sum_{i=1}^{i=k} (m_{pi}/A_{zi})\right]_{nj} \Big/ \left[\sum_{i=1}^{i=k} (m_{pi}/W_{pi}A_{zi})\right]_{nj} = <W_p> + W_p' \qquad (10)$$

For the transverse component of particle cloud velocity, since A_{xi} is not readily determined, U_p is determined by, using the data from, say, the blue beams of the LDV,

$$U_p(t) = \left[\sum_{i=1}^{i=k} (m_{pi}U_{pi}/A_{zi}W_{pi})\right]_{nj} \Big/ \left[\sum_{i=1}^{i=k} (m_{pi}/W_{pi}A_{zi})\right]_{nj} = <U_p> + U_p' \qquad (11)$$

since $A_{zi}W_{pi} = A_{xi}U_{pi}$, and both products are scalars. The PDPA data from the green beam and the LDV data from the blue beam were synchronized. The LDV data from back scattering of the green beam provide a check of the synchronism. The convergence of correlation of particle cloud velocity as Δt_j is reduced in terms of fractions of Δt for a case of $\Delta t = 1/64$ s was demonstrated (van de Wall, 1996).

1.4 Averaging for various instrumentation

This introduction illustrates a common concern of probe dimension or measuring volume in relation to particle size or spacing in the evaluation of measured data. Instrumentation and devices in the following chapters are based on a variety of signals. Different approach toward averaging will be illustrated. Direct measurement of local density is feasible with some optical, acoustic, and nuclear devices, but distribution in particle sizes cannot be determined. These measurements are limited to averages over the length of the beam. In the case of a very dilute suspension, particles of sizes smaller than th wave length may appear transparent to the beam. Sensitivity based on diffuse radiation calls for a thickness many times of the interparticle spacing. At present, the instantaneous averaging theorem may not be available or needed by the nature of the design of some instruments; it remains to be developed for NMRI, for instance. This is in spite of its success as a powerful medical diagnostic tool.

REFERENCES

Barnea, D., Shoham, O., and Taitel, Y., "Flow Pattern Characterization in Two-phase Flow by Electrical Conductance Probe," *International J. of Multiphase Flow, 7*, 387-397 (1980).

Cheng, L., and Soo, S. L., " Charging of Dust Particles by Impact," *J. Appl. Phys., 41*, 585-591 (1970).

Lucht, T. R., and Soo, S. L., "Density and its Fluctuation in Stratified Flow of a Dense Gas-Solid Suspension," *Advanced Powder Technology*, *2* (2), 243-253,1991.

McLaughlin, D. K., and Teiderman, W. G., "Biasing Correction for Individual Realizations of Laser Anemometry Measurements in Turbulent Flows," *Physics of Fluids, 16*, 2082-2088 (1989).

Roco, M. C., *Particulate Two-phase Flow*, Butterworth-Heinemann, Boston, MA (1993).

Saffman, M., "Automatic Calibration of LDA Measurement Volume Size," Spplirf Optics, 26, 2592-2597 (1987).

Soo, S. L., Stukel, J. J. and Hughes, J. M., "Measurement of Mass Flow and Density of Aerosols in Transfer," *Environmental Sci. and Tech., Ind. Eng. Chem., 3*, , 386-393 (1969).

Soo, S. L., "State of Multiphase Instrumentation," *Developments in Theoretical and Applied Mechanics* (T. J. Chung and G. R. Karr, Eds.), Vol. XI. Huntsville, AL, pp. 563-576 (1982).

Soo, S. L., *Particulates and Continuum: Multiphase Fluid Dynamics*, Hemisphere Pub. Corp., NY (1989).

Soo, S. L., "Comparisons of Formulations of Multiphase Flow," *Powder Technology*, *66* (1), 1-7 (1991).

Soo, S. L., Slaughter, M. C., and Plumpe, J. G., "Instrumentation for Flow properties of Gas-Solid Suspensions and Recent Advances," *Particulate Science and Technology, An International Journal*, *12*(1), 1-12 (1994).

Zhu, C., Slaughter, M. C., and Soo, S. L., "Covariance of Density and Velocity Fields of a Gas-Solid Suspension," *Rev. of Sci. Instuments*, *62* (11), 2835-2836 (1991).

van de Wall, R. E., and Soo, S. L., "Measurement of Particle Cloud Density and Velocity using Laser Devices," *Powder Technology, 81*, 269-278 (1994).

van de Wall, R. E., *Measurement of Properties of Gas-solid Suspensions using Phase Doppler Anemometry*, Ph. D. thesis, University of Illinois at Urbana Champaign, Urbana, IL 61801 (1996).

Zhu, C., and Soo, S. L., "A Modified Theory for Electrostatic Probe Measurements of Particle Mass Flows in Dense Gas-Solid Suspensions," *J. Appl. Phys.*, *72*(5), 2060-2062 (1992).

2

Isokinetic Sampling and Cascade Samplers

Chao Zhu

2.1 INTRODUCTION

In the applications of gas-solid flows, measurements of particle mass fluxes, particle concentrations, gas and particle velocities, and particle aerodynamic size distributions are of utmost interest. The local particle mass flux is typically determined using the isokinetic sampling method as the first principle. With the particle velocity determined, the isokinetic sampling can also be used to directly measure the concentrations of airborne particles. For flows with extremely tiny particles such as aerosols, the particle velocity can be approximated as the same as the flow velocity. Otherwise, the particle velocity needs to be measured independently due to the slip effect between phases. In most applications of gas-solid flows, particles are polydispersed. Determination of particle size distribution hence becomes important. One typical instrument for the measurement of particle aerodynamic size distribution of particles is cascade impactor or cascade sampler. In this chapter, basic principles, applications, design and operation considerations of isokinetic sampling and cascade impaction are introduced.

2.1.1 Isokinetic sampling of particle mass flux

The primary standard for direct measurements of local particle mass flux in most gas-solid flows is provided by the isokinetic sampling system. The isokinetic sampling principle requires that the sampling probe which is aligned with the flow (isoaxial) extracts airborne particulates at the sampling velocity matching the original undisturbed local flow velocity (*i.e.*, before the probe's insertion). In other wards, there is no slip between the sampling velocity and the "ambient" flow velocity. When the particulates are collected by a filter or an equivalent particle collector in an isokinetic sampling system, after the initial transition period, the stabilized particle accumulation rate represents the actual particle mass flow rate (or particle mass flux). To eliminate the influence of transient unsteady sampling and the effect of hold-up volume of the sampling system caused by the switching (or starting of the sampling), two samples m_{p1} and m_{p2} over different sampling time duration Δt_1 and Δt_2 should be used (Soo et al., 1969). Hence, the particle mass flux $\dot{m}_p$ is determined from

$$\dot{m}_p = \frac{m_{p1} - m_{p2}}{A(\Delta t_1 - \Delta t_2)} \tag{2.1}$$

where each sampling time duration must be longer than the initial transition period. Equation (2.1) indicates that the measured particle mass flux is mainly time-averaged (also, in part, volume averaged depending upon the effective flow area of the sampling probe A).

2.1.2 Isokinetic sampling of particle concentration

Time-averaged particle mass flux in general can be expressed in terms of the time-averaged particle velocity U_p, time-averaged mass concentration C_p (phase density), and particle diffusive mass flux $\dot{m}_{pd}$ (covariance of particle mass concentration and velocity) as

$$\dot{m}_p = U_p C_p + \dot{m}_{pd} \tag{2.2}$$

However, for a fully developed gas-solid pipe flow, the particle diffusive mass flux is usually negligibly small compared with the particle mass flux (Zhu et al., 1991a). Hence, the isokinetic sampling of a gas-solid suspension flow in principle is able to yield the particle mass concentration provided that the particle velocity can be determined. This principle has served as the most primary method for the calibration of measuring systems on particle mass concentration.

The information of the velocity slip between the particles and gas is important to assist the determination of particle velocity since the isokinetic condition refers to gas phase and, in many cases, only the velocity information of gas phase is available. The general relationship between the particle velocity and gas phase velocity is governed by the momentum equation of particles. For the linear motion of a spherical particle, the governing equation takes the form (Fan and Zhu, 1997)

$$\frac{dU_p}{dt} = \frac{3C_D \rho}{4d_p \rho_p}|U - U_p|(U - U_p) + \frac{1}{2}\frac{\rho}{\rho_p}C_I\frac{d}{dt}(U - U_p) + \frac{\rho}{\rho_p}\frac{DU}{Dt}$$
$$+ \frac{9}{d_p}\frac{\rho}{\rho_p}\left(\frac{\nu}{\pi}\right)^{1/2} C_B \int_{t_0}^{t} \frac{\frac{d}{d\tau}(U - U_p)}{\sqrt{t - \tau}} d\tau + \left(\frac{\rho}{\rho_p} - 1\right) g \tag{2.3}$$

where C_D, C_I and C_B represent the drag, carried mass, and Basset coefficients, respectively. D/Dt denotes for the substantial derivative following the gas flow. Equation (2.3) is a general form of the Basset-Boussinesq-Oseen (BBO) equation. In the practice of aerosol sampling, the particle Reynolds number at terminal velocity is usually within the Stokes regime, which is true for particles less than 60 μm with a typical density of 2000 kg/m^3. In this case, the relaxation time and stopping distance (or inertial range) are very short, less than 100 ms and a couple of centimetres, respectively (Hinds, 1982). Therefore, in the aerosol sampling, it is reasonable to take the particle terminal velocity between the particles and the flow as the relative velocity when the flow moves vertically or just simply assume no-slip between the two.

2.1.3 Development and applications of isokinetic sampling

The study of particle sampling can be dated back to 1940's (Lapple and Shepherd, 1940; Anon, 1941). Despite of the simple appearance of the principle of isokinetic sampling, the practical implement of the method is still far from complete, in part due to the lack of understanding or lack of means to overcome the difficulties such as the determination of flow velocity in the presence of significant amount of particles, the elimination of "intrusive" effect of the sampling probe, the interactions between particles and carrying fluid, the loss of particles to the wall deposition and particle bounce/reentrainment in the sampling tube, the quantification of turbulence effect, the influence of electrostatic charges of particles, the adjustment for sampling of transient or unsteady motions of particulate flows, and the deviation from an anisokinetic sampling.

In the past half century, many significant progresses have been made on the development of isokinetic sampling technique, especially in the area of aerosol sampling. The early stage of the study on isokinetic sampling is represented by Davies (1954), Watson (1954), Levin (1957) and Badzioch (1959). Soo et al. (1969) suggested the use of two different sampling durations to eliminate the influence of transient unsteady sampling and the effect of hold-up volume of the sampling system due to the initial sampling of particles. Studies on directional dependence of isokinetic samplers were carried out by Glauberman (1962), Tufto and Willeke (1982), and Vincent et al. (1986). Deviation in the measurements due to anisokinetic sampling of particulates is also extensively studied (Watson, 1954; Badzioch, S., 1960; Vitols, 1966; Ruping, 1968; Addlesee, 1980; Zhu, 1991; Buerkholz, 1991; Zhu et al., 1997). In order to reduce the error due to the entry effect, the sampling probe is suggested to be sharp-edged and/or with thin-well structures (Whiteley and Reed, 1959; Davis, 1972; Fuchs, 1975). The effects of probe bend and gravity on the isokinetic sampling measurements were investigated by Sansone (1967) and Ruping (1968). The effect of turbulence on particle loss inside the probe entry of aerosol sampling was examined by Vincent et al. (1985) and Wiener et al. (1988). In addition, for not very fine particles, there exists a velocity slip between phases. The effect of velocity slip on isokinetic sampling was explored by Ito et al. (1991) and Zhu et al. (1997). Many references on the introduction of isokinetic sampling principle are available, including Fuchs (1964), Hinds (1982), Vincent (1989), and Allen (1990).

The applications of isokinetic sampling cover but are not limited to the sampling of aerosols such as flu gas in chimney, soots (unburned carbons) from diesel engine exhaust, dusts suspended in the atmosphere, and fumes from various sprayers; measurements of particulate mass fluxes in pneumatic transport pipelines and other particulate pipe flows; solid fuel (also some liquid fuels) distributions in furnaces, engines, and other types of combustors; and calibrations of instruments for the measurements of particle mass concentrations. Isokinetic sampling can also be applied to flows with liquid droplets. In this case, the droplet sample is usually collected by an immiscible liquid (Koo et al., 1992; Zhang and Ishii, 1995).

2.1.4 Cascade impactor for particle sizing

Cascade impactor is a common device for the measurements of particle aerodynamic size (mass) distributions. The basic principle of a cascade impactor is based on the jet impingement. In gas-solid flows, the particle inertia is usually much higher than that of the carrying gaseous media, depending on the mass ratio, relative velocity, and momentum interactions between the two phases. Thus, once a gas-solid flow is forced to detour its course in front of a baffle (for instance, a flat plate), the particles with higher inertia will resist changing their directions with the gas streamlines, leading to an impaction between the particles and baffle (which is known as inertial impaction). Inertial impaction is the underline principle of all inertial impactors developed for the collection and size classification (aerodynamic size) of particulates. To ensure a high level of impaction efficiency of an inertial impactor, the particulate flow is arranged to pass through a nozzle (to form a jet flow) to gain enough inertia before the impaction, and the collection plate is commonly coated with a thin layer of viscous film to trap the colliding particles and prevent those collected particles from bouncing back to the flow stream (particle reentrainment).

It is important to note that a single-stage inertial impactor actually separates impinging particulates into two size ranges: particles larger than the cut-off aerodynamic size of the impactor are removed (collected on the impaction plate) from the particulate-laden stream while those smaller than that cutoff size remain airborne. Therefore, in order to improve the efficiency and accuracy of the aerodynamic size classification of sampled particles, one most common approach is to use a cascade impactor which consists of several single-stage impactors with different cutoff sizes connected in series. When sampling of a polydispersed particulate flow with a wide size distribution, it is a common practice to use a simple particle sizing device such as a settling chamber or cyclone to remove the very coarse particles before a cascade impactor, while the last stage of the cascade impactor is followed by a backup filter to capture all remaining fines.

Multi-stage cascade impactor in principle yields particle mass distribution with respect to aerodynamic size of particles. The combination of mass fraction measured from each cascade stage will form a genuine particle mass distribution if there is a "clean-cut" of the particles collected per stage (*i.e.,* all particles above the cutoff size are collected and all those below go to the next stage). In reality, the collection efficiency of any impaction stage with respect to the aerodynamic size of particles is of S-shape. The cutoff size of a cascade stage is defined as the aerodynamic size at which the collection efficiency is 50%. Usually, there is an overlap of particle size among neighbouring cascade stages, which means that the same sized particles can be found on several collection stages. Due to the overlapping, the actual mass distribution may need to be obtained from a deconvolution process of the measurements. This deconvolution is mainly achieved with the aid of modelling approaches. A good cascade impactor should have a narrow span of the overlap compared to the sizing range of the individual cascade stage.

The original concept of cascade impactor was proposed by May in 1945. The measurements from cascade impactors can be used for the determinations of not only the aerodynamic size distribution of particles but also the chemical compositions of particles. Due to these merits, various design of cascade impactors with different configurations of nozzle jets and collection surfaces have been developed (May, 1945; Ranz and Wong, 1952; Anderson, 1958; Brink, 1958; Lundgren, 1967; Cohen and Montan, 1967; Pilat et al., 1971; Gaussman et al., 1973; Marple and Liu, 1974; May, 1975; Emmerichs and Armbruster, 1981; Lodge and Chan, 1986; Gibson et al., 1987; Marple and Olson, 1995). The May cascade impactor, Anderson sampler, Lundgren impactor, and Mercer impactor are only a few representatives among many practical cascade impactors. The multi-stage number in most cascade impactor is from 4 to 8. The typical particle size range sampled by cascade impactor is from 0.3 μm to 20 μm. Commercial cascade impactors made by companies such as Sierra Instruments, Andersen Samplers, BGI (Casella impactor), and In-Tox Products are available for both industry and laboratory use.

2.2. ISOKINETIC SAMPLING

Isokinetic sampling is for the direct measurements of particulate mass fluxes or concentrations of aerosols (assuming no slip between aerosols and the gas flow) in gas-solid suspension flows. This is accomplished by inserting a thin-wall tube into the particulate suspension flow to draw samples at the isokinetic condition and by passing the collected particles into a sampling train. Typical instruments, influence factors, and design considerations are discussed in the followings.

2.2.1 Principles and instruments

Isokinetic sampling can be defined as the sampling of particles by drawing a particulate suspension into a probe at the same velocity as that of the original undisturbed suspension flow. Thus, the original rate and direction of sampling flow are unchanged by the isokinetic sampling.

The local flow velocity of the fluid in a particle suspension must be determined first in order to match the isokinetic condition. Conventional velocity measuring techniques of a gas stream may face many difficulties when applied to the measurements of gas-solid flows. Pitot static probe is usually not a favorable choice in the flow velocity measurement of a gas-solid suspension due to the particles entering or clogging of the tube. Hot-wire anemometry (HWA) is also not generally applicable for the velocity measurements of gas-solid flows because of the striking of particles (may break the wire) and the uncertain heat transfer environment in the presence of particles. For a very dilute suspension flow, the local gas velocity may be determined by a Shapiro probe or reversed total head tube (see Figure 2.1) originally calibrated against a Pitot static probe in a clean gas stream (Cheng et al., 1970). Hence, the application of a Shapiro

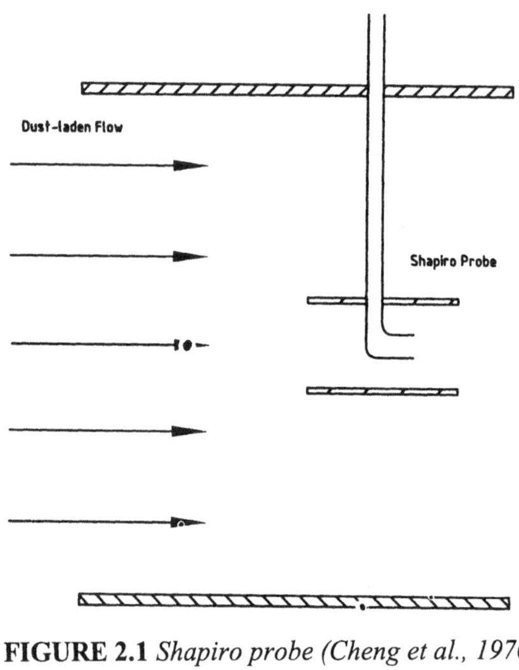

FIGURE 2.1 *Shapiro probe (Cheng et al., 1970).*

probe is always subject to the assumption that the gas velocity is unchanged by the addition of particles into the flow, which is true only for very dilute particulate flows where the effect of turbulence modulation in the presence of particles can be neglected. For a dilute suspension flow, the local flow velocity can be measured using the Laser Doppler Velocimetry (LDV). However, this requires a rather costly optical measurement facility, in addition to an optical window for the measurement (if no optical fibre extension is used). Furthermore, when multiple particles co-exist within the control volume of the measurement simultaneously, the Doppler signal of one particle is undistinguishable from those of other particles in the control volume, yielding unreliable LDV measurements. Therefore, LDV is usually limited to measurements of dilute suspensions only. For general gas-solid suspension flows, the local velocities of both gas and particles can be simultaneously determined by the corona discharge method (Soo et al., 1989; Zhu, 1991). As shown in Figure 2.2a, a negative corona discharge in a particle-laden gas at a high voltage produces ions in air by attachment and modifies electrostatic charges on solid particles suspended in the flow. With the aid of a high speed data acquisition system and an electrostatic probe stationed downstream, the corona discharge probe system utilizes the transient behaviour of a corona discharge to provide a distinction of times of arrival of ions and solid particles with modified charges (see Figure 2.2b). These time intervals yield, from the location of the electric probe to the corona source, the local, instant, and one-dimensional air velocity and particle velocity, respectively. This probe system does not suffer the handicap of other available probe systems in a dense solid suspension flow. For the determination of time-averaged particle velocity, a dual laser beam system based on the cross-correlation technique can be used, as illustrated in Figure 2.3 (Zhu et al., 1991b; Slaughter et al., 1993). This local time-averaged particle velocity is then used for the determination of particle concentration from the isokinetic sampling.

An isokinetic sampling apparatus for the dilute suspension flows is illustrated in Figure 2.4 (Soo et al., 1969). The inlet velocity to the sampling probe is matched by adjusting needle valves 1 and 2 when the switch valve connects the probe to the filter 2. Once matched, two samples via the filter 1 thimbles over two different time durations are taken. Thus, according to Eq. (2.1), the local particle mass flux is obtained. The differential operation of the two samples is aimed at eliminating the effect of the holdup volume of the sampling system. The system in Figure 2.4 may not be suitable for dense suspension measurement because the quick particle build-up in the thimble may lead to a serious deviation from the "isokinetic" velocity matching due to the significant increase in pressure drop over the system. For the measurements in dense gas-solid suspension flows, a modified isokinetic sampling system with cyclone separator may be used, as shown in Figure 2.5 (Zhu, 1991). The cyclone separator functions as first stage particle collector while the filter holder with a cellulose extraction thimble completes the final collection of particles. In

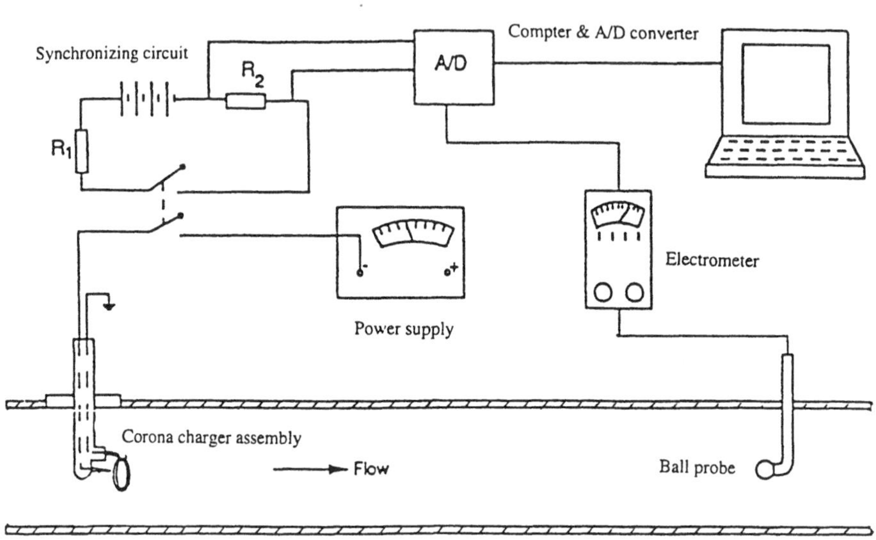

FIGURE 2.2a *Corona charger - ball probe system.*

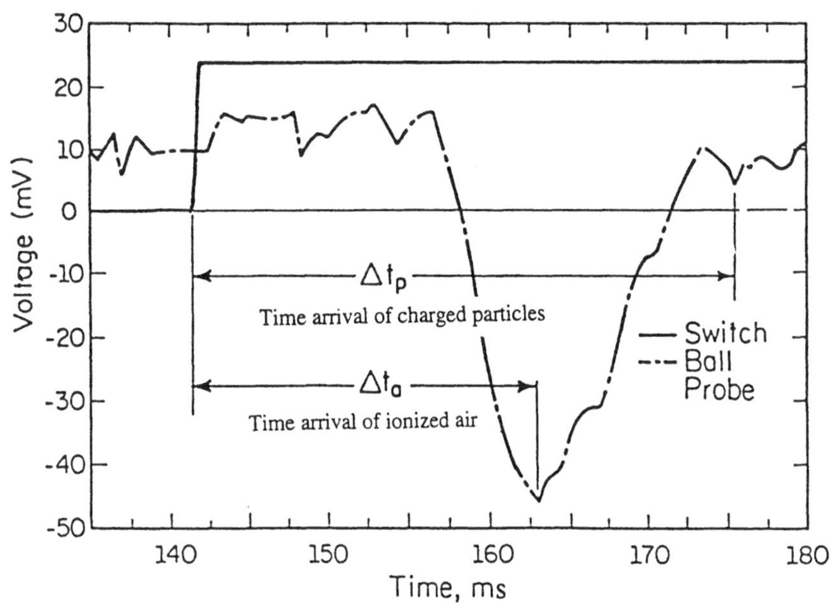

FIGURE 2.2b *Example of phase velocity measurements using corona charger method (Soo et al., 1989).*

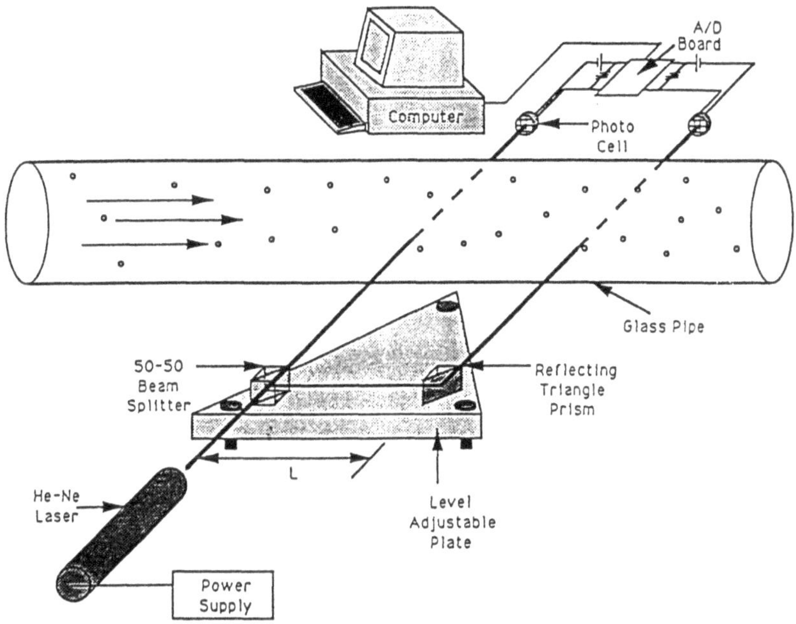

FIGURE 2.3 *Dual-beam optical system.*

addition, a bypass valve is used to adjust the extraction flow rate with a rotameter to match the isokinetic requirement, and a compressed clean air hose is adopted to prevent the particles from entering the sampling tube when it is not in use. Once again, to eliminate the influence of transient unsteady sampling and effect of the holdup volume of the sampling system caused by the switching, two samples over different sampling durations are taken. The difference of the two yields the actual local mass flux.

For the measurements of aerosols such as flying ash in stacks, soots emitted from engines, or dusts in the air, the isokinetic sampling is relatively simple. Since most aerosol suspensions are very dilute and particulates are extremely tiny (typically less than 2 μm), the effect of the holdup volume of the sampling system on the total sampling over a normal sampling duration (from a few minutes to a couple of hours) may be neglected. The local gas velocity can be conveniently determined from a Pitot probe or Shapiro tube (reverse Pitot tube). In addition, the particle velocity in this case is usually regarded the same as that of gas phase (no-slip) so that the mass flux measurements from isokinetic sampling of aerosols can be easily converted into the measurements of particle concentration. An isokinetic sampling train for the measurement of aerosols in hot gas streams is shown in Figure 2.6 (Cadle, 1975). It is noted that a cooling

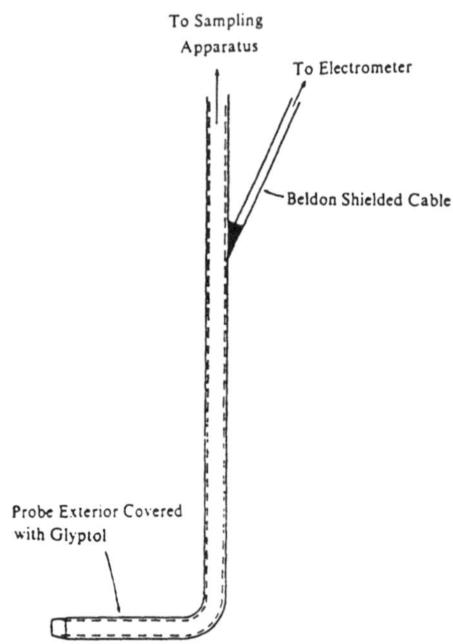

FIGURE 2.4a *Isokinetic sampling probe.*

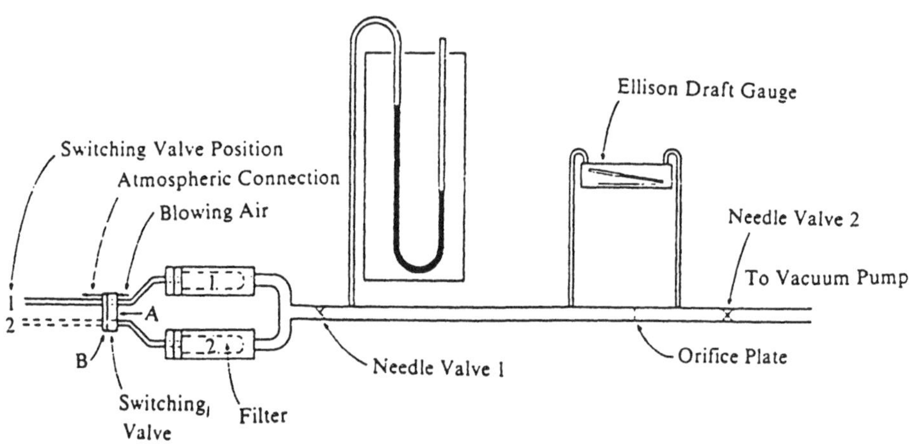

FIGURE 2.4b *Isokinetic sampling apparatus (Soo et al., 1969).*

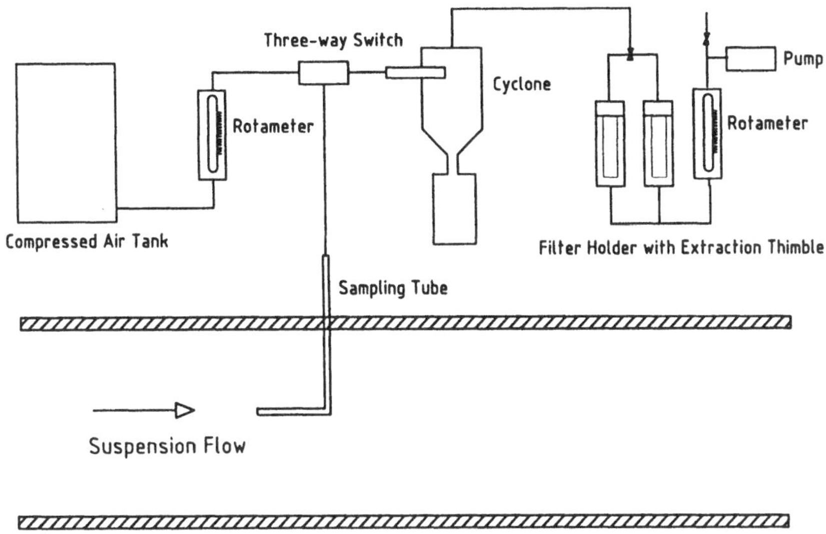

FIGURE 2.5 *Isokinetic sampling system with cyclone separator (Zhu, 1991).*

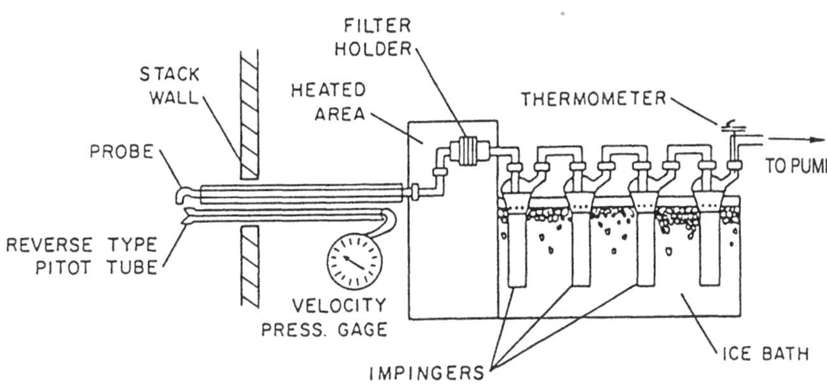

FIGURE 2.6 *Western precipitation isokinetic sampling train (Cadle, 1975).*

system and condenser assembly is used in the sampling train to remove heat, moisture, and other condensibles from the hot, humid, and corrosive gas, protecting the sucking pump and associated meters from contamination and corrosion. In order to adjust the sampling flow rate to accommodate changes in gas velocity, it is convenient to build a velocity sensor such as a Pitot tube into an isokinetic sampler, as exemplified by Figure 2.7 (Boothroyd, 1967; Boubel, 1971; Bohnet, 1978).

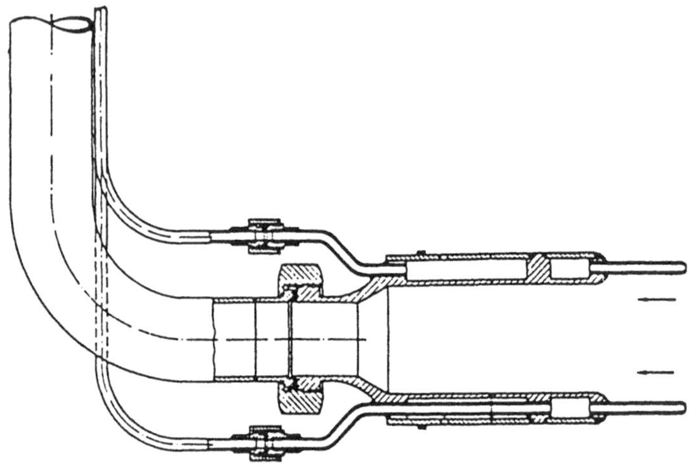

FIGURE 2.7 *Bohnet velocity-sensing isokinetic sampling probe (Bohnet, 1978).*

It is a common practice to combine isokinetic sampler with a particle sizing system, which yields not only the local particle mass flux but also the particle size distribution. For the aerosol measurements, a cascade impactor assembly can be directly connected to an isokinetic sampling probe (see §2.3 for details on cascade impactor). The particle size distribution from a cascade impactor assembly is a mass distribution based on the aerodynamic sizes of particles. If an optical or electrical sizing instrument (e.g., light-scattering photometer; Coulter counter) is used instead, the output will be a particle number distribution with respect to the equivalent particle size defined by the corresponding sizing technique. For the sampling of gas-solid suspensions with larger particles, an isokinetic sampler connected with a series of cyclones may be used.

2.2.2 Anisokinetic sampling

In practice, the isokinetic sampling is closely approached but almost impossible to be rigorously realized. The ansiokinetic sampling refers to the mismatch of velocity, misalignment of the sampling orientation, or both. Many factors contribute to this "anisokinetic" sampling. For example, in the aerosol sampling in stack or in an essentially still air, it is extremely difficult to achieve a precise velocity match. For some confined three-dimensional flows, the matching of the sampling orientation is impossible due to the length restriction of the sampling probe. For the flows with rapid or erratical change in velocity, instantaneous matching of sampling velocity is also nearly impossible. In addition, the insertion of sampling probe inevitably disturbs the original flow field. For the sampling of large or heavy particles or sampling in dense suspensions, the isokinetic condition constantly fails to provide sufficient aerodynamic lift to take the particles through the sampling system. Thus, an anisokinetic sampling (over-sucking) may be operated to keep tracking the particle mass flux in a dense suspension or dilute suspension with large particles. Hence, an investigation of anisokinetic sampling is essential to estimate the departure from the ideal isokinetic sampling.

Anisokinetic sampling can be roughly divided into three different categories, namely, over-sucking (sampling velocity > stream velocity), under-sucking (sampling velocity < stream velocity), and misalignment (probe not aligned with flow), as shown schematically in Figure 2.8. It is interesting to examine the limiting cases of sampling very small or very large particles. Fine particles always follow closely the motion of a gas stream. Therefore, provided that the sampling velocity or flow rate can be determined, the concentration measurement of very small particles during an anisokinetic sampling is the same as that in the mainstream, unaffected by the change of sampling velocity. For very large particles, the great inertia of particles keeps the particle motions unaffected by the sudden change of gas stream near the entry of sampling probe. In other words, the amount of particles entering the probe is nearly the same regardless of the sampling velocity, which means that the mass flux measurement of very large particles is independent of the departure in sampling velocity from the isokinetic condition (Hemeon and Haines, 1954). Therefore, no matter whether the sampling is over-sucking or under sucking, the error from an anisokinetic sampling is resulted mainly due to the particle inertia. The effect of particle inertia is typically characterized by the Stokes number, Stk, which is defined as

$$\mathrm{Stk} = \frac{U_{\mathrm{pt}} U_0}{g D} \tag{2.4}$$

where U_{pt} is the free-falling terminal velocity of the particle. Stokes number may be interpreted as the ratio of the particle stop distance to the characteristic dimension of the flow system. For small particles (e.g., most aerosols) whose Reynolds numbers based on their terminal velocities are within the Stokes regime (say, $\mathrm{Re}_{\mathrm{p}} < 1$), the Stokes number can be calculated by

$$\mathrm{Stk} = \frac{\rho_{\mathrm{p}} d_{\mathrm{p}}^2 U_0}{18 \mu D} \tag{2.5}$$

It should be noted that, due to the different inertia of different sized particles, an anisokinetic sampling of polydispersed particle suspension affects not only the mass of particles in the sample but also the particle size distribution. Sub-sucking

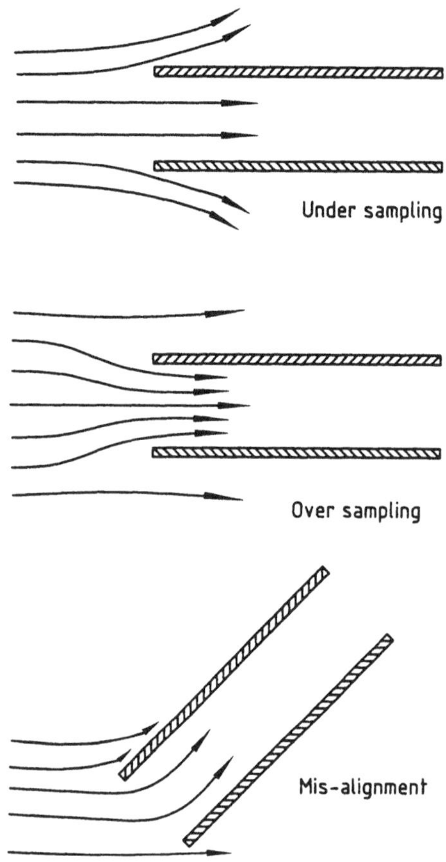

FIGURE 2.8 *Types of anisokinetic sampling.*

sampling results in a skewed size distribution with excess large particles whereas over-sucking sampling leads to a biased size distribution with excess fines.

For aerosol suspensions, the velocity slip between particles and the carrying gas is usually negligibly small compared to the sampling velocity due to the tiny size of aerosols. Measurements from a sampling probe hence can be directly linked to the aerosol concentration. The error in concentration sampling under the over-sucking or under-sucking conditions may be roughly estimated by (Davies, 1968)

$$A_e = \frac{C_s}{C_0} = 1 + \frac{4\,\mathrm{Stk}}{1 + 4\,\mathrm{Stk}}\left(1 - \frac{U_0}{U_s}\right) \tag{2.6}$$

where A_e is known as the aspiration efficiency. Equation (2.6) agrees with the experimental findings for limiting aerosol sampling cases:

a) $A_e = 1$ when $U_0 = 0$ (aerosols suspended in still air)
b) $A_e = 1$ when Stk << 1 (small aerosols)
c) $A_e = U_0 / U_s$ when Stk >> 1 (large particulates)
d) $A_e = 1$ when $U_s = U_0$ (isokinetic sampling)

It is worth pointing out that the isokinetic sampling is not required when very small particles are sampled. A correlation for a better approximation of the aspiration efficiency was proposed by Belyaev and Levin (1972, 1974) as

$$A_e = 1 + \frac{(2 + 0.62\,U_s / U_0)\,\mathrm{Stk}}{1 + (2 + 0.62\,U_s / U_0)\,\mathrm{Stk}} \left(\frac{U_0}{U_s} - 1 \right) \tag{2.7}$$

which is valid for a range of $0.18 < U_s/U_0 < 5$ and $0.18 < \mathrm{Stk} < 2$. Effect of velocity ratio on the aspiration efficiency for an anisokinetic sampling using a thin-wall sampler can also be studied numerically by the use of BBO equation (Eq. (2.3)) and the Lagrangian trajectory modelling approach. A typical result for aerosol sampling is illustrated in Figure 2.9a (Zhu et al., 1997), which is based on a two-dimensional numerical simulation of anisokinetic samplings with the assumption of no-slip in velocity between phases. Figure 2.9a shows that, for the sampling in suspensions moving at very low gas velocities, anisokinetic sampling of coarse particles can cause disastrously large errors. It also illustrates that anisokinetic sampling of submicron particles does not lead to significant error in aspiration efficiency. Although Figure 2.9a further indicates that an isokinetic sampling always ensures a true sampling of particle concentration, it is a biased conclusion from the no-slip assumption which only holds for very fine particles. For particles other than those very fine ones, the assumption of non-slip becomes invalid. The effect of velocity slip on aspiration efficiency is shown in Figure 2.9b, which clearly indicates that even under the isokinetic condition the particle concentration cannot be correctly sampled due to the particle inertia (velocity slip).

When a probe is not aligned with the sampling flow, the effect of probe orientation must be taken into consideration. For very small particles, the efficiency of entry is almost unaffected by the yaw orientation; whereas the efficiency of entry for large particles may follow the cosine relationship (May, 1967). In May's study, the error in efficiency of entry for large particles (> 100 μm) and an yaw angle of 25° is about 10% whereas for 5 μm particles the error is about 7% with an yaw angle of 45° (Allen, 1990). In the area of aerosol sampling, it is widely agreed that the aspiration efficiency in general should be a function of the Stokes number of particles, ratio of sampling velocity to stream velocity, and yaw orientation of sampling probe, which may be expressed in a form (Vincent, 1989)

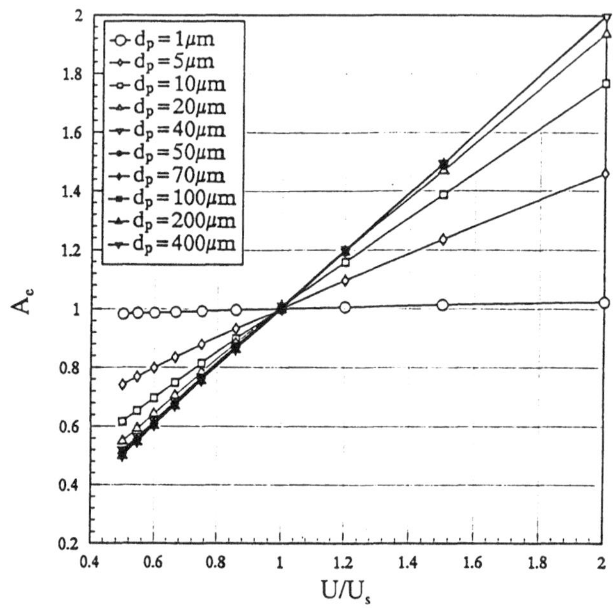

(a) with assumption of no-slip between phase velocities.

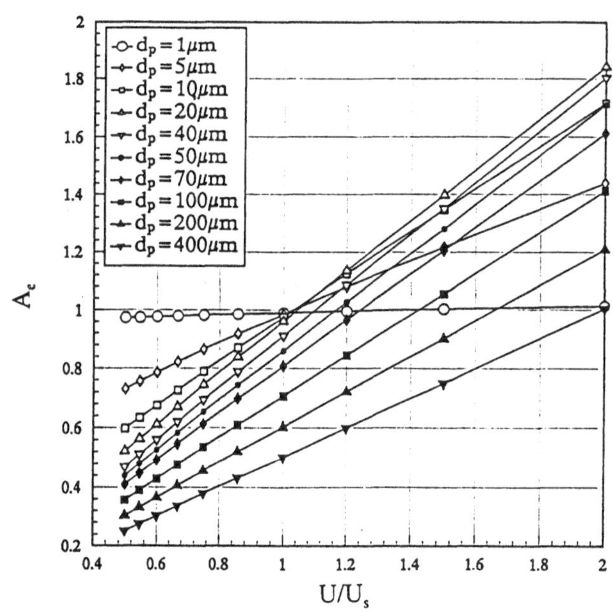

(b) with consideration of slip effect between phase velocities.

FIGURE 2.9 *Aspiration efficiency in anisokinetic sampling (Zhu et al., 1997).*

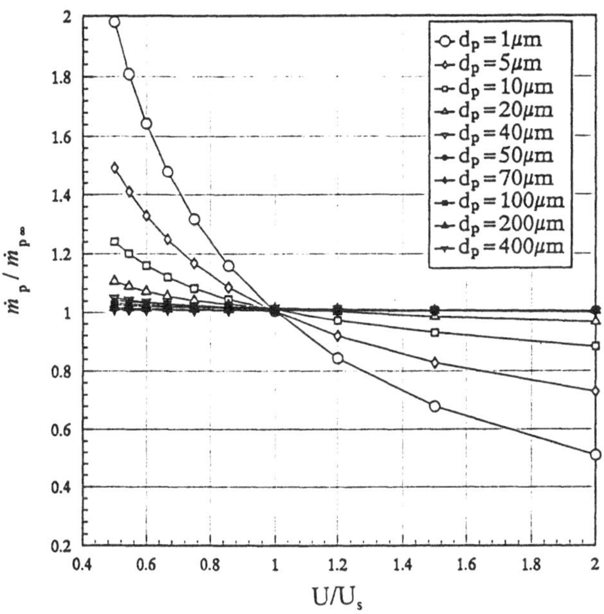

FIGURE 2.10 *Particle mass flux in anisokinetic sampling (Zhu et al., 1997).*

$$A_e = 1 + \left[1 - \frac{1}{1 + G\,\mathrm{Stk}\left(\cos\theta + 4\sqrt{(U_0/U_s)\sin\theta}\right)}\right]\left(\frac{U_0}{U_s}\cos\theta - 1\right) \quad (2.8)$$

where θ is the yaw angle defined by the sampler orientation with respect to the flow, and G is a coefficient with a functional expression of $G = G(D, \theta, U_0/U_s)$. However, the general form of G is yet to be determined.

Particle size (more precisely, particles inertia) plays an important role in an anisokinetic sampling. As stated earlier, the concentration of extremely small particles and mass flux of very large particles can be correctly sampled regardless of the sampling velocity, as exemplified respectively in Figure 2.9a and Figure 2.10 (Zhu et al., 1997). Figure 2.10 further shows that an isokinetic sampling always yields the correct sampling of particle mass flux. In order to yield the concentration of not very fine particles from an isokinetic sampling, the particle velocity must be determined independently. In practice, there can be a significant difference in velocity between the gas and particles, especially for large particles. For instance, for the pneumatic conveyance of glass beads sizing from 100 to 400 μm with carrying air velocity between 8 to 15 m/s, the local particle velocity is about 40 - 60% of the local gas velocity (Zhu, 1991). Therefore, for medium sized particles (say, 5 - 100 μm), anisokinetic sampling in principle provides neither the correct measurements of concentration of particles nor the correct measurements of particle mass flux, as illustrated in Figure 2.9b and Figure 2.10, respectively. This problem is complicated by the coupling of the slip nature (particle inertia) of particles in a carrying flow and the mismatch between the sampling velocity and stream velocity (flowrate and/or orientation). A deterministic model or method for the correction of measurements from anisokinetic sampling of medium sized particles needs to be developed.

2.2.3 Other influencing factors

In a particle sampling process, numerous mechanisms contribute to the error in the particle mass flux measurements. These mechanisms include gravitational sedimentation, impaction on the wall or at the tube bends, wall deposition due to the diffusion of small particles, flow turbulence, surface drag, agglomeration of fine particles, electrostatic charge, stickability of particles to the wall, and flow disturbance by the insertion of the probe, in addition to anisokinetic sampling discussed in §2.2.

2.3 CASCADE IMPACTOR

Cascade impactor is a sampling and size classification instrument for fine particles of aerodynamic size typically ranged from 0.5 - 25 μm. The principle of a cascade impactor is based on the particle inertia in a flow stream. When the stream suddenly makes a right-angle turn, particles must also adjust themselves immediately to make the right-angle turn in order to follow the flow streamlines. In this case, particles with larger inertia may not be able to follow the abrupt turn accordingly and hence impact with the collection surface. The collected particle size mainly depends upon the stream velocity. As shown in Figure2.11, a cascade impactor consists of a series of impaction stages which are arranged in a successive way so that the air stream impact velocity within each stage is progressively increased. An efficient filter or collector is usually used as a final stage to capture all the fines which successfully pass through all the previous stages. Therefore, particles collected at each progressive stages are of smaller size successively. Consequently, the aerodynamics size distribution of the sampled particles can be assessed using a suitable analysis method.

2.3.1 Inertial separation of particles

The particle Reynolds number based on maximum terminal velocity in air (maximum slip velocity of particles suspended in air) can be estimated by

$$C_{\mathrm{D}}\,\mathrm{Re}_{\mathrm{p}}^{2} = \frac{4}{3}\,\frac{(\rho_{\mathrm{p}}-\rho)\,\rho\, g\, d_{\mathrm{p}}^{3}}{\mu^{2}} \tag{2.9}$$

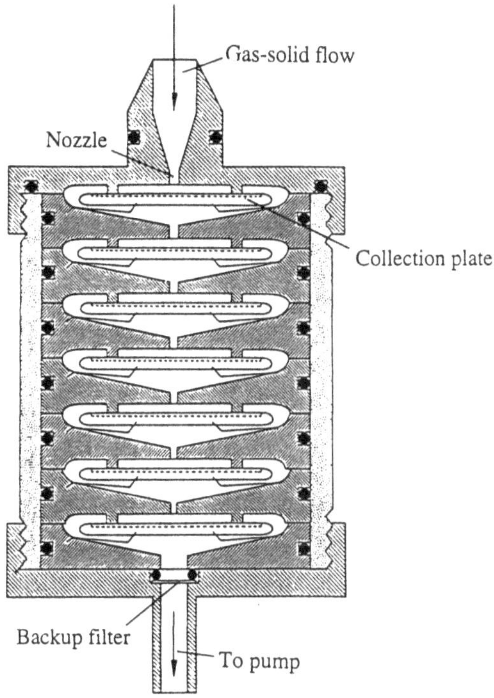

FIGURE 2.11 *Mercer (Lovelace) cascade impactor (Mercer, 1973).*

It is noted that the particles classified by cascade impactors are typically less than 50 μm in size and of material density less than 3000 kg/m^3. With this upper limit, the Re_p is 0.6 which is within the Stokes regime. Thus, in the applications of cascade impactors, the terminal velocity is determined by

$$U_{pt} = \frac{(\rho_p - \rho) g d_p^2}{18 \mu} \tag{2.10}$$

The inertial separation due to the sudden change of the direction of a particle-laden flow stream is characterized by the particle stopping distance. This is defined as the travelling distance of a particle in its forward direction before coming to rest with respect to the surrounding fluid. With Stokes drag, the particle stopping distance of a spherical particle with initial velocity U_0 can be calculated as

$$S = U_0 \tau_S \tag{2.11}$$

where S is the stopping distance, the total distance travelled, or the inertial range; and τ_S is the Stokes relaxation time. For larger or heavier particles with an initial Re_p from 1 to 400, the stopping distance may be estimated using an empirical formulae proposed by Mercer (1973)

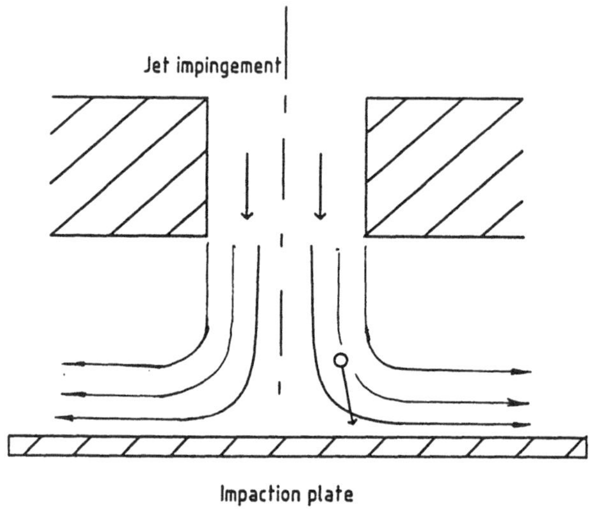

FIGURE 2.12 *Simple single-stage impactor (Hinds, 1982).*

$$S = \frac{\rho_p d_p}{\rho}\left(\mathrm{Re}_p^{\frac{1}{3}} - \sqrt{6}\tan^{-1}\left[\frac{\mathrm{Re}_p^{\frac{1}{3}}}{\sqrt{6}}\right]\right) \tag{2.12}$$

The particle collection efficiency of a single impaction is in general a function of the Stokes number. The importance of the Stokes number in an impaction may be revealed from a simplified analysis of a simple single-stage jet impactor with a rectangular opening, as shown in Figure 2.12 (Hinds, 1982). Due to the symmetry, only half of the impacting flow needs to be analyzed. In this analysis, the particle-laden flow goes through a right-angle impaction. The streamlines of the gas phase are arcs of co-centered quarter circles, forming a right-angle flow tube with an unchanged cross-sectional area which maintains a constant gas velocity. During the quarter-circle turn, particles in the flow are subjected to the action of centrifugal forces which drive them towards the impaction plate. As a first order approximation, the particles are considered to depart their original streamlines with constant radial velocities while traversing the quarter circles. The total departure of a particle towards the impaction plate can be calculated as (Reist, 1993)

$$\delta = \int_0^{\frac{\pi}{2}} U\tau_S \sin\phi \, d\phi = U\tau_S \tag{2.13}$$

It is noted that, when the flow goes though this right-angle impaction, only those particles whose initial flow streamline are within this total departure distance away from the collection plate will be collected by the plate. Hence, the collection efficiency is the ratio of the total departure distance to the flow tube size which is the half of the impactor opening, *i.e.*,

$$\eta = \frac{\delta}{D/2} = \frac{2\,U\,\tau_S}{D} = \mathrm{Stk} \tag{2.14}$$

where Stk is the Stokes number based on the half width of the impactor opening (which is slightly different from the Stk defined by Eq. (2.5)). This is due to the consideration that the streamline of a jet are not strongly affected by the spacing between the nozzle and the impaction plate because the jet of particulates expands only slightly until it reaches within about one jet diameter of the plate. Therefore, the characteristic dimension of an impactor is the half width rather than the spacing between the nozzle and the plate (Hinds, 1982). For impactors with circular openings, D is the diameter of the impactor opening. When dealing with very fine particles (especially submicron particles) or at low pressure conditions, the calculation of Stk needs to be modified to account for the Cunningham slip effect as

$$\mathrm{Stk} = \frac{2\,C_c\,\tau_S\,U}{D} = \frac{C_c\,\rho_p\,d_p^2\,U}{9\,\mu\,D} \tag{2.15}$$

where C_c is the Cunningham slip correction factor, which may be estimated from (Wahi and Liu, 1971)

$$C_c = 1 + \frac{0.163}{P d_p} + \frac{0.0549}{P d_p}\exp(-6.66\,P d_p) \tag{2.16}$$

where P is the static pressure (atm) at the impaction plate; and d_p is the particle diameter (μm).

For a general single-stage impactor, the flow will not make a uniform right turn but in a rather complicated way, neither the particles be monodispersed. For an impactor or an impaction stage, the collection efficiency with respect to particle size or characteristic efficiency curve may be obtained if the individual efficiencies of a series of monodispersed spherical particles can be determined. The determination of this individual efficiency of a given sized particle can be either measured experimentally or estimated theoretically. The theoretical determination relies on particle trajectory modelling in an impingement flow. In this case, it is normally assumed that all the particles which impact with the collecting surface will be captured and remain stuck to that surface. The flow field of the jets with particular nozzle and collecting plate can be calculated by solving the Navier-Stokes equations with the corresponding impactor geometry.

It is a common practice to have the collection efficiency plotted as a function of the root of Stk because, in this way, the term is proportional to the particle diameter.

2.3.2 Typical cascade impactors and applications

Cascade impactors are typically characterized based on the nozzle geometry, number of stages, flow rate capacity, and the range of effective cut-off diameters of the impaction stages. Characteristics of some most commonly used cascade impactors are exemplified in Table 2.1.

Table 2.1 Characteristics of Some Typical Cascade Impactors

Type of cascade impactor	Nozzle type (stage)	No. of stages	Flow rate (lpm)	Range of d_{50} (μm)
Andersen	400 holes	8	28.3	0.4 - 11.0
-	5 annular slits	8	14.2	0.4 - 12
-	400 holes	6	28.3	0.6 - 7.2
Mercer (Lovelace)	single hole	7	0.3	0.5 - 8.5
May (Casella)	single slit	4	17.5	0.4 - 12.4
May 'Ultimate'	single slit	7	5	0.5 - 32
Lippmann (UNICO)	single slit	4	13.4	1.1 - 11.1
Lundgren (Sierra)	9 slits	5	1130	0.5 - 7.2
-	single slit	4	85	0.4 - 13.0

The first cascade impactor was invented by May in 1945. The May cascade impactor is also known as Casella cascade impactor. It consists of four stages with each stage positioned normal to the neighbouring stages, as shown in Figure 2.13. A significant advantage in this design is the easy removal of the collection plates, simply by opening the caps. However, particle loss to the wall deposition becomes a major problem in the accuracy of the measurement using the May cascade impactor. To overcome this weakness, a new design of a seven-stage cascade impactor was proposed by May in 1975, as shown in Figure 2.14, in which the sampling flow makes direct impaction without flow-around over the collection plates in the first three stages while the flow path is kept as short as possible. In the remaining four stages, the flow channels are designed to provide a smoother flow. In this way, the total wall loss is maintained about 1%. This significant reduction of the wall losses makes May to claim the design as "ultimate'.

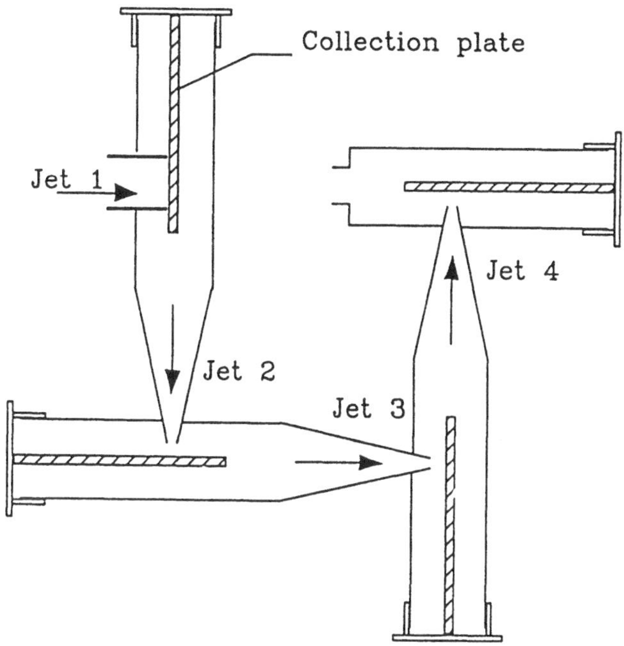

FIGURE 2.13 *May cascade impactor.*

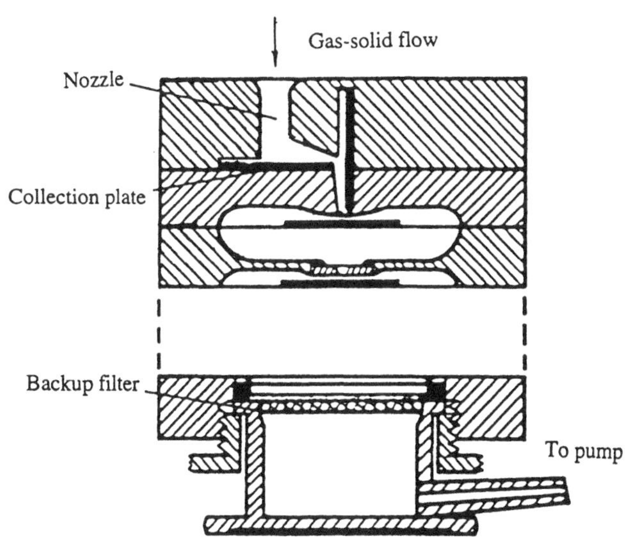

FIGURE 2.14 *May 'ultimate' cascade impactor (Shaw, 1978).*

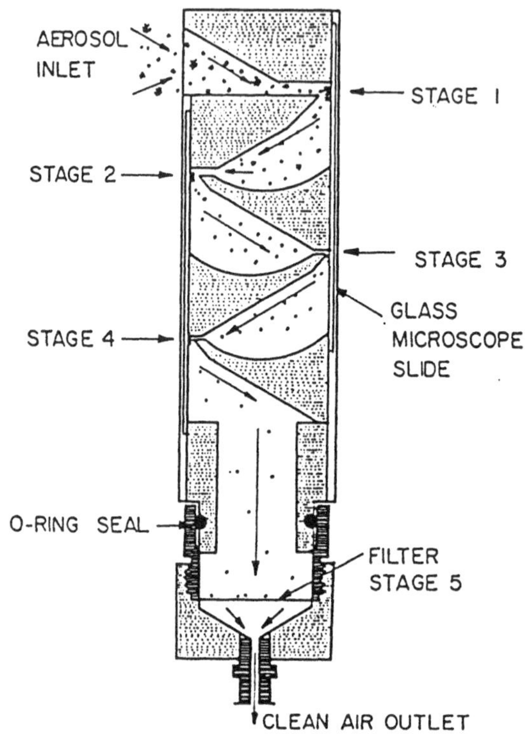

FIGURE 2.15 *Lippmann (UNICO) cascade impactor (Shaw, 1978).*

The operation mechanism of Lippmann (or UNICO) cascade impactor is similar to that of May cascade impactor but is simpler in design and easier to use, as exemplified in Figure 2.15. The UNICO cascade impactor also has the merit of the easy removal of the impaction plates, which is achieved by the implement of a manual slid movement mechanism. The slides can be advanced to provide with multiple collection surfaces with the same stage so that a large number of samples can be acquired. Another interesting cascade impactor which uses externally removable collection cups instead of impaction plates was recently developed by Marple and Olson (1995), as shown in Figure 2.16. This design allows an easy and quick removal of stage deposits from the cascade impactor for analysis without the stage by stage disassembly of the entire impactor.

Lundgren (or Sierra) cascade impactor uses a series of rotating drums instead of flat plates to collect particles, as shown in Figure 2.17. A significant advantage in this design is the easy investigation of time-dependent aerosol distributions and chemical depositions by slowly rotating the collection cylinders. In addition, this arrangement permits a long time sampling and, under steady

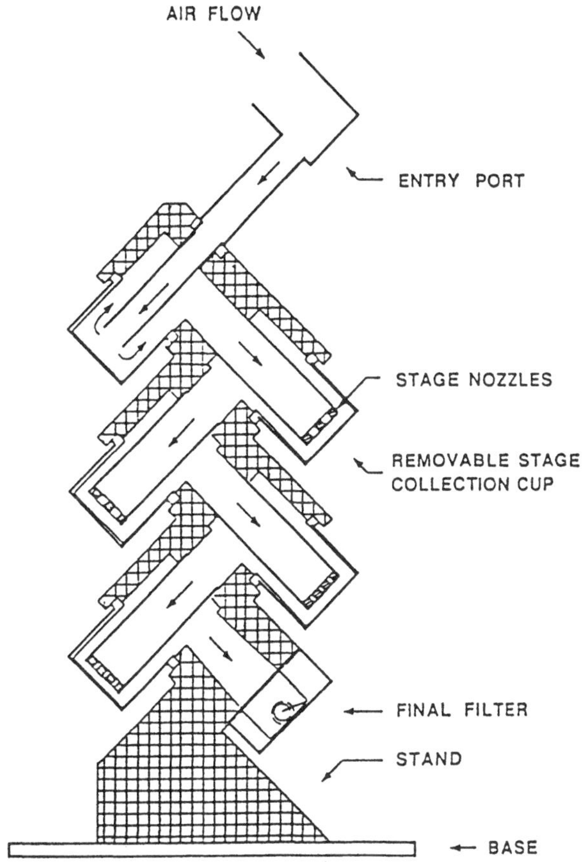

FIGURE 2.16 *Low-loss cascade impactor with stage collection cups (Marple and Olson, 1995).*

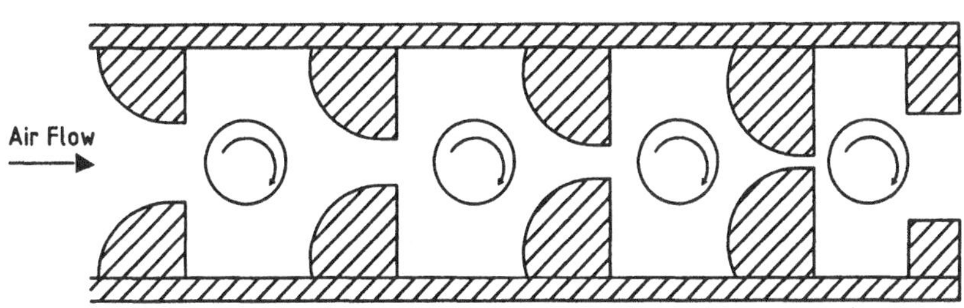

FIGURE 2.17 *Lungren cascade impactor.*

sampling conditions, a uniform deposition of particles on each collection surface. In the applications of Lundgren cascade impactors, it is recommended to use sticky surfaces to prevent the significant "bounce-off" of particles from the collection cylinders.

Mercer (or Lovelace) cascade impactor represents the most commonly used cascade impactors with round single nozzles. As shown in Figure 2.11, seven collection stages are lined up in a series. A membrane filter is used as the final back up filter to collect all remaining fines. The structure of this type of design is simple and compact. However, cascade impactors with single nozzles usually have some disadvantages when dealing with large flow rate sampling. The high sampling velocity leads to not only the large pressure drops across the nozzles but also the severe particle loss due to the increased particle rebounce. To facilitate the aerosol sampling with large flow rate, multijet cascade impactors were developed, as exemplified by the Andersen cascade impactor shown in Figure 2.18. In this design, 400 round nozzles are formed on each jetting plate, providing multiple jets on each collection stage.

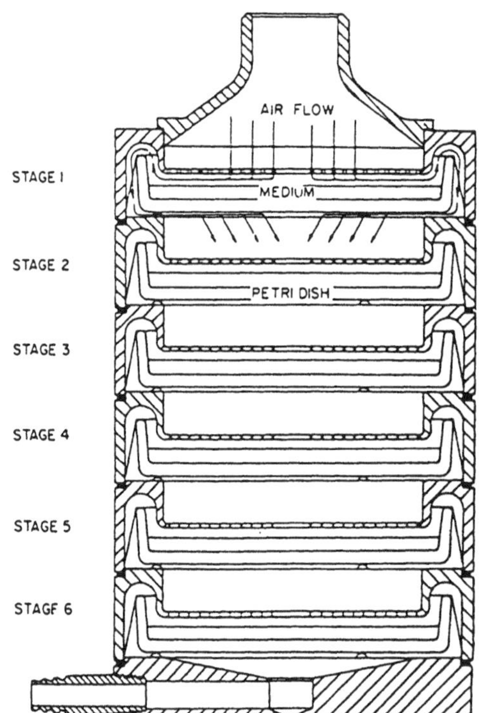

FIGURE 2.18 *Andersen cascade impactor (Shaw, 1978).*

2.3.3 Cut-off size and size analysis

An ideal cascade impactor for particle sizing should consist of a series of stages with each collecting particles larger than a certain size and none smaller. In this way, the distribution of sampled mass on each collection stage and back up filter would directly represent the true mass distribution of particles. In reality, no impaction stage has such an ideal cut-off characteristic. The practical collection efficiency of an impaction stage usually increases monotonically from 0 to 100% over a certain range of particle sizes, as shown in Figure 2.19. The cut-off diameter of an impaction stage is defined as the diameter with a 50% collection efficiency, commonly denoted as d_{50}. Due to the actual cut-off nature of a stage, some oversize particles ($d_p > d_{50}$) fail to be collected while some undersize particles ($d_p > d_{50}$) are captured by the impaction. It is noted that a real impactor will collect the same amount of particles as the ideal stage if the amount of uncollected larger particles matches that of collected smaller particles. With that concern, an effective cut-off diameter (ECD) of an impaction stage is defined as the diameter where the amount of uncollected larger particles equals that of collected smaller particles. In general, ECD is not equal to d_{50} but with a minor difference. The value of ECD depends not only on the size distribution of the particles sampled but also on the characteristics of the impaction stage. In addition, for the same sample, the ECD for a particle number distribution is different from that for a particle mass distribution. For those reasons, in the practical design and analysis of a cascade impactor, d_{50} is much more frequently used than ECD.

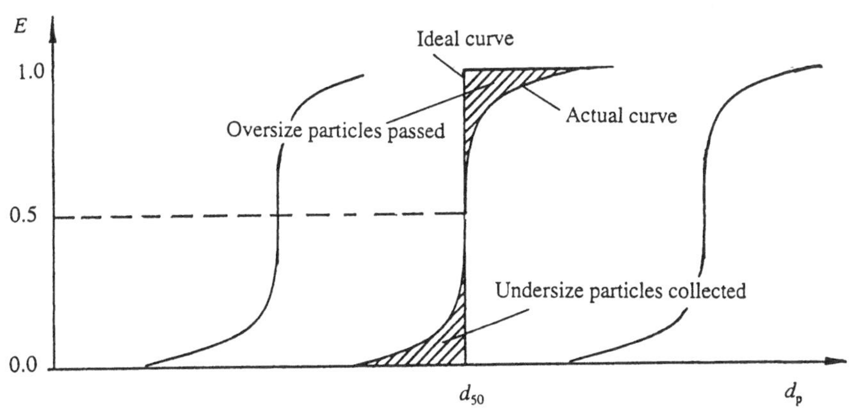

FIGURE 2.19 *Schematic collection efficiencies of a three-stage cascade impactor*

Particle aerodynamic size distributions can be obtained from the measurements of particle mass on each impaction stage of a cascade impactor. However, the analysis is normally not quite straight forward. From Figure 2.19, it is evident that the same sized particles may be collected on several impaction stages instead of on a single stage. This particle size overlapping increases the difficulties of the analysis. Denote $E_i(x)$ as the collection efficiency of particle size x on the ith impaction stage. The actual collection of particle size x on the jth impaction stage may be expressed as

$$K_j(x) = E_j(x)\prod_{i=1}^{j-1}[1 - E_i(x)] \tag{2.17}$$

The mass collected on ith impaction stage m_i can be expressed by

$$m_i = \frac{m_T \int_0^\infty f(x)\, E_i(x)\, h_i(x)\, dx}{\int_0^\infty f(x)\,[1 - E_0(x)]\, dx} \tag{2.18}$$

where $f(x)$ is the particle mass distribution; m_T is the measured total mass concentration, the denominator represents the mass removed by the cascade impactor inlet (stage $i = 0$); and $h(x)$ accounts for the effect of wall loss between stages, which is given by

$$\begin{aligned} h_1(x) &= 1 - WL_1(x) \\ h_i &= h_{i-1}[1 - E_{i-1}(x)] - WL_i(x) \end{aligned} \tag{2.19}$$

where $WL_i(x)$ is the wall loss factor of particle size x between the (i-1)th stage and the ith stage. In practice, $E_i(x)$ and $WL_i(x)$ can be predetermined from the calibration of the ith impaction stage. Thus, from the measured m_i, $f(x)$ can be determined by use of a deconvolution method.

A deconvolution problem in general does not have a unique solution. Instead there are an infinite number of possible solutions that can fit the same set of cascade impactor measurements. It is well recognized that, for most engineering applications, actual particle size distributions can be reasonably represented by a set of log-normal distributions. With this concern, in the following, a deconvolution method (chi-squared method) to extract particle size distributions from cascade impactor data is introduced, which is based on multimodal log-normal size distributions (Dzubay and Hasan, 1990).

Assuming that the effect of wall loss between stages can be neglected, the mass collected (without any measurement error) on the ith impaction stage can be represented by

$$m_i = \frac{m_T \int_0^\infty f(x)\, E_i(x) \prod_{j=0}^{i-1} [\, 1 - E_j(x)\,]\, dx}{\int_0^\infty f(x)\, [\, 1 - E_0(x)\,]\, dx} \tag{2.20}$$

Without loss of generality, let us consider a tri-mode case where the particle size distribution $f(x)$ is a linear combination of three log-normal functions as

$$f(x) = C_1 f_1(x) + C_2 f_2(x) + C_3 f_3(x) \tag{2.21}$$

with

$$C_1 + C_2 + C_3 = 1 \tag{2.22}$$

The kth mode of log-normal function ($k = 1, 2,$ or 3) is expressed by

$$f_k(x) = \frac{1}{\sqrt{2\pi}\, x \ln\sigma_k} \exp\left[- \frac{(\ln x - \ln x_k)^2}{2\,(\ln\sigma_k)^2} \right] \tag{2.23}$$

where x_k and σ are the medium size and geometric standard deviation, respectively. The deconvolution also requires the information of the collection efficiencies of each stage. A commonly used form of the collection efficiency of stage i is given by

$$E_i(x) = \left[1 + \left(\frac{d_{50}}{x} \right)^{p_i} \right]^{-1} \tag{2.24}$$

where p_i is the steepness of the collection efficiency curve. For the last stage (backup filter), $E_N(x) = 1$.

For each mode, three size distribution parameters (C_k, x_k, and σ_k) need to be determined. These can be determined by a nonlinear least squared method (known as chi-squared method) which minimizes χ^2 defined by

$$\chi^2 = \frac{1}{(N - N_p)} \sum_{i=1}^{N} \left(\frac{m_i - M_i}{\delta M_i} \right)^2 \tag{2.25}$$

where M_i is the mass actually measured on stage $_i$ i, δM is the random measurement error in M_i, and N_p is the number of fitted parameters. For uni-, bi-, and tri-mode distributions, $N_p = 2, 5, 8$, respectively. The expectation value of χ is 1. For the cases where no backup filter is used, χ should be evaluated with N-1 instead of N. In order to find the best fitted set of size distribution parameters that yields the minimum χ^2, an interactive approach may be used. This method begins with a set of guessed values for C_k, x_k, and q_k and uses a gradient-expansion algorithm to find a new set of parameters which gives a lower χ^2 (Bevington, 1969). The iteration continues until the relative change in χ^2 reaches an acceptable tolerance, which leads to the best fitted set of parameters. Since each mode is characterized by three parameters, the number of modes which can be fitted depends on the number of data (stages) from the cascade impactor. For instance, for an impactor of five stages and one back-up filter, only six measurements (mass of each stage or back-up filter) are obtained. Hence, the number of modes is limited to two.

2.3.4 Considerations in design and operation

(1) Wall losses and particle bouncing

During the sampling of particles in a cascade impactor, a noticeable part of particles is lost between stages, mainly due to the wall losses (parasitic particle deposition) and particle bounce-off from the collection surfaces. To minimize the wall losses, the wall material should be selected so that it is not subject to the retention of particles. Particle bounce is a major source of error in cascade measurement because the bounce-off particles are reentrained into the stream and lead to not only the biased fractions of particles larger than the cut-off sizes of the following stages but also biased size distribution of the current stage. Particle bounce depends on the impact velocity, particle size and particle composition. The effect of particle bounce can be significant for sampling of solid particles, especially when particles are larger than 6 μm.

To minimize the particle bounce off effect, collection surfaces should also be selected carefully. Common types of impaction surfaces include membrane, fiberglass, silver membrane, Teflon and Nuclepore filter, and brass and stainless steel shim stock. Table 2.2 shows an example of the effect of selection of collection surface on the wall losses (Newton et al., 1990). In Table 2.2, the test aerosols are droplets of 1% CsCl plus 1% uranine. Three types of cascade impactors were used, including Mercer, Sierra Radial Slit Jet (SRSJ), and Lovelace Multi-Jet (LMJ). The occurrence of particle bouncing may be indicated by the presence of excess mass on the back-up filter.

Particle bounce can be effectively controlled by coating an adhesives layer on the collection surface to keep collected particles from bouncing off the plates. Typical coating materials include Antifoam A, Hi-Vac silicone grease, Apiezon® L, viscous oils, Vaseline, and glycerin. The typical thickness of the adhesive

Table 2.2 Effect of collection surfaces on total wall losses (from Newton et al., 1990)

Cascade impactor	Collection surface	Total wall losses
Sierra Radial Slit Jet	Millipore membrane mixed esters of cellulose	19.0
Sierra Radial Slit Jet	Gelman Type A fiberglass filters	9.0
Sierra Radial Slit Jet	Millipore Fluoropore filters	10.2
Sierra Radial Slit Jet	Shim stock (uncoated)	10.3
Sierra Radial Slit Jet	Shim stock coated with Dow Antifoam A	5.2
Lovelace Multi-Jet	Millipore membrane mixed esters of cellulose	25.1
Lovelace Multi-Jet	Gelman Type A fiberglass filters	29.0
Lovelace Multi-Jet	Millipore Fluoropore filters	5.2
Lovelace Multi-Jet	Shim stock (uncoated)	5.2
Lovelace Multi-Jet	Shim stock coated with Dow Antifoam A	4.5
Lovelace Multi-Jet	Flotronics Silver Membrane filter	18.7
Lovelace Multi-Jet	Nuclepore clear, plain, regular filter	11.8
Mercer	Glass cover slip coated with Dow Antifoam A	3.0

layer is ranged from 20 to 100 μm. It is noted that, when sampling with high solid concentration, the adhesive coating becomes less effective as the accumulation of particles on the surface grows over a certain limit (overloading). In addition, the selection of adhesives should be careful to avoid any chemical reactions between the agent and particles, especially when an analysis of chemical composition of sampled particles is required. Another measure to reduce the particle bouncing is to use fiberglass filters or other filter media such as cellulose fiber filters which help to trap the particles in the fiber. Due to the uneven collection with fibers, some analytical techniques such as scanning electron microscope (SEM) or X-ray fluorescence (XRF) may not be applicable to the measurements from the fiber filters.

(2) Pressure drop

Estimation of the pressure drop over a cascade impactor is important for both design and operations of the device. A simple method for the estimation is to assume that the dynamic pressure head of the jet is lost due to turbulence. Hence, the pressure drop in an impaction stage is estimated by (Reist, 1993)

$$\Delta p = p_{\text{up}} - p_{\text{down}} = 2\rho U^2 \frac{p}{p_{\text{down}}} \tag{2.26}$$

where ρ, p, U refer to the density, pressure, and velocity at atmospheric or some reference condition, respectively; and subscripts "up" and "down" refer to the upstream and downstream of the impaction stage.

(3) Sharp cutoff of efficiency curves

It is important to have sharp cutoffs of efficiency curves of each stage of a cascade impactor. In order to produce a steep efficiency curve, the Re in a jet nozzle should be within the range of 500 to 3000. The ratio of the distance between the jet nozzle and the impaction plate to the nozzle diameter or width should be larger than 1.0 for circular nozzles and 1.5 for rectangular nozzles (Hinds, 1982). In the design of multijet impaction stages, the cross-flow parameter should be less than 1.2 (Fang et al., 1991). The cross-flow parameter is defined as $D_n N/4D_c$, where D_n is the nozzle diameter, N is the number of jets per stage, and D_c is the nozzle cluster diameter. The cross-flow parameter indicates the interference between cross-flow and impinging jets in a multijet cascade impactor, which directly affects the collection efficiency of the impactor.

Notations

A	Effective flow area of a sampling probe
A_e	Aspiration efficiency
C_c	Cunningham slip correction
C_D	Drag coefficient
C_I	Carried mass coefficient
C_B	Basset coefficients
D	Pipe diameter or nozzle openness
D_n	Nozzle diameter
D/Dt	Substantial derivative following the gas follow
d_{50}	Cut-off diameter
d_p	Particle diameter
E_i	Collection efficiency of ith impaction stage
f	Particle size (mass) distribution
m_i	mass collected on ith impaction stage
m_p	Total mass of collected particles
$\dot{m}_p$	Particle mass flux
$\dot{m}_{pd}$	Particle diffusive mass flux
N	Number, or number of jets
N_p	Number of fitted parameters

P	Static pressure
Re_p	Particle Reynolds number
S	Stopping distance
Stk	Stokes number
U	Velocity
U_{pt}	Particle terminal (free-falling) velocity
WL	Wall loss factor
x	Particle size

Greek symbols

Δt	Sampling time period
μ	Viscosity
ρ	Density
σ	Geometric standard deviation
τ_S	Stokes relaxation time
χ	Expectation value

Subscripts

0	Free Stream
c	Nozzle cluster
k	k-th mode
p	Particle
s	Sampling or suction
t	Terminal

REFERENCES

Anon, "Sampling of Gas-borne Particles", *Engineering, 152*, 141 (1941).

Addlesee, A. J., "Anisokinetic Sampling of Aerosols at a Slot Intake", *J. Aerosol Sci., 11*, 483 (1980).

Allen, T., *Particle Size Measurement*, 4th edn, Chapman and Hall, New York, 1990.

Anderson, A. A., "New Sampler for the Collection, Sizing, and Enumeration of Viable Airborne Particles", *J. Bacteriol., 76*, 471 (1958).

Badzioch, S., "Collection of Gas-borne Dust Particles by means of an Aspirated Sampling Nozzle", *Br. J. Appl. Phys., 10*, 26 (1959).

Badzioch, S., "Correction for Anisokinetic sampling of Gas-borne Dust Particles", *J. Inst. Fuel, 33*, 106 (1960).

Belyaev, S. P., and Levin, L. M., "Investigation of Aerosol Aspiration by Photographing Particle Tracks under Flash Illumination", *J. Aerosol Sci., 3*, 127 (1972).

Belyaev, S. P., and Levin, L. M., "Techniques for Collection of Representative Aerosol Samples", *J. Aerosol Sci., 5*, 325 (1974).

Bevington, R. P., *Data Reduction and Error Analysis for the Physical Sciences*, McGraw-Hill, New York, 1969.

Bohnet, M., "Particulate Sampling", in W. Strauss (ed.), *Air Pollution Control, Part III: Measuring and Monitoring Air Pollutants*, Wiley, New York, 1978.

Boothroyd, R. G., "An Anemometric Isokinetic Sampling Probe for Aerosols", *J. Sci. Instrum., 44*, 249 (1967).

Boubel, R. W., "A high Volume Stack Sampler", *JAPCA, 21*, 783 (1971).

Brink, J. A., Jr., "Cascade Impactor for Adiabatic Measurements", *Ind. Eng. Chem.*, 50 (1958).

Buerkholz, A., "Untersuchungen zum Messfehler bei nichtisokinetischer Entnahme. Teil 1", *Staub - Reinhaltung der Luft, 51*, 395 (1991).

Cadle, R. D., *The Measurements of Airborne Particles*, John Wiley & Sons, New York, 1975.

Cheng, L., Tung, S. K., and Soo, S. L., "Electrical Measurement of Flow Rate of Pulverized Coal Suspensions", *Trans. ASME, J. Eng. for Power, 92*, 135 (1970).

Cohen, J. J., and Montan, D. N., "Theoretical Considerations, Design and Evaluation of a Cascade Impactor", *Am. Ind. Hyg. Assoc., 28*, 95 (1967).

Davies, C. N., *Dust is Dangerous*, Faber and Faber, London, 1954.

Davies, C. N., "The Entry of Areosols into Sampling Tubes and Heads", *Brit. J. Appl. Phys. (J. Phys. D), 1*, 921 (1968).

Davis, I. H., *Air Sampling Instruments*, 4th edn, Am. Conf. Governmental Industrial Hygienists, 1972.

Dzubay, T. G., and Hasan, H., "Fitting Multimodal Lognormal Size Distributions to Cascade Impactor Data", *Aerosol Sci. and Tech., 13*, 144 (1990).

Emmerichs, M., and Armbruster, L., "Improvement of a Multi-stage Impactor for Determining the Particle Size Distribution of Airborne Dusts", *Silikosebericht Nordrhein-Westfalen, 13*, 111 (1981).

Fan, L.-S., and Zhu, C., *Principles of Gas-Solid Flows*, Cambridge University Press, 1997.

Fang, C. P., Marple, V. A., and Rubow, K. L., *J. Aerosol Sci., 22*, 403 (1991).

Fuchs, N. A., *The Mechanics of Aerosols*, Macmillan, New York, 1964.

Fuchs, N. A., "Sampling of Aerosols", *Atmos. Env., 9*, 697 (1975).

Gibson, H., Vicent, J. H., and Mark, D., "A Personal Inspirable Aerosol Spectrometer for Applications in Occupational Hygiene Research", *Ann. Occup. Hyg., 31*, 463 (1987).

Glauberman, H., "The Directional Dependence of Air Samplers", *Am. Ind. Hyg. Ass. J., 23*, 235 (1962).

Gussman, R. A., Sacca, A. M., and McMahon, N. M., "Design and Calibration of a High Volume Cascade Impactor", *J. Air Poll. Control Assoc., 23* (1973).

Hemeon, W. C. L., and Haines, G. F., "The Magnitude of Errors in Stack Dust Sampling", *Air Repair, 4*, 159 (1954).

Hinds, W. C., *Aerosol Technology: Properties, Behavior, and Measurement of Airborne Particles*, John Wiley & Sons, New York, 1982.

Ito, K., Kobayashi, S., and Tokuda, M., "Mixing Characteristics of a Submerged Jet Measured Using an Isokinetic Sampling Probe", *Metallurgical Transactions B (Precess Metallurgy), 22*, 439 (1991).

Koo, Y. M., Summer, H. R., and Chandler, L. D., "Formation of Immiscible Oil droplets During chemigation I. In-line Dispersion", *Trans. of ASAE, 35*, 1121 (1992).

Lapple, C. E., and Shepherd, C. B., "Calculation of Particle Trajectories", *Ind. Eng. Chem., 32*, 605 (1940).

Levin, L. M., "The Intake of Aerosol Samples", *Ivz. Nauk., SSSR Ser. Geofiz., 7*, 914 (1957).

Lodge, J. P., and Chan, T. L. (eds.), *Cascade Impactor*, American Industrial Hygiene Assoc., Arkron, OH, 1986.

Lundgren, D. A., "An Aerosol Sampler for Determination of Particle Concentration as a Function of Size and Time", *J. Air Poll. Control Assoc., 17*, 225 (1967).

Marple, V. A., and Liu, B. Y. H., "Characteristics of Laminar Jet Impactors", *Env. Sci. Tech., 8*, 648 (1974).

Marple, V. A., and Olson, B. A., "A Low-Loss Cascade Impactor with Stage Collection Cups: Calibration and Pharmaceutical Inhaler Applications", *Aerosol Science and Technology, 22*, 124 (1995).

May, K. R., "The Cascade Impactor: an Instrument for Sampling Coarse Aerosols", *J. Sci. Instrum., 22*, 187 (1945).

May, K. R., *17th Symposium of the Society for General Microbiology*, Imperial College, London, University Press, Cambridge (1967).

May, K. R., "An 'Ultimate' Cascade Impactor for Aerosol Assessment", *J. Aerosol Sci., 6*, 413 (1975).

Mercer, T.T., *Aerosol Technology in Hazard Evaluation*, Academic Press, New York, 1973.

Newton, G. J., Cheng, Y. S., Barr, E. B., and Yeh, H. C., "Effects of Collection Substrates on Performance and Wall Losses in Cascade Impactors", *J. Aerosol Sci., 21 (3)*, 467 (1990).

Ranz, W. E., and Wong, J. B., "Impaction of Dust and Smoke Particles", *Ind. Eng. Chem., 44*, 1371 (1952).

Reist, P. C., *Aerosol Science and Technology*, 2nd ed., McGraw-Hill, New York, 1993.

Pilat, M. J., Ensor, D. S., and Busch, J. C., "Cascade Impactor for Sizing Particles in Emission Sources", *Am. Ind. Hyg. Assoc. J., 32* (1971).

Ruping, G., "The Importance of Isokinetic Suction in Dust Flow Measurement by means of Sampling Probes", *Staub-Reinhalt. Luft (English translation), 28*, 1 (1968).

Sansone, E. B., *Sampling Airbone Solids in Ducts Following a 90 ° Bend*, Ph.D. Thesis, University of Michigan; 1967.

Slaughter, M. C., Zhu, C., and Soo, S. L., "Measurement of Local Statistical Properties of Particle Motion in a Dense Gas Solid Suspension", *Advanced Powder Technology, 4*, 169 (1993).

Soo, S. L., Baker, D. A., Lucht, T. R., and Zhu, C., "A Corona Discharger Probe System for Measuring Phase Velocities in a Dense Suspension", *Rev. Sci. Instrum., 60*, 3475 (1989).

Soo, S. L., Stukel, S. S., and Hughes, J. M., "Measurement of Mass Flow and Density of Aerosols in Transfer", *J. Environmental Sci. and Tech. (Ind. Eng. Chem.), 3*, 386 (1969).

Tufto, P. A., and Willwke, K., "Dependence of Particulates Sampling Efficiency on Inlet Orientation and Flow Velocities", *Am. Ind. Hyg. Ass. J., 43*, 436 (1982).

Vincent, J. H., *Aerosol Sampling: Science and Practice*, John Wiley & Sons, New York, 1989.

Vincent, J. H., Emmett, P. C., and Mark, D., "The Effects of Turbulence on the Entry of Airborne Particles into a Blunt Dust Sampler", *Aerosol Sci. Tech., 4*, 17 (1985).

Vincent, J. H., Stevens, D. C., Mark, D., Marshall, M., and Smith, T. A., "On the Aspiration Characteristics of Large-diameter, Thin-walled Aerosol Sampling Probes at Yaw Orientations with respect to the Wind", *J. Aerosol Sci., 17*, 211 (1986).

Vitols, J. H., "Theoretical Limits of Errors due to Anisokinetic Sampling of Particulate Matter", *JAPCA, 16*, 79 (1966).

Wahi, B. and Liu, B. Y. H., "The mobility of polystyrene latex particles in the transition and the free molecular regimes", *J. Colloid and Interface Sci., 37*, 374 (1971).

Watson, H. H., "Errors due to Anisokinetic Sampling of Aerosols", *Am. Ind. Hyg. ass. J., 15*, 21 (1954).

Whiteley, A. B., and Reed, L. E., "The Effect of Probe Shape on the Accuracy of Sampling flu Gases for Dust Content", *J. Inst. Fuel, 32*, 316 (1959).

Wiener, R. W., Okazaki, K., and Willeke, K., "Influence of Turbulence on Aerosol Sampling Efficiency", *Atmos. Environ., 22*, 917 (1988).

Zhang, G. J., and Ishii, M., "Isokinetic Sampling Probe and Image Processing System for Droplet Size Measurement in Two-phase Flow", *Int. J. of Heat and Mass Transfer, 38*, 2019 (1995).

Zhu, C., *Dynamic Behavior of Unsteady Turbulent Motion in Pipe Flows of Dense Gas-Solid Suspensions*, Ph.D. Thesis, University of Illinois at Urbana-Champaign, 1991.

Zhu, C., Slaughter, M. C., and Soo, S. L., "Covariance of Density and Velocity Fields of a Gas-Solid Suspension", *Rev. Sci. Instrum., 62*, 2835 (1991a).

Zhu, C., Slaughter, M. C., and Soo, S. L., "Measurement of Velocity of Particles in a Dense Suspension by Cross Correlation of Dual Laser Beams", *Rev. Sci. Instrum., 62*, 2036 (1991b).

Zhu, C., Yu, T., and Huang, D., "Numerical Study of Effect of Velocity Slip on Isokinetic/Anisokinetic Sampling of Gas-Solid Flows", *Int. Symp. on Multiphase Fluid, Non-Newtonian Fluid and Physicochemical Fluid Flows '97 Beijing*, Oct. 7-9, Beijing, China, 1997.

3

Electrostatic Measurements

Gerald M. Colver

3.1 INTRODUCTION

The purpose of this chapter is to acquaint the reader with various transducers, probes, sensors, and instruments together with measurement techniques that are used for the detection of electrostatic phenomena in multiphase systems. Both invasive (probes) and noninvasive (coils outside ducts) measurement techniques are discussed. In practice, most experiments in electrostatics are highly specialized utilizing probes fabricated in the laboratory. An emphasis has been placed throughout the chapter on solids-gas systems; however, probe theory and charge measuring techniques are often applicable to related measurements such as charged liquid droplets. A few instruments (atomic force microscope, optical particle counters, laser Doppler tracking devices) are capable of detecting charge interaction at the particle level while most depend on some cumulative electrostatic effect (Faraday cage, particle anemometers, and electrostatic voltmeters).

The fundamental electrical quantities of measurement are electrostatic charge, current (charge transfer rate), voltage (electric potential difference) and particle force resulting from the separation of charge. These quantities are either measured directly by a suitable detector transducer (e.g., an electrometer) or inferred through the measurement of a related quantity such as capacitance or resistance. For purposes of tabulation, electrical data are usually normalized in terms of some geometric factor and expressed as charge density (C/m^3), current density (A/m^2), or electric field strength (V/m).

The chapter begins with the fundamental measurements of resistance, capacitance, charge, and particle force. We proceed with flow measurements with various probes followed by a listing of some commercial electrostatic instruments. Nonelectrosatic measurements in multiphase flow such as the laser-Doppler anemometer, radioactive tracers, and stroboscopic techniques (Polaskowski, et. al, 1995; Soo, 1982) have not been discussed unless in relation to an electrostatic effect.

3.2 ORIGIN OF CHARGE

A common electrostatic effect observed in multiphase systems containing flowing solids is that of *frictional* or *triboelectric* charging caused by particles contacting a solid boundary or by rubbing between the particles themselves. Here the spontaneous transfer of electrons or ions between two dissimilar contacting materials leaves the surfaces oppositely charged following separation. Nonuniform charging of these particles often leads to particle clustering and problems in powder flowability as well as adhesion to walls. *Induction* charging of particles is contact charging that occurs when charge is driven to the surface of a solid or of a conductive liquid by an applied electric field, for example, as in the case of electrified droplet sprays 20-50 μm (Law 1995).

Solid and liquid particle charging is also associated with other phenomena including corona discharge, flame ionization, thermionic emission, radioactive emission, phase change and particle breakup. In some cases, surface-to-surface contact is not necessarily a requirement for charge separation (in contrast to *triboelectric* charging). Charging of particles occurs in normal atmospheres containing about 10^3 ion pairs/cm^3 (from earth's and cosmic radiation) and in controlled environments such as electrostatic precipitators and combustion flames (White, 1963; Lawton and Weinberg, 1969). The overall charging mechanism of the particle may involve several steps including ion diffusion, convection, and some form of ion attachment to the surface from long or short range forces such as image and electronic forces.

Flowing liquid systems containing dielectric or electrolyte solutions (e.g. hydrocarbons) can lead to charging at walls. This is the result of the formation of a double layer of charges having opposite signs at the liquid-solid interface caused by electrochemical reaction (Adamson, 1976). The motion of the liquid subsequently carries away part of the charge furthest from the wall leaving the layer of charge at the wall unaffected (Touchard, 1995). In this way, a large potential difference can build up from the separation of charge by pumping a liquid between two vessels.

3.3 FUNDAMENTAL MEASUREMENTS

3.3.1 Measurement of Bulk Powder Resistivity and Dielectric Constant

The fundamental measurements of dielectric constant and resistivity in multiphase systems follow directly from methods used for solid systems (Curtis, 1915). The material resistivity (or electrical conductivity) together with the permittivity are useful parameters for calculating the charge relaxation time of the material.

3.3.1.1 Measuring bulk resistivity of a powder

The resistivity $\Re$ of a material is based on Ohm's law, which relates the current density J (A/m^2) to the applied electric field strength E (V/m) in the forms

$$J = \frac{E}{\Re} \quad or \quad I = \frac{V}{R} \quad \text{and} \quad R = \Re\,\frac{L}{A} \tag{3.1}$$

R (Ohms) is the measured resistance of the sample over its length L (in the direction of a the electric field) and A is the current carrying cross-section area. The dimensions of $\Re$ are reported as Ohm-meter (Ω-m). A material following Equation 3.1 is said to be Ohmic; whereas, a material following a non-liner power law such as $V \sim I^n$ in the current-voltage characteristic is non-Ohmic. Non-Ohmic behavior has been discussed by Lampert and Mark (1970), Lacharme (1978), and Kingery (1976).

An alternative representation of the volume resistivity $\Re$ is its reciprocal or *electrical conductivity* $\sigma = \Re^{-1}$. For isotropic samples, the material resistivity (or conductivity) is independent of the direction of the applied field. The resistivity depends primarily on the material temperature and is independent of the size of the sample (Weast, 1970).

The measurement of bulk resistivity of a powder includes volume and surface conduction mechanisms. It is generally not possible to separate out the two effects so that the effective powder resistivity, either the volume or surface resistivity, for dielectric and insulating particles such as glass depends on such factors as the presence of surface impurities and the relative humidity. For clean metal powders, the volume resistivity will dominate conduction in a bed of particles; whereas, the presence of a surface oxide film can dominate conduction via the contact resistance for only lightly compacted powders.

When, Equation 3.1 is applied to a packed bed of powder using the standard apparatus in Figure 3-1 an *effective* resistivity will be measured (i.e.

not the material resistivity). For high bulk resistivity powders (> 10^7 Ohm-cm: for example, fly ash (Bickelhaupt, 1975)), the standard code ASME/ANSI (1973) recommends an electronically controlled environment of temperature and relative humidity. Base tests are conducted at 300°K and 5 % relative humidity at potentials 90 % of the breakdown voltage. Both positive and negative electrodes are porous to permit diffusion of humidified gas into the sample and to help increase particle-electrode contact. The outer guard-ring electrode confines the test region to a uniform electric field away from the outer edges of the electrode where strong field effects distort the flow if current.

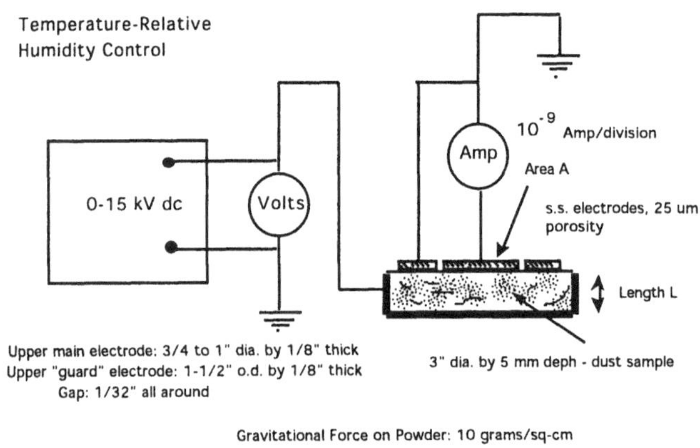

FIGURE 3-1 *Measuring the effective powder resistivity by ASME/ANSI PTC 28, 1973.*

A test standard for measuring the volume and surface resistance of solid samples is ASTM D-257 (1983) in which various guarded electrode configurations are described. The resistivity of the sample can also be determined using unsteady measurement of capacitance and rate-of-change of voltage. The recommended voltage for solid samples in the range 10^{12} to 10^{17} Ω is 500±5 V depending on the circuit. The measurement of volume resistivity above 10^{19} Ω-m is of doubtful validity with commonly used apparatus. A typical commercial unit for volume and surface resistivity measurement can accommodate sample sheet sizes of 64 to 102 mm with thicknesses up to 3 mm using voltages to 1000 V. Volume resistivity measurement up to 10^{18} Ω-m (for samples 0.1 cm thick) and surface resistivity up to 10^{18} Ω/sq with ASTM standards are claimed (Keithley, 1996).

The reproducibility of powder resistivity measurement depends to some extent on the user since compaction can alter the particle contacts. For low resistivity/high current (~100 A) measurements used in metal sintering processes, high compaction pressures to 700 MP reduce the bulk resistance of copper by an order of magnitude (Weissler 1978). Compaction also affects particle stacking and elastic and plastic deformation. For electrostatic powder coating, the electrodes may be submerged in the test powder to simulate more closely the conditions of deposition (Misev 1991; Corbett 1974). Powders having a low bulk resistivity, 10^9-10^{11} Ohm-cm, can be used successfully with electrostatic guns only for small particles (~5 μm) due to charge and particle loss while resistivities > 10^{14} Ohm-cm are desirable for use with larger particles.

Low resistivity measurements of spherical coke and irregular graphite particles for both packed and fluidized states were reported by Graham and Harvey (1965) utilizing two pairs of 0.75 inch graphite electrodes (unguarded) mounted flush with the walls of a 2 inch I.D. column or a pair of graphite electrodes (1 inch exposure) mounted vertically in either of two rectangular columns (1.75 in. x 3 in. and 8 in. x 4 in.).

A fluidized bed utilizing a guard electrode for measuring a high resistivity bulk powder such as glass was used by Colver (1977). This approach has some interesting features including a forced supply of humidified gas to condition the particles. Fluidization also allows for convenient measurement of voidage and for gravity force compaction (unhindered settling) of the bed. A small leakage gas is provided to the packed bed to control ambient conditions (e.g. to exclude oxygen) or for control of ion mobility (e.g. moisture deposition). The *average* bed voidage α_c (fraction of gas volume = 1- fraction of solids volume) is determined by sighting the test level of powder h through scales mounted on either end of the bed with the relation

$$1-\alpha_c=\alpha_d=\frac{M_d}{\rho_d A_b h} \tag{3.2}$$

where M_d is the mass of powder in the bed, ρ_d is the material (solids) density, and A_b is the bed cross-section.

In one study Colver (1980) finds an empirical relation for the effective bulk resistivity of glass powder (3-M Superbrite) to vary with the percent relative humidity (R.H.) and particle diameter d (μm) at room temperature as

$$\Re_b\,(\Omega - cm) = \begin{cases} 325\,(d)^{5.12} \exp\,(-0.188xR.H) & (d \geq 65\ \mu m) \\ 6.22x10^{11} \exp\,(-0.188xR.H) & (d \leq 65\ \mu m) \end{cases} \qquad (3.3)$$

Equations 3.3 show the interesting result that the bed resistivity takes on a *pseudo-continuum* behavior for particles smaller than 65 μm.

3.3.1.2 Measurement error in resistivity

The *maximum* propagated uncertainty in the evaluation of the resistance R(I, V) using the chain rule for differentiation and Ohm's law, Equation 3.1, is

$$\begin{aligned} \Delta R_{max} &= \left| \frac{\partial R}{\partial I} \Delta I \right| + \left| \frac{\partial R}{\partial V} \Delta V \right| \\ &= \left| \frac{R}{I} \Delta I \right| + \left| \frac{R}{V} \Delta V \right| \end{aligned} \qquad (3.4)$$

in which ΔI and ΔV are the experimental uncertainties (assumed independent) in the measurement of the current and voltage respectively (e.g., uncertainties from precision or unknown bias errors such as instrument resolution, variability from calibration due to extraneous drift, intrinsic error in the instrument calibration source). Equation 3.4 is reported in ASTM D-257 (1983) and by Northrop 1997 as the *limiting instrument error.* If errors from the two variables current and voltage are independent and assumed to cancel as a result of multiple readings taken over many samples, then the propagated uncertainty (error) will be reduced in value as (Kennedy and Neville, 1976)

$$\begin{aligned} \Delta R_{rms} &= \sqrt{\left(\frac{\partial R}{\partial I} \Delta I \right)^2 + \left(\frac{\partial R}{\partial V} \Delta V \right)^2} \\ &= \sqrt{\left(\frac{R}{I} \Delta I \right)^2 + \left(\frac{R}{V} \Delta V \right)^2} \end{aligned} \qquad (3.5)$$

For example, if the overall percentage uncertainty in current due to readability and indicated error is ±5 % and that due to voltage is ±3 %, then ΔR_{max}=±0.08 R and ΔR_{rms}=±0.058 R, the latter being smaller in magnitude.

3.3.1.3 Surface resistivity

Another useful resistivity measurement for solids and powders is the *surface resistivity* (its reciprocal is *surface conductivity*). Surface resistivity for large specimens is measured either directly by guarded surface contact electrodes or indirectly by transient *RC* response with typical source voltages of 200 to 1 kV (Takahashi 1995, Curtis 1915). The current is usually assumed to be distributed entirely over the surface of the sample by a conductive film such as adsorbed water molecules. An explanation for the surface conduction in silicate glass is that alkali metal ions react with adsorbed water by the process of ion exchange forming metallic hydroxides such as NaOH which in turn reacts with water to form mobile ions on the surface (Doremus, 1973). This process accounts in part for the dissolution of glass in water. The surface resistivity as an electrical property has no intrinsic relationship to the volume resistivity. The measurement of surface resistivity of a solid sample is discussed in ASTM D-257 (1983). Various electrode setups are mentioned along with their circuits for different sample geometries (rectangular and cylindrical). Commercial meters designed for the measurement of surface resistivity are also available (Keithley, 1996; Monore, 1997; Trek, 1998).

The corresponding Ohmic relationship to that of Equation 3.1 relating the surface current density J_s (A/m), electric field strength E (V/m), and the surface resistivity γ is,

$$J_s = \frac{E}{\gamma} \qquad and \qquad R_s = \gamma \frac{L}{W} = \gamma \qquad (L = W) \tag{3.6}$$

in which R_s is the measured surface resistance of the sample over surface length L (in the direction of the electric field) and W is the surface width perpendicular to the flow of the current. Since the surface resistivity is independent of the dimensions of L and W, no loss of generality results if one takes L=W (i.e. a surface square) in reporting γ. The dimensions of γ are Ohms/square (Ω/sq). The reciprocal unit in SI is the Siemen-sq (S-sq).

The surface resistivity of clean glass in air can be very high, of the order 10^{14} Ohm/square or larger (Morey, 1954). In the case of glass it is possible to increase the surface resistivity with a water repellent which serves to prevent the formation of a continuous layer of water (Holland, 1966). With powders this a common practice using silicone compounds.

In contrast to surface resistivity, the volume resistivity of glass and ceramics is controlled largely by its composition. For example, the conductivity of sodium silicate glass increases in direct proportion to the sodium ion concentration (Kingery et. al., 1976). Electronic conduction is also possible in some glasses. Special electrodes may be required for *dc* measurements taken over extended periods of time with alkali-containing materials such as some glasses to replenish ions being stored or removed at the source electrode, thereby producing electrode polarization. This problem can be circumvented by incorporating an *ac* measurement. A similar polarization problem occurs in the application *dc* resistance probes to electrolytes.

3.3.1.4 Packed bed models of resistivity for conduction probes

A model for the bulk effective resistivity $\Re_b$ of a dilute suspension (disperse phase) of noninteracting conducting spheres (not necessarily mono-dispersed) of material resistivity $\Re_d$ and void fraction α_d suspended in a continuous medium of material resistivity $\Re_c$ was derived by Maxwell (1954). His result is

$$\Re_b = \Re_c \frac{2\,\Re_d + \Re_c + \alpha_d(\Re_d - \Re_c)}{2\,\Re_d + \Re_c - 2\alpha_d(\Re_d - \Re_c)} \qquad \alpha_d << 1 \tag{3.7}$$

which satisfies both the lower and upper voidage limits α_d at 0 and 1 respectively. It is implied that that the particles are suspended uniformly and are stationary, or if moving, they transport no net charge. A packing constraint applies to the upper limit of the void fraction such as $\alpha_d = \pi/6$ for a cubic array.

Holm (1967) identifies the *contact* resistance between particles of clean metal to be the result of current constriction at the point of contact. This "geometric constriction" together with the volume and surface resistivities integrated over the remaining volume and surface of a particle constitute the total resistance measured between two contacts located at the poles of the particle. In addition, if a thin film exists between the particle contacts, the tunnel effect provides a current independent of the film resistivity.

By integrating the uniform surface resistivity γ over the surface of a spherical particle, Johnson and Melcher (1975) give the total resistance of a single particle of radius *r* with the small contact "cap" radius *a* at opposite poles of the particle through which the current enters and leaves as

$$R_{sur}(\Omega) = \frac{\gamma}{\pi}\ln\left(\frac{2r}{a}\right) \qquad \frac{a}{r} << 1 \quad (\textit{single particle})$$

$$\Re_{sur}(\Omega - m) = \frac{2r\gamma}{\pi}\ln\left(\frac{2r}{a}\right) \qquad \frac{a}{r} << 1 \quad (\textit{cubic array}) \tag{3.8}$$

The second equation is the bulk effective resistivity due to particle surface resistivity for a cubic array of mono-dispersed particles with the direction of the electric field aligned with the poles and volume conduction neglected. The constriction resistance is included in the integration of Equations 3.8. These equations are a weak function of the particle geometry.

Holm (1967) gives the resistance for the volume resistivity of a single particle of radius r measured between opposite "pole caps" of diameter a and uniform material resistivity $\mathfrak{R}$ as [1]

$$R_{vol}(\Omega) = \frac{\mathfrak{R}}{\pi}\left(\frac{1}{a} - \frac{1}{2r}\right) \approx \frac{\mathfrak{R}}{\pi a} \qquad \frac{a}{r} << 1 \quad \textit{(single particle)} \qquad (3.9)$$

Using the cubic model, the bulk effective resistivity due to volume resistivity of the particle is

$$\mathfrak{R}_{vol}(\Omega\text{–}m) = \frac{2\mathfrak{R}r}{\pi}\left(\frac{1}{a} - \frac{1}{2r}\right) \approx \frac{2\mathfrak{R}r}{\pi a} \qquad \frac{a}{r} << 1 \quad \textit{(cubic array)} \qquad (3.10)$$

We can give an approximate relationship for the bulk bed resistance including the effects of surface and volume resistivities of the particles for the case of negligible continuum conductivity $\mathfrak{R}_d << \mathfrak{R}_c$ by employing Equations 3.8 (lower) and 3.10 in a parallel circuit as

$$\mathfrak{R}_b(\Omega - m) = \frac{\frac{2r\gamma}{\pi}\ln\left(\frac{2r}{a}\right)}{\frac{a\gamma}{\mathfrak{R}}\ln\left(\frac{2r}{a}\right) + 1} \qquad \frac{a}{r} << 1 \quad \left(\textit{cubic array}, \alpha_d = \frac{\pi}{6}\right) \qquad (3.11)$$

Equation 3.11 can be used as a first estimate for the bulk resistivity of a packed cubic array of spherical particles. Equation 3.11 also indicates that the surface resistance of a particle is controlling over the conduction process for the condition that

---

[1] The volume resistance of a sphere of uniform resistivity has also been considered by Smythe (1968).

$$\frac{a\gamma}{\Re}\ln\left(\frac{2r}{a}\right) << 1 \quad \textit{(surface dominated conduction, } a/r<<1) \tag{3.12}$$

Moslehi's (1983) result, accounting for surface-volume field gives

$$\frac{r\gamma}{2\Re} << 1 \quad \textit{(surface dominated conduction, finite } a) \tag{3.13}$$

Moslehi and Self (1983) and Moslehi (1983) solved the combined problem of surface and volume resistances in terms of Legendre polynomials. They show that the two resistances combine in parallel.

A difficulty in applying Equation 3.11 is the lack of detailed information for the particle contact "cap" size *a* which in turn results from elasticity and the local deformation of the contact points. Additionally, there is the likely possibility of multiple contacts at the cap in a bed of particles and that the positions of the caps are not aligned as poles in the direction of the field. The most frequently utilized contact theory to handle the local deformation for interparticle forces is the Hertz problem (Timoshenko 1951) combined with the assumption of a cut-off or breakdown voltage across points of contact (e.g. Mclean, 1977, Dietz and Melcher, 1978A). This cut-off is taken as the limiting breakdown field strength near the point of the contact (usually 3×10^{6} V/m) so that a Schottky type of partial breakdown is expected. Ikazaki and Kamamura (1984) also considered both surface and volume conduction in their conduction/force model of spherical particles again adding the two effects in parallel. The volume resistance of a sphere of uniform resistivity has also been considered by Smythe (1968).

3.3.1.5 Packed bed models of permittivity for capacitance probes

The experimental setups for measuring powder resistivity in Figures 3-1 can also be used to measure the *effective* dielectric constant of the bulk powder. The procedure is to measure the capacitance with (′) and without the powder. Using the definition of capacitance $C = \varepsilon A/x$ for parallel plates of area A and separation distance x gives

$$\frac{C'}{C} = \frac{\varepsilon_b}{\varepsilon_c} = \frac{\kappa_b}{\kappa_c} \tag{3.14}$$

where the bulk permittivity and bulk dielectric constant are ε_b and κ_c respectively, and the continuous phase permittivity is ε_c with dielectric constant κ_c.

Using the analogy between electrical conductivity and permittivity (Smythe 1968), Maxwell's result Equation 3.7 can be transformed into an *effective* dielectric constant κ_b for a dilute suspension of spheres having dielectric constant κ_d (permittivity ε_d) suspended in a continuous medium of dielectric constant κ_c (permittivity ε_c)

$$\kappa_b = \kappa_c \frac{\kappa_d + 2\kappa_c + 2\alpha_d(\kappa_d - \kappa_c)}{\kappa_d + 2\kappa_c - \alpha_d(\kappa_d - \kappa_c)} \qquad \alpha_d << 1 \qquad (3.15)$$

The same restrictions apply to Equation 3.15 as Equation 3.7.

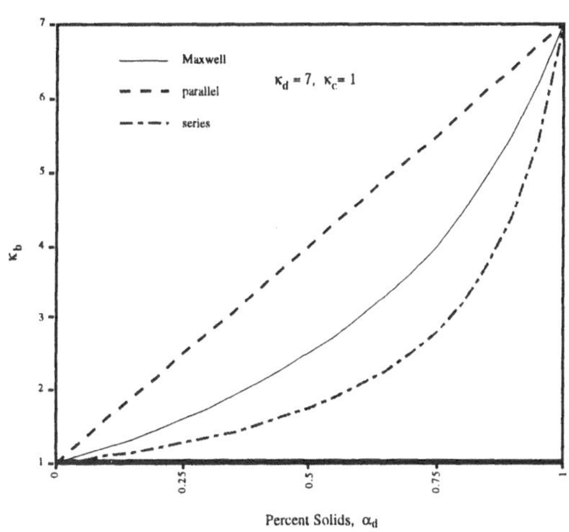

FIGURE 3-2 *Effective bulk dielectric constant for different models, κ_d = 7.0, κ_c=1.*

In considering capacitance probes, various models for the dielectric constant of a packed bed were reviewed by Jones (1979) and for fluidized beds Louge (1997). Jones includes the effects of frequency, voidage, lossy dielectrics, and conductive disperse (particles) and continuous phases. For example, Equation 3.15 represents an intermediate case between the parallel and series arrangements of two capacitors each with it own dielectric material.

$$\kappa_{parall} = \alpha_d \kappa_d + \alpha_c \kappa_c \qquad \frac{1}{\kappa_{series}} = \frac{\alpha_d}{\kappa_d} + \frac{\alpha_c}{\kappa_c} \qquad (3.16)$$

series *parallel*

where $\alpha_d + \alpha_c = 1$. Figure 3-2 shows Equations 3.15-3.16 with $\kappa_d = 7.0$ (glass) and $\kappa_c = 1$ (air). Jones (1979) extends the dielectric model of Equation 3.15 for capacitance probes to include Ohmic conduction of the particulate [2] and continuum phases by replacing the permittivity of each phase by their complex quantities.

3.3.1.6 Measuring effective dielectric constant (permittivity) of a powder

An experimental setup for measuring the *effective* dielectric constant κ_b (or permittivity ε_b) of a bulk powder is shown in Figure 3-3 following Masuda et. al, 1995. The capacitance meter is a Wheatstone bridge driven by an oscillator having either two or four legs with the unknown capacitance being one of the legs. Measurements are taken with and without the test powder at the same electrode spacing. The setup is similar to that used to measure the effective bulk powder resistivity (Figures 3-1). To avoid edge effects a guard ring is employed. The electrode spacing used by Masuda et. al. was 3 mm. Equation 3.14 is used to determine the unknown dielectric constant or permittivity of the bulk powder. Masuda et. al. found the Maxwell result of Equation 3.15 gave good agreement for their powders.

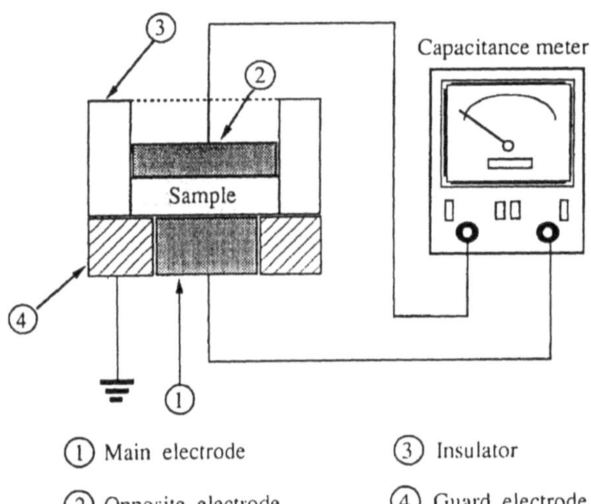

FIGURE 3-3 *Experimental setup for measuring the effective dielectric constituent of a powder (Masuda et. al. 1995).*

[2] Particles remain noncontacting, i.e. restricted to volume conduction mechanism only.

3.3.2 Measurement of Charge

3.3.2.1 Electrostatic charge, its origin and magnitude

A review of charging mechanisms is given by Lapple 1970. Common sources for charge production in multiphase systems includes triboelectric (frictional) charging and charge separation relating to applied electric fields including induction and field charging of surfaces by ions. At high temperatures thermionic emission can become important (Soo, 1967). The combined effect of triboelectric charging Q_{el} and induction charging of small cubes of ebonite and sand sliding down an inclined grounded metal plate in the presence of an applied electric field E_o was given by Gill and Alfrey (1949) according to

$$Q = \pm K_1 E_o + Q_{el} \tag{3.17}$$

in which K_1 is a constant. The appropriate sign for each term in Equation 3.17 must be considered. Equation 3.17 shows that an electric field can alter or even reverse the sign of triboelectrically charged materials. Figure 3-4 shows example charge magnitudes using equations taken from the literature and assumed physical properties. Some representative charge magnitudes encountered in various industrial processes are listed in Table 3.1.

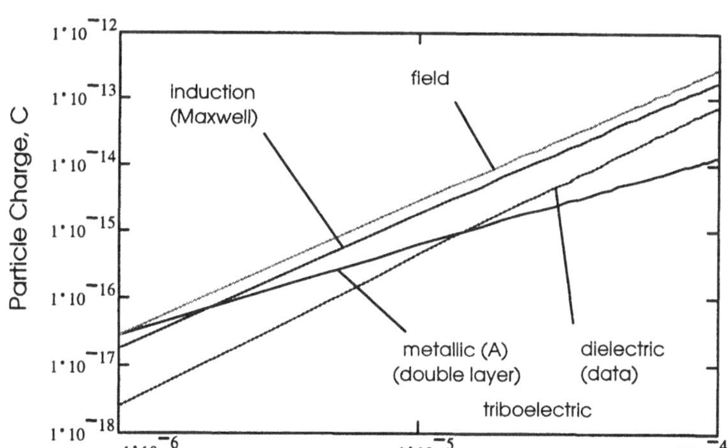

FIGURE 3-4 *Particle charge calculations; dielectric charge from Figure 3-6, (upper charge limit).*

TABLE 3.1 Charge-to-mass ratio of organic powders for industrial operations (Glor, 1988; Hassler, 1978; Tucholski and Colver, 1993A, Greaves and Makin, 1980)

Operation	Charge/mass ratio [C/kg]
Sieving	$10^{-11} \rightarrow 10^{-9}$
Pouring	$10^{-9} \rightarrow 10^{-7}$
Scroll Feed Transfer	$10^{-8} \rightarrow 10^{-6}$
Grinding	$10^{-7} \rightarrow 10^{-6}$
Micronizing	$10^{-7} \rightarrow 10^{-4}$
Pneumatic Transfer	$10^{-6} \rightarrow 10^{-4}$
Fluidized Bed (bubbling)	$10^{-6} \rightarrow 10^{-5}$
Fluidized Bed (freeboard)	$10^{-6} \rightarrow 10^{-5}$
Aerosol Spray Can	$10^{-5} \rightarrow 10^{-4}$
Fog: Supersonic Nozzle	$10^{-4} \rightarrow 10^{-3}$

3.3.2.2 Contact and zeta potentials of particles

Charge transfer is related to the *contact potential* developed between unlike materials. The difference in *work function* between the two materials gives an indication of the tendency for spontaneous exchange of electrons or ions which can lead to charging of the materials following separation. To measure the contact or surface potential of a powder requires a special setup and is more difficult than the corresponding measurement for a solid since the effective permittivity of the packed powder is required. A typical experiment is shown in Figure 3-5 following Masuda (1995). The upper gold electrode is oscillated over a distance of 2.5 mm at 1.2 Hz. The measured effective contact potential difference V_c (~ 1 V) of the powder depends on the contact potential $V_{c/Au}$ relative to the gold electrode and is corrected for the presence of any charge ρ_q distributed through the powder according to the equation

$$V_c = V_{c/Au} + \frac{\rho_q}{2\varepsilon} h^2 \qquad (3.18)$$

where V_c is the measured voltage to null the circuit current.

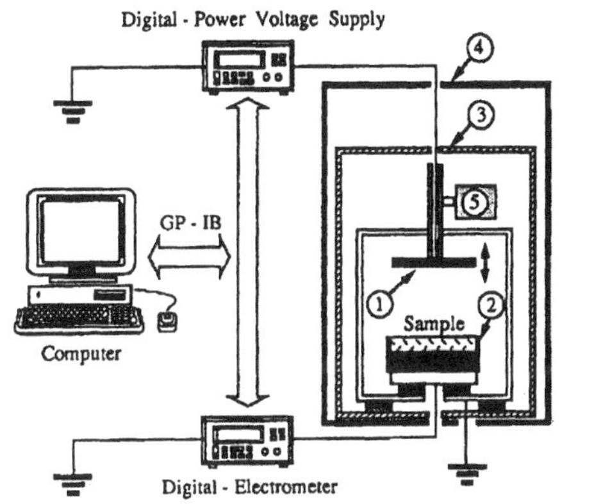

FIGURE 3-5 *Experiments setup for measuring contact potential difference (Masuda et. al. 1995).*

3.3.2.3 Triboelectric charging

Triboelectric or frictional charging of contacting materials is the result of differences in surface potentials or Fermi level (Lapple 1970) as well as the physical nature of the contact, which helps explain the wide variations of experimental charge reported in Table 3.1. Glor (1988) emphasized that a particular operation can readily dominate the charging process. Castle and Schein (1995) give a macroscopic based model of charging for two contacting insulating spheres (toner and carrier particles) to explain the observed linear relationship between the toner mass-to-charge ratio and the ratio of toner mass to carrier mass. Their model satisfies both the low and high surface state density limits. Environmental effects such monolayers of adsorbed gases can have a significant effect on charging between surfaces. For example, Debeau (1944) demonstrated that adsorbed air, oxygen, and nitrogen on nickel first reduced the charge on quartz and sodium chloride particles for pressures from 760 to 30 mm Hg and then increased charge below 0.1 mm Hg. He also found that the presence of water vapor and hydrogen reduced charge for these materials. Charge magnitude can be associated with the following variables (Montgomery, 1959): *material properties* (chemical composition, crystallinity, strain, size and shape of object); *ambient conditions* (temperature, pressure of atmosphere, gas adsorbtion, electromagnetic fields); *mechanical variables* (type of contact, rubbing, impacting, rolling , duration, force, relative velocity).

The driving force for charge transfer in conductors is the Fermi level difference between two materials. However, other factors will contribute to the charge transfer and charge reflux (back discharge on separation of materials). We see that conditions conducive to charging in multiphase systems include Fermi level difference, high speed flow, large forces, and low humidity. One rule for significant charging is that at least one of the materials should have a bulk resistivity > 10^7 Ω-m.

3.3.2.4 The triboelectric series

The tendency for insulating and conducting materials to charge positively relative to one another is the basis for the "triboelectric series." Sources of triboelectric tables of materials with references can be found in Schein (1992), Cross (1987), Lapple (1970) and Montgomery (1959). It is not surprising to find differences in the ranking of the same polymer material from different sources because of the sensitivity of the measurements to chemical differences in the material and possible variations in testing. For this reason, knowledge of the manufacturer of a particular material may be relevant in placing an item into the series. A typical triboelectric series is shown in Table 3.2.

TABLE 3.2 An example triboelectric series

Material & Relative Charge			
Quartz	**(positive)**		
Epoxy resins		Nickel, Copper	
Glass		Brass, Silver	
Mica		Cellulose acetate	
Nylon		Polystyrene	
Wool		Silicone	
Aluminum		PVC (Vinyl)	
Paper		Polypropylene	
Steel		Polyethylene	
Hard Rubber		Teflon	**(negative)**

3.3.2.5 Charge relaxation in a powder

Haenen (1975) found that when the decay of surface charge was not exponential, the time constant was not a good measure of the actual decay time. In the case of a bed of particles, charge transport can take place between particles when one particle contacts another in the presence of an electric (self) field. For an isolated spherical particle of uniform surface resistivity γ placed in an electric field, the relaxation time constant τ for adjustment of charge over the surface of the particle was given by as Johnson and Melcher (1975)

$$\tau = \frac{\gamma(\kappa_p + 2)\varepsilon_o d}{4} \tag{3.19}$$

They calculate a range of τ values ($10^{-3} \rightarrow 100$ s) corresponding to increasing relative humidity $0 \rightarrow 100$ % with sand particles (d=0.5 mm, κ_p=3.3) and assumed values of the contact cap radius (r/a= 10 and 10^4).

An alternative expression for the charge relaxation time can be approximated for a multiphase system by an extension of continuum theory of solids. Applying conservation of charge $\nabla \cdot \boldsymbol{J} + \partial \rho_q / \partial t = 0$, the equation of Poisson $\nabla \cdot \boldsymbol{E} = \rho_q / \varepsilon$ and Ohm's law $\boldsymbol{J} = \boldsymbol{E} / \Re$ and solving for the charge density $\rho_q(t)$ in terms of the initial charge ρ_{qo} leads to

$$\rho_q(t) = \rho_{qo} \exp\left[-\frac{t}{\tau}\right], \qquad \tau = \Re \varepsilon \tag{3.20}$$

Masuda et. al. (1995) employ this equation to infer the time constant for a packed beds of polymer powder.

Equation 3-20 enjoys general validity for charge decay without regard to charge conditions at the boundaries of the medium. The *continuum* assumption for a multiphase system implies that the three defining equations (conservation of charge, Poisson's, Ohm's law) remain valid over length scales for which the variation in the quantities of electric field strength, current density, bulk resistivity, and charge density are large in comparison to the size of a particle and the characteristic voidage dimension.

It is natural to extend Equation 3-20 to an isotropic-homogenous multiphase system using an appropriate mixture resistivity and permittivity such as Equations 3.7 and 3.15 as

$$\Re \rightarrow \Re_b, \ \varepsilon \rightarrow \varepsilon_b \left(= \kappa_b \varepsilon_o\right), \ \tau \rightarrow \Re_b \, \kappa_b \varepsilon_o \tag{3.21}$$

Colver (1980) observed experimentally that the bulk resistivity of a packed bed of spheres less than about 65 μm takes on a *pseudocontinuum* behavior.

3.3.2.6 Preparation of powders for charge measurement and storage

As noted above, triboelectric charge depends on many factors including the surface conditions, crystallinity, material, and shape. When making charge measurements, ridding the surface of impurities is important. For example, Kunkel (1950) conditioned the surface of amorphous and crystalline quartz by washing the powder in dilute hydrochloric acid and then rinsing in distilled water. Next, each powder was dried at 120 °C for several hours before being stored in a closed glass jar. Even after the treatment, Kunkel found that the quartz had to be baked out at 500 °C to rid it of all surface water (one or two molecular layers). Clearly, powders that oxidize such as sulfur or carbon cannot successfully be baked out in air.

Bulk storage of powder materials containing charge has been discussed by Lapple, (1970). He points out that charge can be lost by charge relaxation and self corona discharge if the material charge is sufficiently high and the packing is dense (i.e., a powder in a container). Unipolar (monopolar or containing charge of one sign) and bipolar (containing charge of opposite signs) charge on powders are susceptible to charge degradation over time as a result of charge migration to ground and recombination of bipolar charge respectively. Another source of storage charge loss is neutralization of charge by free atmospheric ions. All of this simply means that storage and compaction of powders should be avoided prior to a charge measurement.

3.3.2.7 Charge measurement of powders

It is common practice to report triboelectric charge on a charge-to-mass basis when using the Faraday cage method since *net* charge and mass are easily measured. However, triboelectric and induction charging are more closely related to the surface area of a particle (e.g. Castle and Schein, 1995) rather than to volume (mass) so that the particle diameter should also be reported along with the charge if possible. Even so, it is unlikely during the processing that irregular or even spherical particles will receive uniform surface charge density so that some disparity will remain when relating charge to the particle's *total* surface area.

Figure 3-6 illustrates the disparity based on the assumption that the surface charge is uniformly distributed over the surface of a particle for a number of organic powder data charged in a laboratory scale pneumatic transport experiment (Glor, 1988). A disparity would also be expected if the charge were replotted based on the charge being uniformly distributed throughout the mass of the particle. Glor calculated an *ideal* conversion from uniform surface-charge to charge -to-mass ratio (dashed line in Figure 3-6) assuming a uniform surface charge of magnitude q_s (=2.7x10^{-5} C/m^2) large enough to cover his data

$$q_v = q_s\left(\frac{6}{\rho_d d}\right)$$
$$= q_s S \qquad (3.22)$$

where q_v (C/m^3) is the charge to mass ratio, ρ_d (=10^3 kg/m^3) is the material density, d is the particle diameter and S is the specific surface area $=6/(\rho_d\, d)$. The dashed line shows that the assumed surface charge q_s is too large to represent the actual data and that the (logarithmic) slope =1 is too large (a better slope ≈ 0.75). This means that the ideal assumption of constant surface charge is not valid over the range of particle diameters tested.

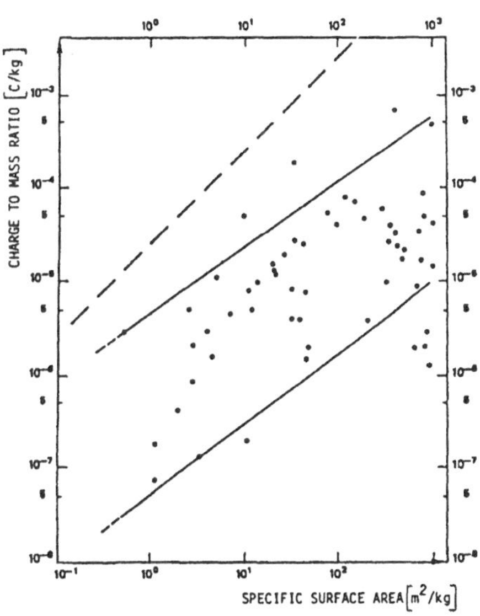

FIGURE 3-6 *Charge-to-mass ratio versus specific surface area for organic powders (Glor 1988) - dashed line is the relationship for uniform surface charge.*

3.3.2.8 "Closed" Faraday cage for charge measurement

Powder charge data are often reported on a mass basis (Coulomb/kg) by the Faraday cage (or cup) method. A sample of aerosol or powder is measured for its charge and then weighed (Haus and Melcher, 1989). The appeal of the Faraday cage method is its simplicity and accuracy (the *net* charge being automatically integrated). When reasonable care is exercised in handling samples, the Faraday cage method can provide accurate and reproducible results for powders as well as other charged materials such a fibers or webbing. The collection cage may be no more than a conducting container insulated from ground by a piece of Teflon and placed inside a larger container which in turn is grounded in order to shield the cup from the surroundings and extraneous capacitances such as the experimenter. A typical setup is illustrated in Figure 3-7.

As noted above, the Faraday cage gives a direct measure of the *net* charge on the sample placed in the cage. The cage reading is based on Gauss's law applied to an isolated conducting enclosure so that any *net* charge (sum of positive and negative charges) within the cage container will exactly induce an equal magnitude of *net* charge on the (outside) walls of the container, which can then be detected as a voltage or charge reading using a high input impedance measuring device such as a commercial electrometer (the electrometer is configured to indicate charge directly, Figure 3-7). The high input impedance of the electrometer (~10^{14} Ohms) ensures that any charge lost to ground through the electrometer is negligible.

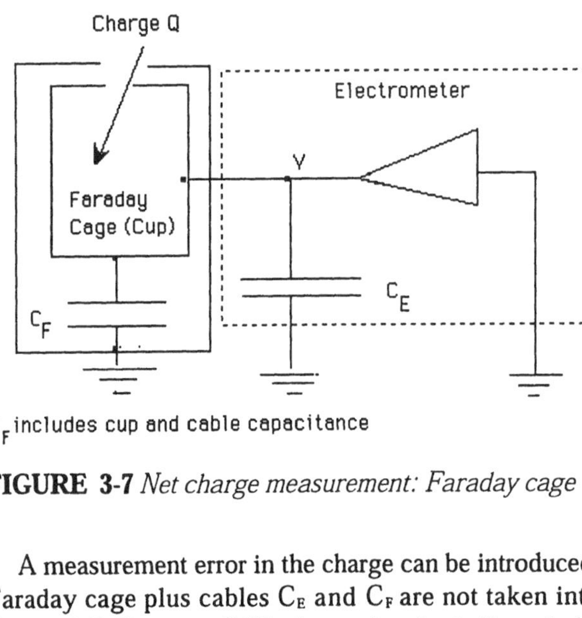

$$Q_{act} = VC_E\left[1 + \frac{C_F}{C_E}\right] = Q_E\left[1 + \frac{C_F}{C_E}\right] \qquad (3.23)$$

FIGURE 3-7 *Net charge measurement: Faraday cage (cup) with electrometer.*

A measurement error in the charge can be introduced if the capacitance of the Faraday cage plus cables C_E and C_F are not taken into account. Referring to Figure 3-7, Equation 3.23 shows that the indicated charge on the electrometer Q_E must be corrected to the actual charge Q_E to account for the low voltage reading V from the distributed charge on capacitors C_E and C_F. A second charge measurement error results if a significant fraction of the electric flux lines from the measured charge do not terminate on the cage as a result of an opening (e.g. the open cage method). Such an error can be corrected theoretically or detected experimentally.

As with any charge measuring system care should be exercised so that extraneous charge is neither gained nor lost during handling of the sample as in "spooning" the test powder. Because of the likelihood of extraneous contact, it is desirable to sample a multiphase flow system directly through a probe that is itself a part of the Faraday cage. An example of one such Faraday cage designed for easy disassembly and weighing of the powder is shown in Figure 3-8 (Left) (Tucholski and Colver, 1993). A second design used for sampling fly ash in a precipitator experiments is shown in Figure 3-8 (Right) (Kobashi, 1978). The double insulation design (inner and outer tube) is for the purpose of reducing signal to noise ratio.

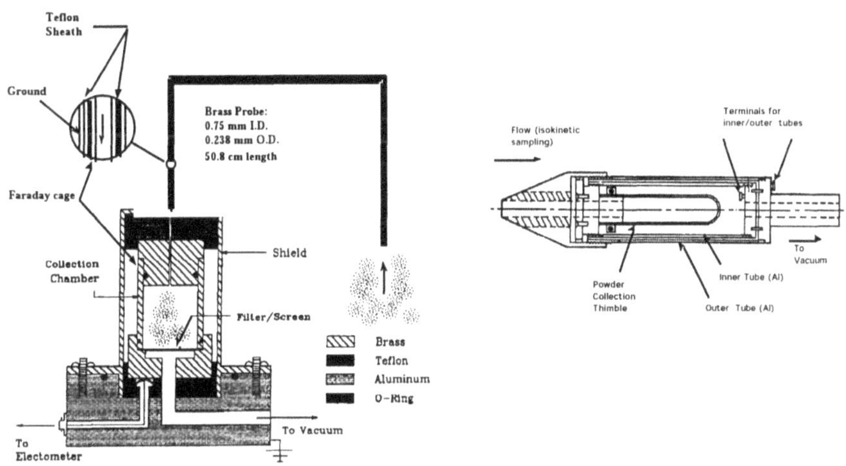

FIGURE 3-8 *Left: powder flow sampling Faraday cage (Tucholski and Colver 1993); Right: Isokinetic sampling Faraday cage (Kobashi 1978)*

A double shielded Faraday cage was also used by Tardos et. al., (1984) for sampling the continuous flow of monodispersed aerosol of 1.049 μm polystyrene latex particles (std dev = 0.0587 μm). The total particles N_p were captured on a filter over time t at currents of 10^{-13}A. The double shielding was required for these small current.

A variation of the Faraday cage method utilizing several grids was used by Greaves and Makin (1980) to capture an aerosol spray. The collection system consisted of a series of 1 to 5 parallel grids (4 proved satisfactory) measuring 30 x 30 cm^2 enclosed inside and contacting a five-sided metal cage. One side of the cage was open to the spray and the opposite side perforated for passage of the flow (the grids were situated perpendicular to the flow). Each steel grid contained multiple 2.5 mm holes on 4 mm centers, which projected a 35 %

opening. The complete box assembly was placed inside a grounded metal duct that was exhausted by a fan. The aerosol to be measured was sprayed downstream and collected by the grids. The charge on the grids system was measured directly either by an electrometer set to the charge mode or else to the voltmeter mode (peak volts were about 3 V) as an electrostatic voltmeter with readings taken across a 0.047 μF capacitor connecting the grid box and ground. They measured maximum aerosol charge of 3×10^{-5} C/kg or typically (1.8→4.7) $\times10^{-7}$ C over about 5 seconds.

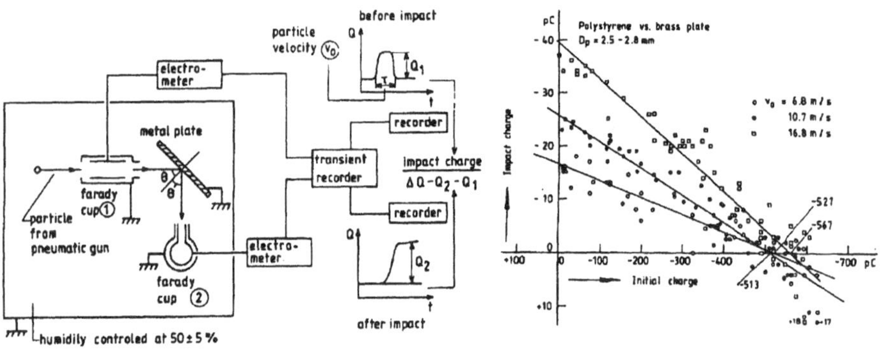

FIGURE 3-9 *Charge acquired by 2.5 mm polystyrene particles impacting a brass plate at different velocities (Yamamoto and Scarlett 1986).*

3.3.2.9 "Open" Faraday cage & ring probe methods

A variation of the closed Faraday cage is the open Faraday cage used by Yamamoto and Scarlett (1986) for measuring the initial charge and charge transfer to large moving particles before impact with a metal plate, Figure 3-9 (Left). Typical results are shown in Figure 3-9 (Right) for 2.4-2.8 mm polystyrene particles striking a brass plate at different velocities. Matsuyama and Yamamoto (1995) extended the method to successive collisions between a Teflon sphere and two brass plates. They find a relationship for the charge transferred ΔQ to a 3.2 mm Teflon sphere for a single impact having an approach angle of 60° and particle velocity of 11.4 m/s based on the initial charge as

$$\Delta Q(pC) = -0.058\, Q_{initial}(pC) - 72 \qquad (3.24)$$

If the open cage is replaced by a simple conducting ring, then the flux leakage from a charged particle will be large and a theoretical or experimental correction must be made that relates the peak voltage of the signal to the particle charge. The theory for a point charge Q passing along the center of a ring at velocity u as detected by voltage on an oscilloscope of resistance R was developed by Gajewski (1984). He gives the equation for peak voltage V_{max} and charge Q in terms of ring capacitance C_r, oscilloscope capacitance C, and ring diameter D as

$$V_{max} = \pm \frac{2\sqrt{3}}{9} \frac{C_r C}{C_r + C} \frac{R}{\pi \varepsilon_o D^2} Qu$$

$$= \pm cQu \tag{3.25}$$

where c is a calibration constant.

An improvement to reduce external electrical noise is to employ two concentric rings and a voltage follower circuit connected to the outer ring that acts as a guard ring (Vercoulen et. al., 1992). With the two ring system, one before and one after particle impact with a metal plate, Vercoulen (1995) measured the triboelectric charge transfer of 2-4 mm coated glass particles of nylon, silane, alginate, and gold coatings in atmospheres of air nitrogen argon and helium. He found charging in the range of 1 to 60 pC per particle contact.

Ally and Klinzing (1985) inferred the charge transfer from particle impacts using nickel electrodes amplified with an electrometer in a vertical pipe flow (0.0254 m diameter, copper, Plexiglas, and glass). The particles were copper, Plexiglas, glass beads, and crushed glass. They recorded the voltage output in time to infer particle contact time and charge transfer at 5 locations along the pipe. They measured charge to mass ratios in the range 5×10^{-6} to 0.022 C/kg depending on the particle and pipe materials and the relative humidity (increasing humidity decreased charge due to increasing particle conductivy). They noted increased pressure drop in the flow due to the electrostatic effects in the pipe observing that glass-copper combinations had the greatest effect.

3.3.2.10 Charge measurement by particle mobility (electrostatic precipitation)

Monodispersed polystyrene latex particles 1.049 μm in diameter (std dev = 0.0587 μm) were captured utilizing a radial flow parallel-plate mobility analyzer (Tardos et. al. 1984). The mobility of the particles was determined from measurements of the collection efficiency of the analyzer by sampling particle number density for the inlet and exit flows (Figure 3-10). The principle was fundamentally that of electrostatic precipitation. The particles were charged by a corona discharge. The particles capture efficiency in the mobility analyzer was varied with the applied voltage (0→4 kV) from which the mobility distribution was back calculated. A knowledge of the mobility K_p gives the particle charge Q from the relation (K= 0.7 →1 for the Cunningham correction, Brodkey 1967, d<$3\,\mu$m)

$$Q = \frac{3\pi\mu d K_p}{1 + \frac{2K\lambda}{d}} \tag{3.26}$$

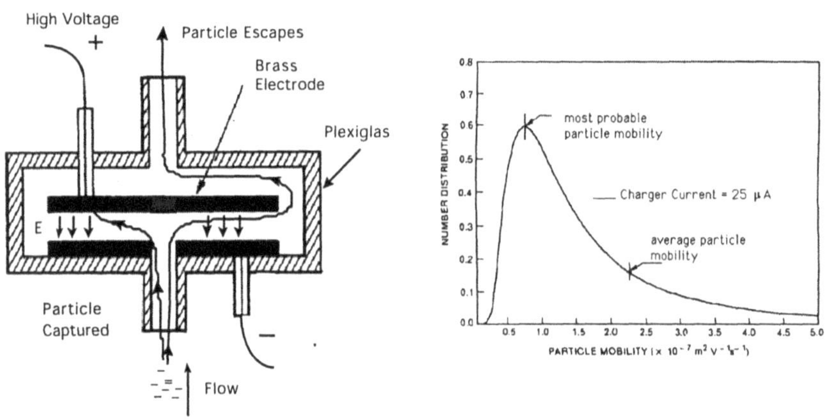

FIGURE 3-10 *Particle mobility analyzer and mobility distribution (Tardos et. al. 1984).*

The number average particle charge was found to be in the range (2.84-3.22)x10^{-17} C (peaking at 3.41x10^{-17} C) as the corona charging current was varied form 10 to 34 μA. A charge distribution over the monodispersed particles was observed due to effects related to corona charging. From their mobility distribution, the authors find a most probable particle mobility of about 0.7x10^{-7} $m^2V^{-1}s^{-1}$ and an average mobility of 2.25x10^{-7} $m^2V^{-1}s^{-1}$ for a corona current of 25 μA (Figure 3-10, right).

A unique self-contained probe for measuring bipolar (positive and negative) charge was designed for in-situ measurement of fly ash by Self et. al. 1979 based on particle mobility (electrostatic precipitation). Fly ash particles were sampled in the range 2-10 μm (duct mass loading of 2-10 g/m^3). A cyclone section could be added to the tip of the probe to remove large particles (>20 μm). Bipolar charge-to-mass ratios of ± (10-50) μC/g were reported.

A simple method for determining the average bipolar charge in a flowing suspension was described by Kobashi (1978) in which precipitated fly ash particles of one sign are first collected on one plate (ground side) of a parallel plate capacitor and then the particles of the other sign collected on the same plate by reversing the sign of the power supply. The particles are cleaned from the plate and weighed, and the experiment is rerun with the power supply reversed (the ground side is the same). An electrometer on the ground side plate measures the total powder charge of each sign. This method gives an average charge-to-mass ratio for each sign in the particle distribution.

3.3.2.11 Faraday cage method applied to fluidized beds and suspensions.

A clever method for measuring particle charge in a fluidized bed was utilized by Tardos and Pfeffer 1980 in which they sampled 2 mm porcelain particles out the bottom of a tapered bed. They also measured charging in the bed with a ball probe. A decrease in charge was apparent over ranges of increasing relative humidities 21 % to 42 %. An interesting effect observed was a peak in the charge with increasing superficial velocity that was attributed to charged fines leaving the bed. They present their charge data in terms particle surface area.

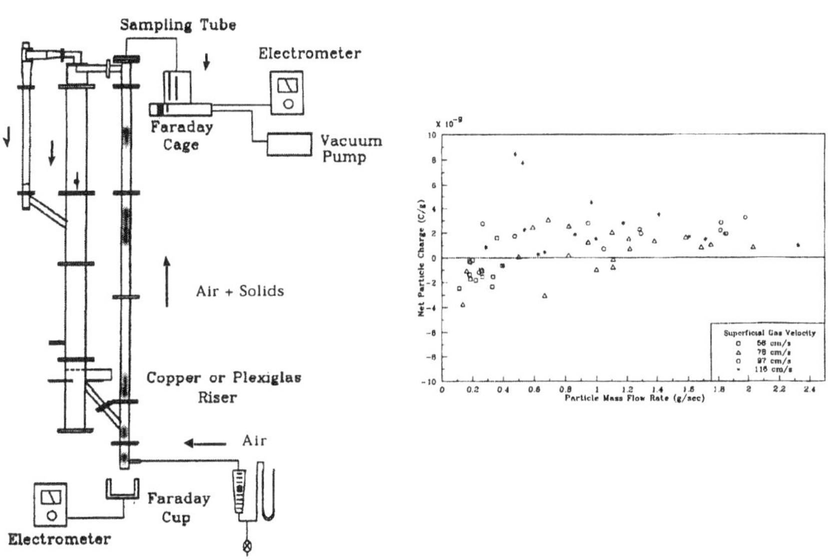

FIGURE 3-11 *Circulating fluidized bed (Tucholski and Colver 1993).*

Figure 3-11 shows the experimental setup for measuring particle charge in the freeboard of a circulating fluidized bed fabricated from either copper or

Plexiglas tube of 2.54 cm I.D. (Tucholski and Colver, 1993A). The particles were 44-75 μm glass spheres. The *net* charge on the particles was evaluated with the Faraday cage shown in Figure 3-11 (right) positioned to sample particles at the top of the copper or Plexiglas riser. They also sampled particles from the bottom of the bed into a second Faraday cage. A vacuum assist was employed on the upper Faraday cage. In a related study, Tucholski and Colver (1993B) employed a high voltage parallel plate arrangement at the top of the same bed riser to separate out negatively charged pyrite from positively charged carbonaceous material in pulverized coal.

In evaluating the *net* charge distribution from a suspension of different particle size ranges, Fasso et. al. (1982) sampled 30-55 μm glass beads with a Faraday cage utilizing a Schmitt trigger in the bed freeboard of a 9.52 cm Plexiglas bed. At a superficial velocity of 5.45 cm/s they fit the mean particle charge and mean particle diameter to an equation

$$Q(fC) = 0.0030\left[\frac{d(\mu m)}{2}\right]^{1.4} \qquad (3.27)$$

They found a decreasing charge to mass ratio from 88.9 to 69.8 fC as the superficial air velocity was increased from 5.45 to 8.40 cm/s.

A double shielded Faraday cage was used by Tardos et. al. (1984) for sampling 1.049 μm aerosol of polystyrene latex. Charge and current magnitudes were of the order 10^{-17} C per particle and 10^{-13}A respectively. They find that the particle charge Q is related to the current I by the relation

$$Q = \frac{I}{n_p \dot{V}ol} = \frac{I\,t}{N_p} \qquad (3.28)$$

3.3.2.12 Charge measurement on single particles

The open and closed Faraday cages and the ring probe method are well suited for charge measurement of individual particles (v. Figure 3-9; Yamamoto and Scarlett, 1986). The challenge is to get the sample into the cage (or ring) so that little or no charge is transferred in the process.

Various other techniques have been employed for measuring the charge on single particles. Inculet et. al. (1983) investigated corona charging for a single 3.175 mm metal particle by first suspending the particle from a nylon string, charging it by corona discharge, and then discharging it through an electrometer to measure the charge (presumably the charge mode of the electrometer was used). They measured charge magnitudes in the range 1 to 4 nC depending on the distance of the corona source from the particle. Rhim and Rulison (1996) levitated individual charged droplets with electric fields primarily for studying surface tension and viscosity. They utilized the simple relationship Q=mg/E to determine the charge *Q*, where *m* is the mass of the particle, *E* is the electric field strength, and *g* is gravity.

Colver (1976) studied inductive charging of individual copper spheres 388-1315 μm in an electric field by measuring the current in the external circuit for single particles oscillating between high voltage parallel plates. Particle charge was in the range 3 to 100 pC with electric fields covering 6 to 60 kV/cm. He also utilized levitation by an electric field to infer charge on individual particles (irregular and spherical shapes) of different materials (coal, lime, glass, copper) as well as the effect of relative humidity.

3.3.2.13 Bipolar charged suspensions

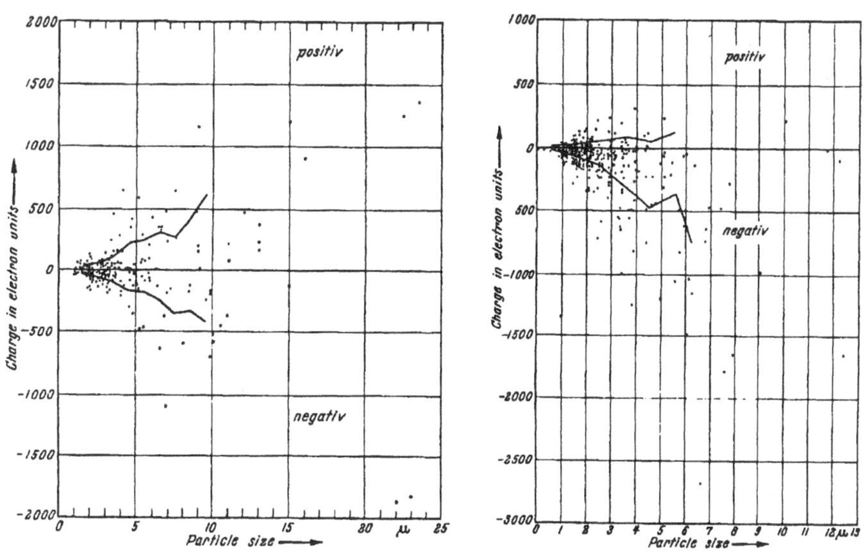

FIGURE 3-12 *Crushed quartz; (left), symmetrical charging quartz against quartz; (right), unsymmetrical charging quartz against platinum (Loeb 1958; Kunkel 1950).*

A disadvantage of the Faraday cage method for powders and aerosols containing both positive and negative charge (bipolar charge) is that only *net* charge is indicated. The problem is to identify both charge and size distribution. Charge spectrometer systems will often separate particles during free-fall or pneumatic transport (laminar) utilizing an electric field directed perpendicular to the flow. Turner and Balasubramanian (1976) used this technique to determine the charge distribution of glass particles of means size 49, 69, and 83 μm in an electric field of 75 kV/m. An electrometer was used to detect the charge in each cage with the quantity of particles determined by weight. They fit linear curves to the cumulative charge distribution on log-normal plots. A self-contained probe for measuring bipolar (positive and negative) charge was used by Self et. al. 1979 for in-situ measurement of fly ash.

Symmetrical charging of a powder is illustrated in Figure 3-12 from the data of Kunkel, (Loeb, 1958; Kunkel, 1950) using a total of 1500 crushed quartz particles (only 1 in 5 particles is plotted). The solid lines represents averages in the numbers of equivalent elementary electrons. Various experimenters have measured +/- charge distributions in powders but invariably at the cost of a more complex measuring system. Kunkel utilized strobe photographs to detect particle diameter by application of Stokes' law and charge-to-mass ratio for individual particles in an electric field.

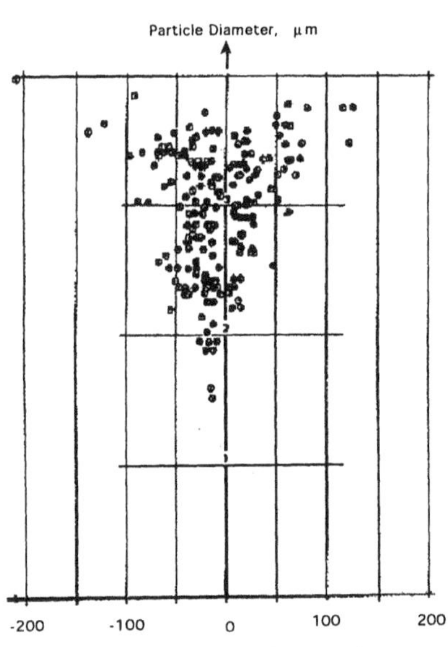

FIGURE 3-13 *Crushed quartz; e=1.602x10^{-19}C (Hassler 1978).*

Figure 3-13 shows bipolar charged particles that were produced by crushing quartz to sizes smaller that 4 μm (Hassler, 1978). An optically based dust chamber was used to photograph and evaluate the paths of individual particles falling through a sedimentation tube. A symmetrical voltage sawtooth was used to produced both positive and negative direction electric fields of equal duration, which when applied to the falling particles produced prescribed paths. With the application of Stoke's drag law, both the particle size and the charge were determined. Note that the same particles collectively placed in a Faraday cage would register only a slight negative *net* charge while in fact the charge has a wide range of positive and negative values. The 1 μm particles in this study having 5 elementary charges corresponds to about $3x10^{-4}$ C/kg. In the same study, Hassler reported charges on fogs of particles generated in supersonic nozzles of $1x10^{-4} \rightarrow 17x10^{-4}$ C/kg (droplets were in the range 1-15 μm). The charging of water droplets was increased by limiting the conductivity of the water by distillation and deionization to about 0.6 μS/cm (S = Siemens; ordinary tap water is about 300 μS/cm).

Various configurations of charge spectrometers used to detect charged toner particles ~10 μm were reviewed by Schein (1992). The basic theory for detection of charge is to determine particle trajectory and location of deposition on a surface in an electric field. The quantity of powder is then measured (e.g. by counting particles) as a function of its position y from which the sign and charge magnitude of the fractional sample is determined. If in addition the powder has a particle size distribution then a computerized microscope system is employed to measure particle diameter at each position.

The so called E-SPART system can determine number (mass) of particles in real time up to 100 s^{-1}. With this system, bipolar charge-to-mass ratios of 0 to ± 20 μC/g for 0.4 to 20 μm toner particles have been measured using a combination of acoustic excitation to determine aerodynamic particle diameter and electric field migration velocity to determine particle charge (Mazumder et. al., 1991).

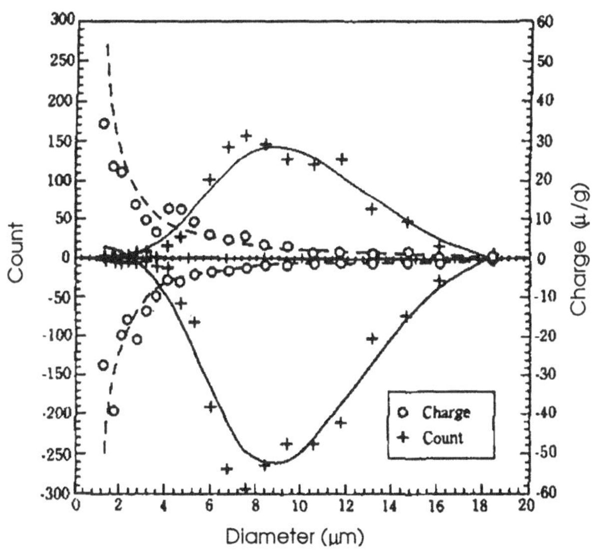

FIGURE 3-3 *Bipolar charge and particle count with aerodynamic particle diameter using E-SPART system of paint powder after tribocharging (Mazumder 1993)*

Figure 3-14 shows bipolar charge and particle count plotted against aerodynamic particle diameter for a paint powder with the E-SPART system after tribocharging (Mazumder, 1993). The investigators considered optimization of the Q/m ratio of toner for electrophotographic imaging. The particle velocity component in the direction of the electric and acoustic fields is measured by laser-Doppler-velocimeter (LDV). The acoustic field is driven at

1 kHz (2.0 to 20.0 μm particles) or 24 kHz (0.4 to 4.0 μm particles) by a speaker and is monitored with a microphone. The *dc* electric field strength is varied form $2x10^3$ to $2x10^5$ V/m (50 to 5,000 V over 21.4 mm) depending on the charge on the particles. Particles are transported at constant velocity through the test cell at right angles to the fields by a small forced convection flow. A limitation of the system was found for particles having charge > 20 μC/g because of electrostatic attraction to the metal walls of the test cell.

A sinusiodal electric field was used by Mizuno and Otsuka (1984) in combination with optical tracking of charged submicron particles (< 1 μm) in a vertical gas flow experiment to determine charge-to-radius ratio. The optical technique utilized a TV monitor and lens system with three different slit configurations used to view particle motion. The field amplitude was E=1.1 kV/cm between parallel plates with 5 mm separation. Incense smoke of mean diameter 0.8-1.0 μm and Di-octyl phthalate particles of mean diameter 0.5 μm were charged in a boxer charger utilizing an *ac* field. Particle diameter was determined by sedimentation velocity in gravity. They obtained charge values of $2.3x10^{-18}$ C for the Di-octyl phthalate particles and $6.4x10^{-18}$ C for the smoke particles.

Suitable particle drag laws must be considered for large particles in free-fall. (>50 μm) or field forces where one encounters Re_d> 0.5. Various drag equations covering the standard drag curve are given by Clift et. al. (1978). The terminal velocity of free-falling particles can be determined graphically or analytically with the appropriate drag equation (Kunii and Levenspiel, 1991). For particles small enough to be affected by the mean free path of the gas (d<3 μm), the Cunningham correction factor and Brownian motion must be considered in the formulation of particle drag (Brodkey, 1967).

For larger particles 50-75 μm, Ban and co-workers used the so-called laser Phase Doppler Particle Analyzer (PDPA) (Ban et. al, 1994). The method utilizes the measurement of particle velocity, particle diameter (for spherical particles), and number density by Doppler shift in light scattered (particle velocity) and phase change (particle diameter) from an argon ion laser. The measurement of particle trajectory from electric field deflection (100-200 kV/m) is carried out in a test volume of cross-section 10x14 cm^2 from which individual particle charge is calculated. Stokes drag is utilized for particle Reynolds number less than 2. They also find that particle charge increases with particle diameter and particle velocity, decreases with relative humidity, and increases or decreases with temperature depending on the material (Schaefer et. al., 1994).

3.3.3 Measurement of Particle Force

3.3.3.1 Particle force equations

The various electrostatic forces acting between particles, particles and surfaces, and liquid interfaces in the presence of electric fields having been the subject of numerous theoretical and experimental investigations. While the fundamental force mechanisms between materials have been identified (Lapple 1970; Krupp 1967; Adamson 1976), there remains practical limitations to their application because of the uncertainty of detailed descriptions at contact points such as the number and size of asperities, close contact separation distance and contact area, presence of films, and gas breakdown from electric fields. Complications arise from the presence of other permanent forces such as van der Walls and contact electronic forces or if there is a distribution of particle sizes. Dielectrophoretic effects resulting from field gradients and dielectric present yet another electrostatic force factor (Jones, 1995).

In contrast to contacting particles, an isolated particle of charge Q that interacts with a local field E is simply evaluated from the relationship $f_p = Q \cdot E$. Such electrostatic forces for isolated particle can be simply evaluated by levitation, i.e., utilizing gravitational forces and steady-state experiments (Colver, 1976; Jones, 1986, Tombs and Jones, 1990, 1993). Charged droplets have been levitated for the studies of surface tension and viscosity (Rhim and Rulison, 1996).

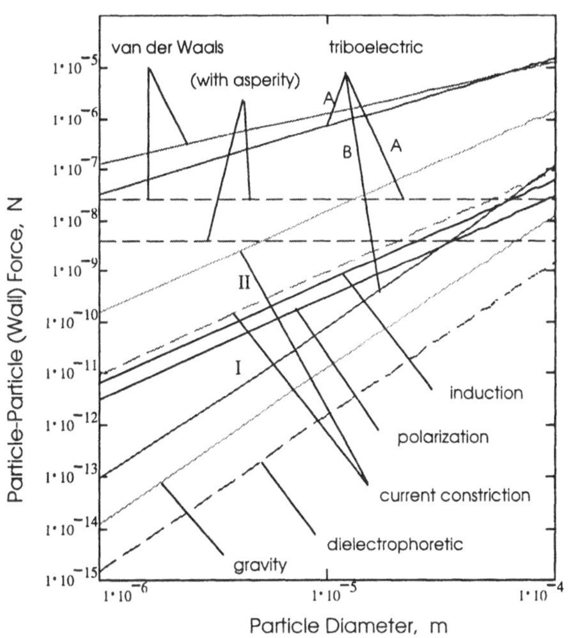

FIGURE 3-15
Particle-particle (wall) forces, (calculated).

Figure 3-15 shows plots of various equations given in the literature for particles sizes 1 → 100 μm using the specified assumptions of particle spacing, asperity diameter, electric field strength etc. The calculations represent only an estimate of the probable force based on the values assumed for each equation. Most notable in Figure 3-15 is the observation that gravitation forces for small particles can be readily overcome by virtually all of the electrical forces depending on the particular assumptions made for the field strength and field gradient, contact potential, etc. The van der Waals force theoretically can dominate electrostatic forces for a perfectly smooth small sphere; however, the presence of a 0.2 μm surface asperity dramatically reduces van der Waals forces to a small constant value that is more easily overcome by other electrostatic forces, even for particles smaller than 100 μm.

Brunton and Rozelaar (1968) noted a 5 fold drop-off in current constriction and contact resistance forces at the contact point between a conducting sphere and a conducting flat surface as the roughness of the polished metal surfaces was increased from $2\text{x}10^{-6}$ to $60\text{x}10^{-6}$ inches (0.05 to 1.5 μm) centerline average distance. They also observed an increase in force with an increase in the relative dielectric constant of the intervening material. Martin et. al. (1991) using shear experiments argued that deformation of the contact cannot be ignored but that in the case of current constriction and contact resistance forces, the Hertzian model of deformation at the contact leads to unrealistically large interparticle forces by up to an order of magnitude. Using a commercial Nanotest device, they report individual force measurements on glass spheres (1.9 and 3.7 mm diameters) in contact with a polished stainless steel surface to be intermediate between limited field and elastic cap models (Martin et. al. 1994). Robinson and Jones (1982) also utilized a shear apparatus with applied electric field to demonstrate a near linear relationship between the yield shear and the applied electric field strength for 600-710 μm sand and spherical beads.

The *dc* field force equations can be expressed in the form

$$F_{dc}=K_2\pi\varepsilon_o d^2 E_{\max}^{2-\beta} E_o^{\beta} \qquad (3.29)$$

where $1 \leq \beta \leq 2$ for a conducting spherical particle in contact with a conducting wall or another particle. The limit $\beta \sim 1$ implies field limited breakdown with no deformation at the point of contact (McLean, 1977, Dietz,1978A); whereas, β=2 applies to a point contact (Lebedev and Skal'skaya, 1962). Intermediate values of β between 1 and 2 are found for a Hertzian deformation of the contact area with the field included (McLean, 1977, β=1.54). Colver (1980) reported an experimental value of β less than unity (β=0.656) for glass particles at the interface of an electric suspension. For conducting spherical particles in contact with a conducting flat surface Colver (1976) reports $\beta=2$, and $K_2 = 1.24 \rightarrow 1.33$ with an average value 1.29.

Colver (1976) used the method of images to evaluate both charged and uncharged particle-wall forces for a noncontacting conducting sphere in the presence of a uniform electric field. He utilized levitation, circuit current measurement, and manipulation of spheres to confirm the Maxwell charge constant. Stoy (1995) gives an expression for the force between two uncharged touching dielectric spheres in a parallel electric field. Jones (1986) used levitation and the method of images for the case of two and three contacting spheres comprising chains of particles having dipole and quadrupole moments in an external electric field. He also considered the inversion method for the case of dipole forces with strong electric fields on small biological cells, droplets, and bubbles, (Jones, 1987). Tombs (1995) considered the theoretical force on a lossy dielectric sphere with finite conductivity and dielectric constant near a ground plane in an electric field.

3.3.3.2 Particle force with ac fields

The *dc* force model of Equation 3.29 can be extended approximately to include low frequency *ac* fields for an array of semi-insulating particles using lumped capacitance. Colver and Wang, 1994, show approximately that

$$F_{ac/dc} = \frac{K_2 \pi \varepsilon_o d^2 E_{max}^{2-\beta} E_o^{\beta}}{1 + \left(\tau_c \omega\right)^2} \quad \text{where} \quad \tau_c = \frac{R_c R_p \left(C_c + C_p\right)}{R_p + R_c} \tag{3.30}$$

which satisfies the *dc* limit $\omega \rightarrow 0$ of Equation 3.29 and τ_c is the stated time constant.

3.3.3.3 Force measurement

Cohesive forces between a bed of fly ash particles with an applied electric field were measured by McLean (1977) (diameter not given) using a free electrode attached to a mechanical balance. These stresses were measured normal to the electrode. The device was similar to a guarded electrode used to measure particle resistivity. He observed cohesive forces in the range 0-200 N/m^2 that varied in proportion to the electric field strength. Colver (1980) utilized two experimental methods (interfacial forces and stringers) to evaluate the force on a packed beds of 62 μm glass spheres with applied electric fields. Fitting his data he showed that the dimensionless interparticle force is F/F_g ~

$1.84+5.95\times10^{-3}\,E^{0.66}/\rho_d d$ (SI units). Robinson and Jones (1980, 1981) observed that the angle of repose of a packed bed increased monotomically up to 90 ° with applied electric fields up to 3 kV/cm in the bed. They observed a similar field dependence for glass spheres of 75-125 µm and 600-710 µm but almost no effect on 1500-2000 µm particles.

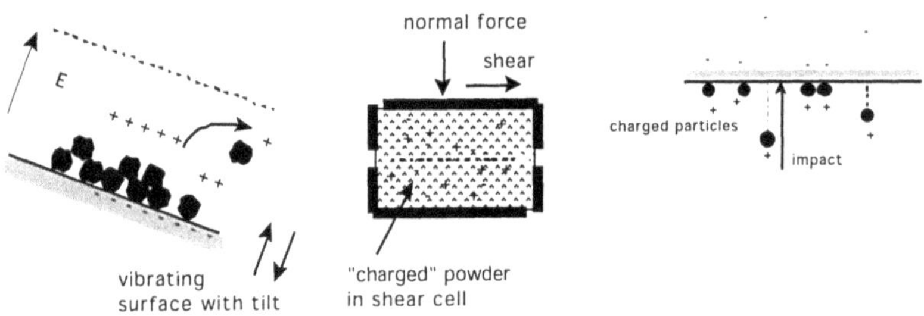

FIGURE 3-16 *Experimental configurations used for the measurement of powder forces (vibration-shear cell-impaction) with electrostatic effect added.*

Various experiments have been devised to measure forces in bulk powders such as the bulk shear stress (Miyanami, Terashita, and Yano, 1983) and forces between individual particles using vibration, impaction (sudden acceleration), and rolling friction (Jimbo and Yamazaki, 1983). Figure 3-16 shows how such techniques can be extended to include powders already charged or powders subjected to an applied electric field. Ikazaki and Kamamura (1984) measured tension force versus time with a load cell to indicate the attraction force of individual glass hemispheres (5 and 10 mm diameter) and for talc powder (4.2 µm) under the influence of an applied voltage. For an electric field strength of 10^5 V/m, they measured increasing forces to 4.5×10^{-2} N as the temperature was increased from room to 400 °C. They considered both surface and bulk conduction in their force model.

3.3.3.4 Agglomeration of particles

The agglomeration of submicron particles (0.67 µm) by diffusion and coulomb attachment forces with an isolated unipolar charged micron sized target particle (4-7 µm) was shown by Nakajima et. al., (1996) to be significantly enhanced by a factor up to 10^5 over Browian motion alone if the target particle was either charged or oscillated (800 Hz). In the experiment, a single target particle was highly charged (70-80 fC) by a plasma, evaporated to test size and suspended by quadrupole electrodes, and then exposed to a suspension of micron size particles. An *ac* electric field controlled the vibration amplitudes of the target particle in the range 128 to about 1280 μm. Charge on the submicron particles was eliminated by increasing the conductivity of the particles with a chemical agent during its generation or by charge neutralization using bipolar charge from an *ac* corona charger. The authors claimed errors of at most 50 nm in the size measurement of the 4-7 μm target particles (vibrating) using laser Doppler anemometer fringes. The measured change in diameter of the target particle over time was used to determine the capture rate of submicron particles.

Kobashi (1978), studied agglomeration of charged fly ash particles (size distribution typically 0.5→15 μm) in an *ac* electric field of amplitude ~ 14.6 kV. Particles were charged by a corona wire (- 9 kV) at concentrations of 10^4 to 10^5 cm^{-3}. Agglomeration was achieved by the difference in relative velocity between different particle sizes. The particle size distribution (assumed to be log-normal) was determined following agglomeration using particle settling time as determined by the measurement of laser beam attenuation over time and, in a second continuous flow experiment, a cascade impactor. An electron microscope was use to check the cascade impactor results. Agglomeration by bipolar charging was found to be superior to that of unipolar charge on the particles.

Liu and Colver (1991) used the EPS (Electric Particulate Suspension) to capture flowing fly ash particles (mean diameter 2.4 μm) in an electric field on charged moving copper spheres of diameters 74-88, 104-125, 125-147 μm. The copper target particles were directed at right angles to the flow of fly ash. The copper spheres were inductively charged to 10^{-13} C in an electric field of strength 10.9 kV/cm while the charge on the fly ash was not controlled. With 104-125 μm copper spheres moving at a speed of 117 cm/s (electric field of strength 10.9 kV/cm) and the fly ash particle velocities in the range 3 to 19 cm/s at concentrations of the order 10^5 cm^{-3}, typical capture ratios were found to be 13.2 to 75.33 particles of fly ash per copper sphere. They determined the quantity of fine particles captured on the target particles by measuring the concentration of fly ash flowing into and out of the test section.

3.3.3.5 Particle diffusion

Colver and Howell (1980) used the electrostatic EPS (Electric Particulate Suspension) to measure diffusion of spherical copper spheres (74-88 and 125-147 μm) along a copper parallel plate duct having a 1 cm separation distance. The particles were dynamically suspended in the duct by inductive charging from an applied electric field of strength ~10^5 V/m while the concentration of particles (~ 10^8 m^{-3}) was scanned along the duct using attenuation of a laser beam. The mass flux of diffusing particles was measured by collecting the particles flowing from the end of the duct. Self-diffusion coefficients found were of the order 10^{-3} m^2/s.

3.3.3.6 Particle-wall drag

Using a similar EPS setup, Cotroneo and Colver (1978) evaluated drag and wall friction effects of flowing 85 μm copper particles suspended by an electric field. This study was unique in that the solids flow was investigated below the saltation velocity of the particles. They considered various particle forces including electrostatic, gravity, drag, lift and particle-wall interactions. The Reynolds number based on air flow in the duct was 0 →1870 while the particle Reynolds number based on relative flow was less than 13. They calculated particle angular rotation speeds in the range 57 to 2735 rad/s. In a later study Colver and Sarhan (1996) extended this study into the turbulent regime for Reynolds number 1,300 to 12,550. A correlation for the apparent particle drag was found (Sarhan, 1989).

3.3.3.7 Atomic force measurement

Long and short-range forces at the molecular level between surfaces have been directly measured using the *surface forces microbalance* (SFM) first developed by Israelachvili and Tabor and Klein. The SFM utilizes the simultaneous measurement of separation distance and surface force. Electrostatic repulsion forces and van der Walls forces between surfaces in liquids are discussed by Israelachvili and McGuiggan (1988).

The applications of the SFM include force measurement between surfaces in liquid and vapor, adhesion between similar or dissimilar materials, contact deformation, wetting and capillary condensation, viscosity in thin films, forces between surfactant and polymer-coated surfaces, and surface chemistry. Fluid-electrolyte interactions between conductive surfaces can also be measured [Smith, et. al., 1988]. A typical microforce of 10 nN can be detected over separation distances to a resolution of 0.1 nm with optical interoferometry between reflective surfaces. With electrostatic forces, relatively large separation are measured 1-100 nm, whereas, short range forces such as van der Waals forces take place over distances of less than 3.0 nm. Ultrasmooth and electrically conductive surfaces can be formed by the deposition of a metal film (40 nm thickness) such as Pt on a smooth substrate of mica [Smith, et. al., 1988]. The separation distance between the two surfaces is controlled by a combination of mechanically reduced motions in the micron range using a stepping motor and a piezoelectric tube having an expansion coefficient of about 1 nm/V. The force between the surfaces is determined by a cantilever deflection with spring constant ~100 N/m. A review of the subject of "scanning probe microscopy," which can provide images and surface topography at the atomic scale, has been given by Lawrence et. al. (1996).

3.4 PROBES AND SENSORS

3.4.1 Capacitance Probes

Capacitance probes can be used with flowing suspensions as well as in fluidized beds to evaluate voidage and particle concentration through Equations 3.14 and 3.15 together with the relationship between voidage, particle volume. and concentration given by

$$\begin{aligned}\alpha_d &= 1 - V_p n_d \\ &= 1 - \frac{\pi d^3}{6} n_d \qquad \textit{(spheres)}\end{aligned} \tag{3.31}$$

Figure 3-17 shows two common designs used for probing fluidized beds. Example dimensions for the electrode diameter are 0.8 mm diameter at a separation distance of 0.25 mm for the needle-type probe (left) and 6 mm x 1.6 mm at a separation distance of 1 mm for the plate-type electrodes (right), each with a base diameter of 3 mm. The needle-type probe might be calibrated for use with glass bead diameters of 0.6 to 0.7 mm and the plate-type for bead diameters of 2 to 3 mm diameter. Smaller particle diameters can also be used.

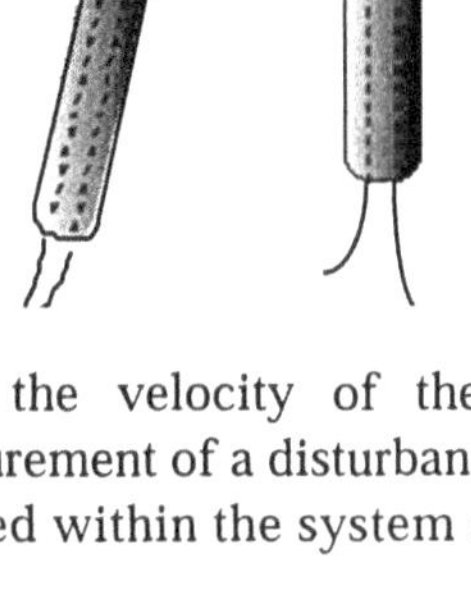

FIGURE 3-17 *Capacitance probes: needle-type (left); plate-type (right).*

With two capacitance probes positioned at a known separation distance (as in a fluidized bed), the velocity of the solids phase interface can be determined by measurement of a disturbance such as a bubble. Small probes can be designed to be used within the system or mounted flush with the walls and may include a second measurement system such as a optical fiber. Reviews of various capacitance probes used in fluidization are given by Geldart and Kelsey (1971), Dutta and Wen (1979), Grace and Baeyens (1986), and Louge (1997).

A simple example of an ideal parallel plate capacitor will demonstrate the capacitance method. A capacitance probe driven by an *ac* source can detect a disturbance signal from any one of the fundamental parameters ε, A, x as illustrated by the equation for a parallel plate capacitor

$$C = \frac{Q}{\Phi} = \varepsilon \frac{A}{x} \tag{3.32}$$

where ε is the permittivity of the material between the plates, A is a plate area (one side), and x is the separation distance between the plates. To sense a change in bed voidage, a measurement of the effective bulk permittivity ε_b would utilized, Equations 3.14 and 3.15

Figure 3-18 shows a modulated output signal Δe(t) generated by a step pulse of magnitude Δε(t) in the capacitance of one arm of a simple *ac* bridge circuit assumed initially to be in balance and driven by a peak *ac* voltage V_o at carrier radial frequency ω_c.[3] This signal approximates the passage of a bubble over a single capacitance probe submerged in a fluidized bed if the probe is small compared to the dimensions of the bubble. The output Δe(t) is shown as the raw signal would appear on an oscilloscope. The bridge-capacitor relationship for fixed A and x of the probe gives the output signal Δe(t) corresponding to the capacitance change ΔC as

$$\Delta e = \left| \frac{\frac{\Delta C}{C}}{\left(2\frac{\Delta C}{C} - 4\right)} \right| V_o \sin(\omega_c t) \approx \frac{\Delta \varepsilon_b}{4\varepsilon_b} V_o \sin(\omega_c t) \qquad \left(\frac{\Delta \varepsilon_b}{2\varepsilon_b} << 1\right) \tag{3.33}$$

For subsequent demodulation and a good signal profile recovery, the period of the *ac* carrier frequency T = 1/f = 2π/ω should be small compared to the duration of the signal (i.e., less than 1 s), the ratio being 1:10 in this example.

[3] In practice, two of the dummy capacitors C could equally be replaced by resistors in adjacent arms of the bridge with provision made for balancing the bridge initially.

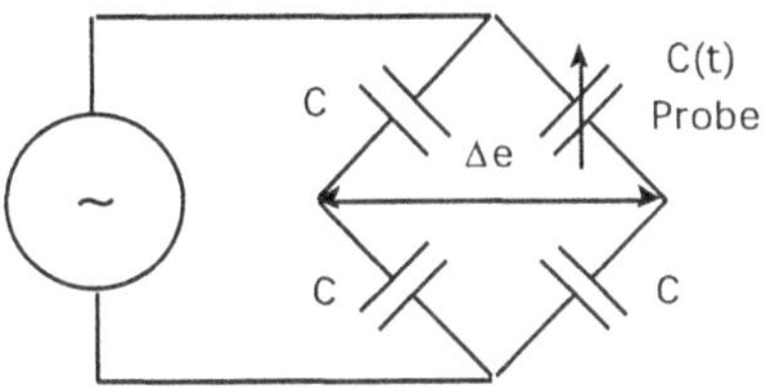

Bridge Circuit and Capacitance Probe (transducer)

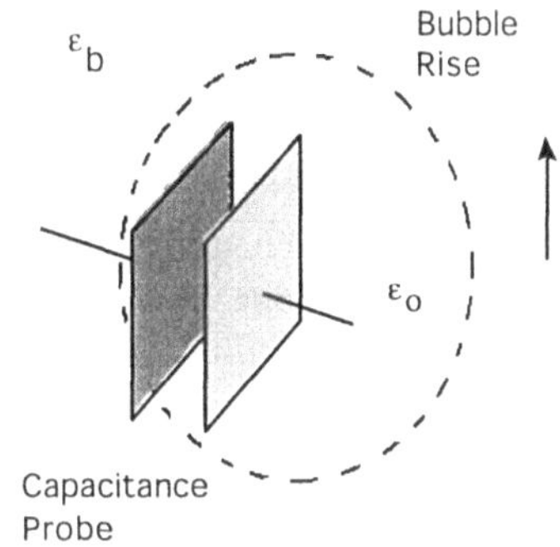

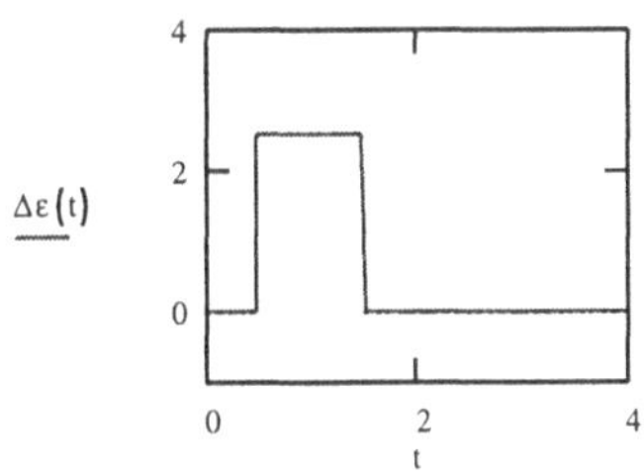

Input signal $\Delta\varepsilon(t) = \Delta\varepsilon_b(t)$ to probe for large moving bubble.

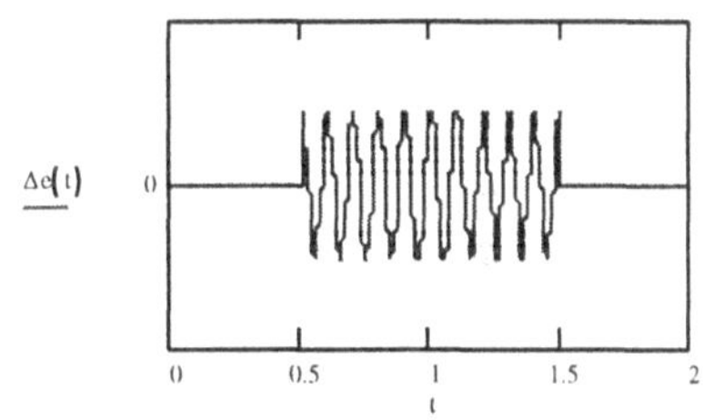

Output signal on oscilloscope

FIGURE 3-18 *Modulated bridge circuit for use with capacitance probe.*

In practice, the output from the capacitance circuit can be sent to an integrating DVM or suitable filter to smooth out fluctuations in the signal from the fluidized bed. Calibration of the probe for gas-solid suspensions can be carried out at the extreme points of either an empty or packed bed where the voidage α_d is know or can be calculated. Intermediate calibration points can be found by expanding a powder in a gas fluidized bed. Larger bed expansions (e.g. 10 %) are most readily obtained using small particles (< 100 μm). For liquid-solid suspensions, both fluidization and sedimentation can be utilized as techniques for calibration.

3.4.2 Current Probes

The current probe is one of the simplest of the multiphase electrostatic detectors requiring only the measurement of *dc* current. Typical probe data will range from nA to μA for dilute multiphase flows and pA (or larger) for fluidized beds. Current probes are similar in principle to the well known Langmuir probe used to measure ion flux and electron and ion concentration and electron temperature in flames and MHD plasmas (Clements et. al. 1978; Lawton and Weinberg, 1969). Other types of current probes used in ionized fields are discussed by Cross 1987. Current probes used in multiphase flows can generally be categorized as being either "external" or "internal" flow types.

Tardos and Pfeffer 1980 utilized an external ball current probe to investigate charging in fluidized beds. They note a disparity between the particle charge in the bed and the current measured with increasing superficial velocity, which they attributed to the distinction between current and particle charge measurement.

Internal probes may or may not be designed to capture all of the particles entering the probe. Similar to ion flux probes, these probes can give the particle number density of a charged suspension, the mass flux, or the charge-to-mass ratio of the particles depending on the application and theory. Calibration of the current probe is dilute systems is desirable if quantitative results are desired for the charge. Fasso et. al. (1982) devised a cylindrical precipitator type of current probe with 100 and 300 V dc sources to attract particles to the probe surface. The probe detected particle concentration when both the charge and particle velocity were known. Both the mass and current flows are sampled (the mass flow rate being sampled isokinetically). The charge-to-mass ratio of the particles is determined by dividing the current by the mass flow rate.

To interpret the probe current quantitatively in terms of the suspension (number density, particle charge etc.) some assumption must be made regarding the actual charge transferred during a particle-surface encounter with the probe. The simplest assumption for current probe theory is that all previously acquired charge on the particle is transferred during a probe encounter. In practice this might be the case if the particles were highly charged by an insulating surface and were themselves good conductors and subsequently gave up the majority of their charge to another good conducting surface (the probe). However, such assumptions always raises questions as to interpretation of the current so that calibration of the probe is preferred.

As noted above, and depending on the electronics, generally there is no net current transfer in the probe by a charged particle unless electrical contact is made through physical contact or through gas phase discharge (Colver, 1976). A unique case is the induction type probe, which does not depend on direct electrical contact with the particles but generates a surface or "image" charge on the probe. The resulting signal is then interpreted electronically (Nieh, et. al., 1986).

3.4.3 Potential Probes

Early potential probes were used to indicate static buildup in fluidized beds (Ciborowski, 1962). More recent investigators of charge effects in fluidized beds were reported by Tardos and Pfeffer (1980), Bafrnec et. al. (1993), Fujino et. al. (1985), and Borland and Geldart (1971/72). As is the case for current probes, a potential probe may not give an adequate characterization of the actual charge residing on the particles. The simplest potential probes are spheres suspended by a wire in a fluidized bed. Kisel'nikov et. al. (1967) considered the buildup of potential on a submersed spherical probe over times of a few minutes noting three regimes of charging in fluidized beds covering voltage ranges from 100 to over 1000 V.

High impedance potential probes can be categorized as: (1) passive probes that provide a voltage reading relative to a reference ground by virtue of their capacitance and associated charge, and (2) active or nulling probes that match the local spatial potential being measured using feedback control. Electrostatic voltmeters of the contacting and non-contacting types respectively are used as the instruments of readout for these probes.

Other commercially available high voltage contacting probes (typically 40 kV with input impedance $10^9\ \Omega$ are based on a simple voltage dividing circuit (built in the probe itself) and cannot be used to sense a spatial potential from an electric field existing in air since such probes rely on a current flow and no charge source is provided with the field. The ideal potential probe would detect the voltage at a specified location in a fluidized bed without itself becoming charged by the particles so as to indicate the local potential in the absence of the probe.

We can demonstrate the maximum predicted voltage (at the centerline) resulting from a net free charge assumed to be distributed uniformly throughout a cylindrical air fluidized bed of diameter D containing particles of a uniform charge density ρ A small diameter (assumed initially uncharged) wire used as a potential probe situated along the axis of a would indicate a potential difference from a grounded outer wall ($\Phi_{wall}=0$) as

$$\Phi_{probe} = \frac{3\left(1-\alpha_d\right) Q\, D^2}{8\pi\, \kappa_b \varepsilon_0 d^3} = \frac{\rho_q D^2}{16\, \kappa_b \varepsilon_0} = \frac{9.60x10^{-4}(0.0254)^2}{16x3.5x\left(8.854x10^{-12}\right)} \approx 1.25\, kV \tag{3.34}$$

The *ideal* probe potential Φ_{probe} is seen to be proportional to the individual particle charge Q in the bed, or equivalently, to the particle charge density ρ_q

existing throughout the bed. The calculated value of 1.25 k*V* is based here on a nominal value for the particle charge in an air fluidized bed of $9.60\text{x}10^{-4}$ C/m^3 (sampled from packed bed of *net* charge density $8\text{x}10^{-7}$ C/kg of microspheres, Table 3.2), a fluidized bed voidage of 0.5, a bed diameter of 2.54 cm, and a particle dielectric constant of 7.0 used in Equation 3.15 to calculate the effective fluidized bed bulk dielectric constant of 3.5.

If the probe wire is charged, then the centerline potential indicated by the probe is not simply related to the volume charge density in the bed ρ_q (C/m^3) but also depends on the charge acquired by the probe σ_q (surface charge density C/m^2) according to the relation

$$\Phi_{probe} = \frac{\rho_q\left(D^2 - D_w^2\right) + 2\left(\rho_q D_w^2 - 4\,\sigma_q D_w\right)\ln\left(D_w / D\right)}{16\,\kappa_b \varepsilon_0} \tag{3.35}$$

where D_w and D are the diameters of the wire and the bed respectively. However, Equation 3.34 and 3.35 reduce to the same potential for the condition $D_w << D$ so that the indicated probe potential still reflects the actual potential in the bed for probes of small diameters. This simple example shows that submersion probes should generally be small in comparison to other characteristic dimensions of the system such as the bed diameter to measure spatial potential but not so small as to emit corona (Equation 3.36).

The above example shows that even small bed diameters can produce large voltages from the net charge distributed over the bed. Increasing the bed diameter in our example to 10 cm would theoretically raise the probe potential to 19.4 k*V* at its center. Such high potentials cannot necessarily be sustained on bare wires of small diameter without charge leakage or corona discharge so that elevated readings should be ex examined. For example, by incorporating Peek's formula for corona onset in a coaxial system into an equation for the surface potential of our probe wire at 19.4 kV one obtains [4]

$$\Phi_{probe}(kV) = \frac{31x0.72}{2}\,D_w(cm)\,\ln\left(\frac{D}{D_w}\right)\left(1 + \frac{0.308}{\sqrt{D_w(cm)/2}}\right) \tag{3.36}$$

(Corona onset potential)

[4] Peek's *ac* formula is approximately correct for *dc* fields, (Cobine, 1958).

for the critical potential at normal air conditions of temperature and pressure. For a bed diameter of D=10 cm, Equation 3.36 gives the minimum permitted wire diameter D_w of 2.54 mm for corona onset in our problem. The predicted corona onset voltage is valid only over portions of the probe extending above the surface of the fluidized bed since Peek's formula applies to a coaxial cylinder suspended in air.

A type of null-current potential probe was devised by Bailey et. al. (1995) where it was assumed that the probe tip attains the local potential under conditions that the corona current from the probe was balanced to zero. His probe was constructed from small carbon fiber electrodes (single and double) measuring 1 μm diameter by 1 mm length mounted on 3 mm glass tubing. The probe produced a near-zero corona current of (-)2×10^{-11} A at an offset (error) voltage of (-)1.125 kV with respect to the local space potential being measured. The maximum offset current was maintained electronically through feedback control. The probe was designed to map the spatial potential above charged powders up to (-)30 kV without adding significant charge to the powder from the leakage corona itself.

3.4.4 Resistance Probes

Flow disturbances in gas-liquid and bulk powder systems can be detected by local variations of the bulk resistivity. Resistivity probes were briefly discussed by Dutta and Wen 1979. Hayakawa et. al. (1964) utilized dissimilar resistance properties of coke particles in a fluidized bed to track their rate of mixing. Conducting and non-conducting coke particles were used (-65+100 mesh). The two types of coke particles were prepared at 600 and 900°C. The variation of voltage drop was detected across a pair of platinum ring electrodes of 1/8 in diameter, spaced 1/8 in apart and mounted on a 1/8 in ceramic rod. Current through the electrodes was also detected by a microammeter to measure bed resistivity. Typical currents were in the range 0-150 μA. For example, they found that the characteristic time constant for mixing was independent of the inventory ratio of conducting to non-conducting coke.

The variation of the bed voidage in a gas fluidized bed was used by Colver (1981, 1996) and Donahoe and Colver 1984 to track single injected air bubbles, fast moving waves, a rising plastic float, and falling objects for both *ac* and *dc* electric fields. The rise velocity of the bubble slows from about 25 cm/s to 10.5 cm/sec near a the surface of the bed. A fast compression wave moving at about 270 cm/s was detected just ahead of the bubble.

The interpretation of the output signal as being purely resistive is complicated by the effect of charge transport through bulk convection of particles and by a transient capacitance effect due to the variation in the effective dielectric constant between the electrodes, the latter effect depending on the size and placement of the electrodes (Colver 1996). An additional conductivity effect is that of particle charging in the nose and wake regions of the bubble as suggested by Boland and Geldart [5] (1971, 1972), particularly for larger particles (> 200 μm). Such complications are of little consequence if the objective of the measurement is to mark a replicated disturbance.

A simple resistance probe consisting of two conducting electrodes in contact with the powder of bulk resistance R_b can be utilized with a *dc* source of voltage V connected in series with an external resistance R_e across which a voltage difference is monitored. The model can be extended using Figure 3-19 to include the effects of electrode capacitance C and inductance L. Ignoring the probe inductance, which is usually small for parallel electrodes, the differential equation describing the output signal Φ is given in Figure 3-19. For a resistance dominated signal (rapid decay of the capacitance signal) we require that the time constant be small in comparison to the characteristic time of the signal $\tau_e \ll \Delta t_{signal}$ so that the output signal follows the bulk resistivity signal with time. This condition can be achieved by design of the electrodes making C small and/or small R_e. [6] For example, for the system of Colver and Donahe, $\tau_e = 1.1 \times 10^{-4}$ s (Colver, 1996). However, reducing R_e in relation to R_b will reduce the output signal magnitude as can be seen through the defining equations for the steady-state signal, $\Phi \approx (R_e/R_b)V$.

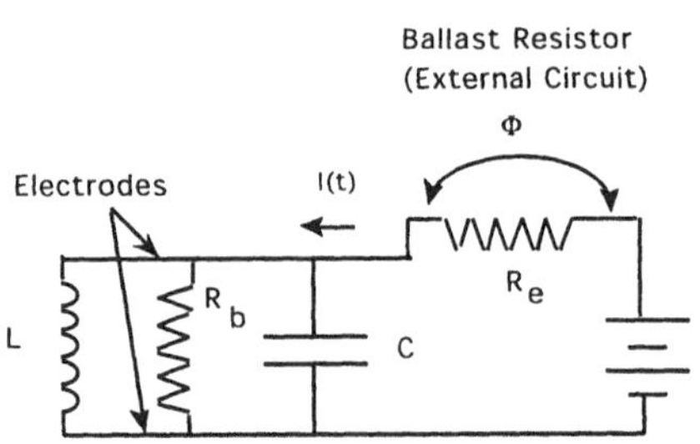

$$\frac{d\Phi}{dt} + \frac{\Phi}{\tau_e} = \frac{V}{\tau_b}$$

where $\tau_b = R_b C$

$$\tau_e = \frac{R_b R_e}{R_b + R_e} C \sim R_e C, \qquad R_b \gg R_e$$

FIGURE 3-19 *Simple electrode model and equations for detecting a bulk resistance change.*

Resistance probes can also be used in liquids to detect the void fraction of bubbles (Lamarre and Melville, 1992). Such probes respond to the difference in conductivity of the liquid-gas phases. Direct-current probes are simple and

---

[5] They used the current and voltage modes of an electrometer with insulated electrodes in the wall of a Perspex bed to track the passage of bubbles -- the interpretation of their signals was speculative.

[6] C must be calculated or measured; the usual case for powders is $R_b \gg R_e$.

inexpensive but also suffer from many problems such as electrochemical attack, polarization, and sensitivity of the measurement to liquid resistivity (Teyssedou et. al. 1988). Polarization occurs when the ions of the electrolyte in water are attracted to electrode of opposite sign creating a back emf that diminishes the current. The ionic interface problem is similar to that encountered in the measurement of the bulk resistivity of solids (see 3.3.1.3). In both cases the solution is to use an *ac* source for the resistance measurement

The simple electrode model in Figure 3-19 illustrates the *ac* impedance theory following Lamarre and Melville (1992). The capacitive Z_c and resistive Z_R (due to the electrolyte) impedances are in parallel with the inductance being ignored. The following definitions and limitations together with Equations 3.7 and 3.15 give the complex impedance Z of a liquid-gas probe:

$$\mathbf{Z} = \frac{\mathbf{Z}_c\mathbf{Z}_R}{\mathbf{Z}_c + \mathbf{Z}_R} = \left(\frac{\mathrm{kx}}{\mathrm{A}}\Re_c\right)\frac{(1+\alpha_d/2)}{(1-\alpha_d)}\frac{(1-j2\pi f\,\Re_c\varepsilon_c)}{\left[1+(2\pi f\,\Re_c)^2\right]}$$

$$\approx \left(\frac{\mathrm{kx}}{\mathrm{A}}\Re_c\right)\frac{(1+\alpha_d/2)}{(1-\alpha_d)}, \qquad (2\pi f\,\Re_c\varepsilon_c) << 1$$

where

$$\mathbf{Z}_c = \frac{1}{j2\pi f\,C}, \quad C = \varepsilon_c\frac{A}{kx}, \quad \mathbf{Z}_R = R_b = \Re_c\frac{kx}{A}$$

$$\Re_d \to \infty, \quad \kappa_c \to \infty \tag{3.37}$$

and where k is a geometric correction for a non-parallel plate electrode. It has been assumed that the resistivity of the gas phase $\Re_d$ and the permittivity of the liquid phase κ_c are large in comparison to their counterpart phases. As a numerical example, taking the resistivity and permittivity of water $\Re_c = 200$ Ω-m and $\varepsilon_c = 7.1\text{x}10^{-10}$ F/m respectively gives the design exciting frequency f << 1 MHz for a resistive dominated impedance probe. A capacitance probe results when f >> 1 MHz. Equation 3.37 shows that the resistance probe is sensitivity to the voidage α_d from spherical bubbles.

3.4.5 Particle Velocity Probes (anemometers) and sensors

Particle velocity measurement in multiphase systems can be approximately categorized as *invasive* (probes, impaction devices) and *noninvasive* (rings, coils, optical beams). Yan (1996) has further detailed particle velocity measurement according to: (1) Doppler methods: laser and microwave; (2) Cross-correlations methods: capacitance, electrodynamic, acoustic or radiometric sensors; and (3) Spatial filtering: optical, microwave, capacitance or namic sensors.

In multiphase flows, charge induction from either linear motion or fluctuations of charged flowing particles can be used to detect particle motion using electromagnetic radiation and the fundamental laws of Ampere and Faraday. Velocity is subsequently determined directly by time of flight measurement or indirectly by field intensity measurement. Naturally occurring charge of 10^{-13} C on particles is usually sufficient for detection (Nieh *et. al.* 1986).

A time of flight experiment was devised by Soo et. al. (1989) utilizing a sudden release of ions emitted from a negative corona discharge as a tracer to detect both solids and gas phase velocities. The method is similar to an ion flow anemometer utilizing released ion pulses to measure wind velocity (Asano, 1995). Yet another method involves deflection of ions across a duct (Castle and Sewell, 1975). Soo et. al. used 210-230 μm glass spheres in a dense-gas-solid suspension in air have solids loading 6 kg/kg of air and pipe Reynolds number 34,000. Tests were run with and without particles. The ion pulse was generated in the phases by a negative corona discharge during initial breakdown. Time of flight times were of the order 300 ms giving velocities of 7- 15 m/s for their system.

Figure 3-20 shows the time of flight probe of Nieh, et. al. (1986) for which the particle velocity in a dilute pipe flow is measured. The probe concept is similar to that used by Yamamoto and Scarlett (1986) who measured single particle velocity. Axial phase velocities were evaluated in the radial direction of a pipe. The probe and the associated electronics convert an induced charge on the inner brass tube to a square pulse when a single charged particle enters and leaves the probe. The time of flight of the particle is then determined from the pulse time and knowledge of the length of the inner probe of 5d (d = inner probe diameter, 2.5 mm).

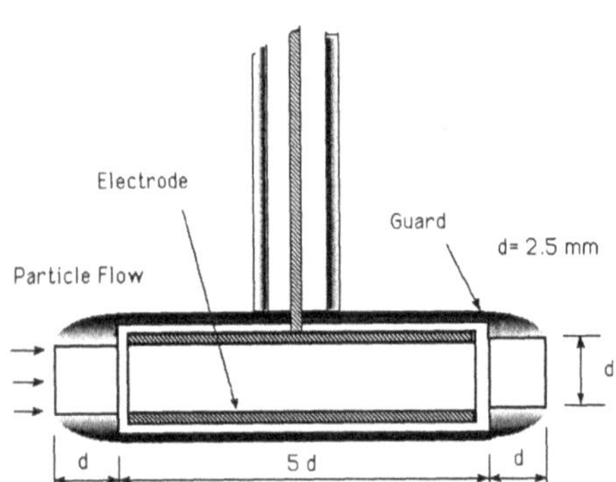

FIGURE 3-20 *Electrostatic induction probe (Nieh, et. al., 1986).*

The electronics utilized a charge sensing op-amp (operational amplifier) with rapid signal decay (circuit RC = 0.11 or 0.64 ms) that converted the charge disturbances at either end of the probe end into separate voltage spikes. The two spikes were inverted as positive signals and shaped into a square pulse using a pair of Schmitt triggers placed in parallel with two start/stop flip-flop circuits. The time of the square pulse representing the time of flight over distance 5 d was chopped by a 10 MHz oscillator. This gave a high resolution over a typical ms time of flight through the probe. An example particle time of flight time was 1.2805 ms.

An alternative and indirect approach to measuring particle velocity u_d using particle charge is to measure the mass flux and the effective density of the dispersed phase (particles) and apply the relationship

$$\dot{m}_d = \rho_d \alpha_d u_d \qquad (3.38)$$

The effective density $\rho_d \alpha_d$ of the dispersed phase is determined by a capacitance measurement of the solids voidage α_d (with ρ_d known).

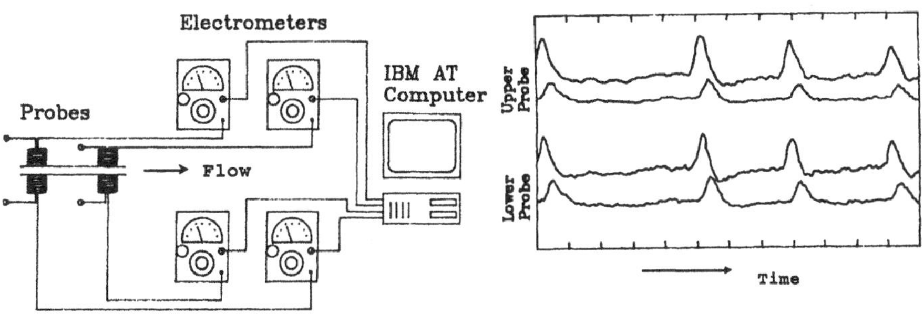

FIGURE 3-21 *Computer output signals (right) from two pairs of coils(left) showing a correlation of signals for 450 μm glass particles in PVC pipe at upstream and downstream locations (Klinzing et. al. 1987).*

Klinzing et. al. (1987) describe two systems in which the detectors are either two separate coils placed perpendicular to the flow (external to the pipe, Figure 3-21) at known separation distance or, alternatively, a single coil of wire wrapped around 0.0254 and 0.0508 m insulating PVC pipe (not shown). The first method gives a direct particle velocity without a calibration curve. In this setup two coils (one terminal of each grounded) were spaced at 0.34 m along the pipe and cross-correlation of signals was applied at the two locations. Similar signal patterns from fluctuations in the particles flow are assumed to persist along the pipe for some distance, Figure 3-21 (right). The time of flight of particles is measured from the known separation distance of the coils and the time measured between like signals (upper or lower pairs). Electrometers were used to amplify the signals. Approximate particle velocities of 5 - 18 m/s corresponded to gas velocities of about 5 - 22 m/s for 270 μm methyl methacrylate particles in a 0.0254 m horizontal pipe (the particle velocity lagging the gas velocity). They found that the particle charge increased pressure drop in vertical pipe flow for a solids loading of 100 kg solids/kg air (Ally and Klinzing, 1985, Zaltash et. al. 1988).

In a second system, Klinzing et. al. utilized a single coil to detect the wave action associated with accelerating /decelerating particles undergoing collisions with the walls and with themselves. The particles were 450 μm and 270 μm crushed glass beads and 270 μm methyl methacrylate particles. Coil currents (one end grounded) of 0.5 to 1.5 μA were detected with an electrometer corresponding to gas velocities of 7 to 10 m/s for constant solids flow rates of 0.020 and 0.029 kg/s. This system was found to be limited in usefulness since the particle velocity was not identified and calibration curves were be required for different flow configurations.

Zaltash et. al. 1988 describe a method for measuring vertical flow particle velocity in a pipe (Plexiglas and copper) utilizing two aluminum cylinders 0.0508 m in length with an internal diameter of 0.0254 m (1 in) spaced at 0.610 m (2 ft). The two small sections of cylinder constituted part of the pipe wall of 0.0254 m (1 in). Cross-correlation was used to interpret particle bombardment and charge transfer to the probes using an electrometer to detect the signals. A pair of optical probes was used to compare the flow rates, again using cross-correlation. Solid flow rates were 0.011-0.025 kg/s, gas velocities were 3.5-15.5 m/s, particle sizes were 18.8-446.3 μm (Plexiglas and copper), and the relative humidity was varied from 19.5 % to 61.6 %. Good agreement was reported for the two types of probes (electrostatic and optical).

A single ring-type capacitance sensor was used by Yan (1996) to measure particle velocity in a metal pneumatic pipeline. The grounded pipe constituted the second electrode of the capacitor (the ring being insulated from the pipe). This setup was more sensitive to particles located near the wall than located at the centerline. The method utilizes spatial filtering of the noise produced by the bulk flow of particles moving inside the ring. The ring measured 53 mm in diameter and 2 mm in axial length and was flush with the inner diameter of the pipe. Yan found that the bandwidth of the frequency response of the fast Fourier transform (FFT) produced by the flow was related to the dimensions of the ring in the axial direction and the mean particle velocity. A linear relationship is given between the measured bandwidth from 0 to 5.0 kHz and the mean particle velocity from 0 to 45 m/s. The method is sensitive to particle size but not the material. He gives data for cement and two grades of coal.

3.5 INSTRUMENTATION

3.5.1 Electrostatic Voltmeters, Fieldmeters, and Electrometers

Electrostatic fieldmeters, voltmeters, and electrometers have been discussed by Cross (1987), Glor (1988), Horenstein (1995), Pratt (1997), and Northrop (1997). *Contacting* electrometers by definition are electrostatic devices with a high input resistance (>10^{14} Ω) and come in two basic designs, high voltage capacitance based input and low voltage low current solid state input. In the conventional capacitance based electrometer movement, a potential differences from 100 V to 100 kV, corresponding to charge loadings of 0.01μC to 1 μC or capacitance change of 100 fF to 10 fF (f =10^{-15}), is measured by a mechanical force utilizing a capacitance-spring arrangement. By comparison, a modern low current solid state electrometer has typical input ranges as follows: potential difference of 10 μV to 100 V, current of 1 fA to 300 mA, charge of 1 fC to 10 μC, and resistance of 1 to 100 TΩ (T=10^{12}).

3.5.1.1 Contacting electrometer

A simple *contacting* high voltage electrometer is based on charge induced electrostatic torque of a variable capacitor. The high voltage source to be measured is placed in electrical contact with the rotating plate of the capacitor with the remaining plate held at ground potential. The torque on the capacitor varies as the square of the applied voltage while the restraining torsion spring is linear so that a calibration of the device results in a non-linear scale having reduced resolution at low voltages. This meter responds to both *ac* or *dc* voltage and has a true rms output (i.e., gives correct RMS for any wave shape). At radio frequencies, the reactance of the instrument must be considered. As noted previously, potentials from a few hundred volts to 100 kV or more can be measured with a typical input capacitance of 10 to 225 pF.

3.5.1.2 Noncontacting fieldmeter and voltmeter

Electrostatic fieldmeters and voltmeters are *noncontacting* devices used to measure high voltages to 100 kV or more of charged surfaces. Alternatively, the electric field strength from a charged suspension or a packed bed of powder can be measured directly if the instrument is properly employed. Low voltage electrostatic millivoltmeters in the range ± 0.010 to ± 10 V are designed to measure contact potentials of materials (see section on contact potential). Surface resolutions down to a few mm can be achieved with suitable probes.

Ideal meters have a high input impedance and low capacitance to avoid draining the source charge.

In a typical commercial *noncontacting* electrostatic *voltmeter*, the housing on the sensing probe is driven to the potential of the source so that the intervening electric field is nulled. The readout is given in volts at a specified distance. A nulled voltmeter cannot detect the intervening electric field strength between the test surface and the meter since only potential is indicated. Surface potentials from 0.01 to 200 kV can be measured by design of such instruments. Zacher and Williams (1995) discuss an *ac* feedback electrostatic voltmeter utilizing low voltage nulling and claim high accuracy and low cost.

Electrostatic *fieldmeters* detect the electric field strength at the sensing electrode and are calibrated to give the voltage of a charged surface at a known separation distance of the sensing electrode (usually limited to a few cm) or else to indicate electric field strength directly (if used properly). The surface voltage readout is volts/m (reading) multiplied by the gap distance (measured) when used within the specified calibration distance of the instrument. If the insertion of the *grounded* fieldmeter meter itself disturbs the electric field strength to be measured (by bringing ground potential close to the surface) then a theoretical correction to the measured electric field strength is required.

A hand-held electrostatic fieldmeter (assumed to be grounded) indicating the field lines with and without a grounded parallel plate is shown in Figure 3-22 (left). Field uniformity is improved near the test region by the addition of the ground plate that serves to null the electric field (E=0, from induced charged on plate) in the region behind the test surface for the case of either a charged insulator or conductor. This aids in the interpretation of the reading and extends the calibration distance of the instrument. A small opening for the electrode is provided to sense the field. [7] To stabilize the instrument, the sensing electrode can be vibrated to produce an *ac* signal (typical of commercial fieldmeters). Phase detection provides for the determination of positive or negative polarity of the measured field. In a dusty or corrosive environment, gas purging may be incorporated to protect the meter. In the electrostatic *field mill* design, either a rotating or tuning-fork shutter is used to produce an *ac* electric field at the sensing electrode. which is proportional to the electric field (Horenstein, 1995). Castle et. al. 1988 designed a self-purging electrostatic *field mill* for measuring the field strength of fly ash in an industrial electrostatic precipitator.

[7] In practice, the sensing electrode is placed behind a small hole of about 1 mm diameter.

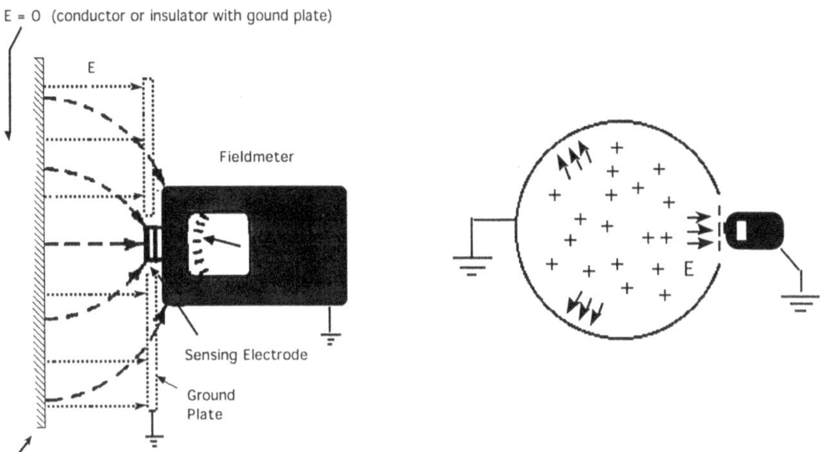

FIGURE 3-22 *Left: Non-contacting electrostatic fieldmeter, reads surface potential with/without ground plate; electric field behind charged test surface is zero when ground plate is added; Right, fieldmeter correctly mounted to read a uniform electric field strength E from charged suspension inside the grounded tube.*

Figure 3-22 (right) shows the correct placement of the electrostatic fieldmeter so that the electric field strength E can be read directly from the meter. In this application, the surface potential is not meaningful.

Two limiting electrostatic conditions of interest in potential measurement are constant voltage and constant charge flat surfaces. In both cases a uniform electric field is desired between the source and the sensing electrode so that the calibration of the instrument is retained over a wider range of separation distances (Blitshteyn, 1984).

Firstly, for a flat test object that is electrically isolated (charged dielectric in Figure 3-22, left) the charge remains constant while the surface voltage is altered by the presence of the meter due to the capacitance change as seen by the test object. The charge on the test surface remains constant. The unknown surface charge density q_s can be calculated using Gauss's law for E=0 behind the surface. The grounded metal plate on the meter situated at a distance L from the surface serves to exactly balance the charge on the test surface, so that [8]

$$q_s = \frac{V_m \varepsilon_o}{L}, \qquad V_{so} = \frac{C_s}{C_{so}} V_m \approx \frac{C_{so} + C_m}{C_{so}} V_m \qquad \text{constant charge} \tag{3.39}$$

[8] This equation should not be used for a constant voltage source measurement since in this case the charge on the surface will vary when the meter (and its ground plate) are inserted.

where the meter reading is V_m. The desired voltage-to-ground of the test surface V_{so} (without the meter) is then calculated from q_s. Its deviation from V_m depends on the ratio of capacitance-to-ground of the object before C_{so} and after C_s the introduction of the fieldmeter (C_m is the capacitance between the object and the grounded fieldmeter). Equations 3.39 are approximate, but demonstrate that the original voltage is greater than that indicated by the fieldmeter reading.

For a constant voltage source, the charge on the test surface will be altered by the capacitance of the meter; however, the voltage of the test surface can be correctly read from the meter (reading in V/m multiplied by the separation distance in m). For closely spaced parallel plates, the original charge density on the test surface is calculated by the relation

$$q_s = \frac{V_m \varepsilon_o}{L}\frac{C_{so}}{C_s} \approx \frac{V_m \varepsilon_o}{L}\left(\frac{C_{so}}{C_{so} + C_m}\right) \quad \text{constant voltage} \tag{3.40}$$

Equation 3.40 shows that the charge density on the test surface is increased by the presence of the meter (and ground plate) and must be reduced by the ratio of capacitance before and after introduction of the fieldmeter.

The insertion of a grounded (or ungrounded) instrument near the charged surface to be measured alters the surface potential for an electrically insulated surface or else alters the charge for a surface held at constant voltage. It should be noted that the meter reading (V/m) multiplied by the separation distance (m) gives the correct surface potential with the meter inserted since by its calibration.

3.5.1.3 Contacting voltmeters

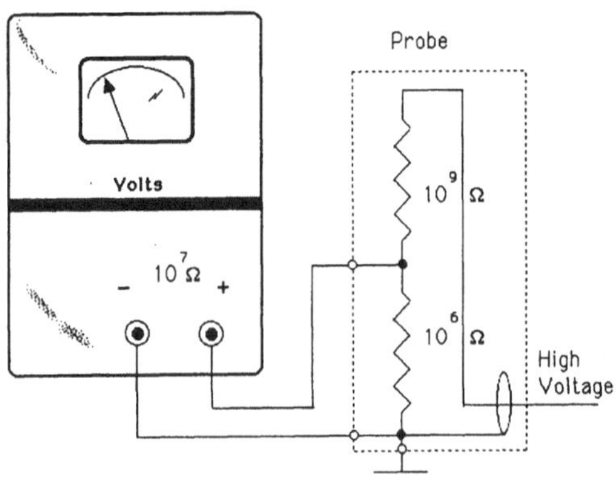

FIGURE 3-23 *Voltage divider circuit in high voltage probe connected to voltmeter.*

A *contacting* high voltage probe connected to a multimeter or similar laboratory instrument is fundamentally different from that of a an electrostatic *noncontacting* voltmeter described above since a sustainable current source is required at the measured potential. The probe voltage reading will be correct providing that the current drawn is not excessive in loading the test circuit. Laboratory voltmeters (or multimeters) typically have an input impedance of 10 MΩ and are designed to measure circuits but are not suitable for the measurement of spatial potential within an air gap. When such a voltmeter is connected to a *contacting* 40 kV high voltage probe (a dividing circuit is built in the probe, Figure 3-23), the input impedance is increased to about $10^9\,\Omega$ as seen at the probe tip. This impedance will be much less than that of the air gap (infinity). The solution is a high impedance instrument such as the electrostatic fieldmeter or voltmeter described previously.

3.6 OTHER MEASUREMENTS

3.6.1 Tomography

An established method in tomography is that of Magnetic Resonance Imaging (MRI) in which a magnetic field is used to detect and visualize a multiphase system (Hill and Kakalios, 1996). A recent development is electrical tomography utilizing capacitance, resistance, inductance or triboelectric sensing (Williams and Beck, 1995; Plaskowski et. al., 1995; Beck and Williams, 1996). Non-electrical tomography includes ultrasonic, optical and emission. These non-intrusive techniques give both temporal (real time) and spatial variations of bed voidage and bubble activity in fluidized beds. The Electrical Capacitance Tomography (ETC) system consists of the sensing electrodes, sensor electronics, transputer network, host pc, and display (Kuhn, et. al., 1994). Visualization in two and three-dimensional give information on bubble shape, size, and coalescence and bubble rise velocity (Wang. et. al., 1996). The dynamic behavior of a fluidized bed is considered from cross-correlation in either space or time. ECT is also being applied in chaos analysis to consider scaleup of fluidized beds. The volume being surveyed is limited to areas of the bed covered by the electrodes. A temporal resolution of less than 0.015 s is possible. At the present time, bed diameters of a few centimeters (10-28 cm) are utilizable by ECT with diameters that may reach the meter range (Kuhn, et. al., 1994).

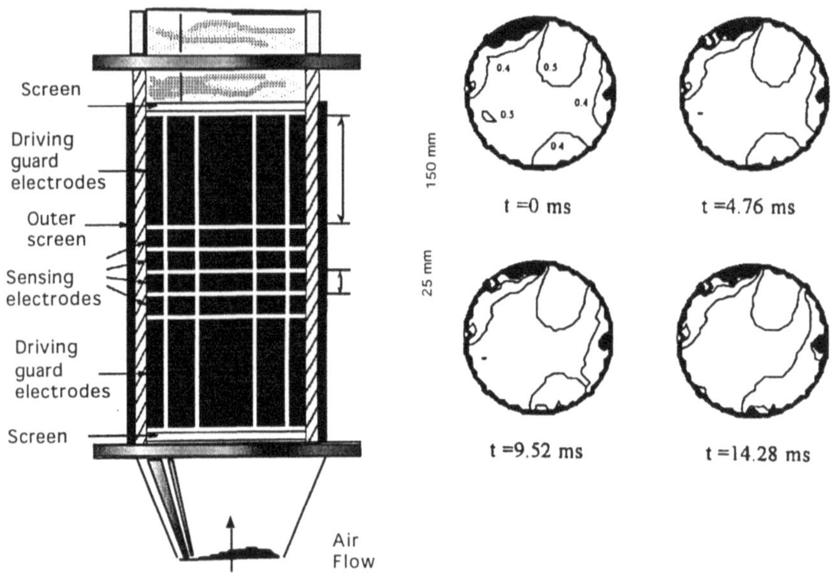

FIGURE 3-24 *Left: Four plane capacitance sensor with driven guard electrodes located near the distributor in fluidized bed (distances in mm); Adapted from Beck et. al.(1995); Right: Bubbles (dark) attached to a wall of bed in time sequence (see also Wang et. al. 1996)*

With ETC, capacitance measurements are made in the horizontal plane by 8 to 12 electrodes placed circumferentially around the bed with an outer shield kept at zero-potential to shield the sensing electrodes, Figures 3-24. Phase velocity can be detected by analyzing the time lag of a disturbances such as the tip of a rising bubble at different planes (4 planes are shown). A single plane gives a two-dimensional image of the voidage across a slice of the bed evaluated in real time as shown in Figure 3-24 (right) for a 15 cm bed. A three-dimensional construction of the bed is made possible using multiple layers of electrodes (Wang. et. al., 1996). Guard electrodes above and below the electrodes are used to overcome fringing of the field. The time between images is 4.76 ms with frame rates of 100 (two plane) to 200 per second (single plane). The spatial resolution of ECT is smallest near the center of the bed so that small bubbles may go undetected in this region.

The capacitance tomography scheme consists of sampling at low voltages (15 V) and high frequencies (1.25 MHz) all combinations of successive pairs of electrodes around the bed (Figure 3-46, left). The displacement current is sensed for each electrode pair, which is assumed proportional to the permittivity of the material between the electrodes. The wave length of the electric field is much greater than the diameter of the bed so that any queried process of fluidization is a solution to the average electrostatic field equations. The system is calibrated for both an empty bed and an all-solids compacted bed to establish the extremes, with the void fraction being characterized as intermediate grey levels as individual pixels on a monitor. An image is reconstructed using back projections to calculate the spatial permittivity (Plaskowski et. al., 1995). This procedure is successful if the relative permittivity of the solid is somewhat larger than that of the air (ratio: 3-6). To improve the resolution of an 8 electrode system (~0.3 fF), a 12 or 16 electrode can be employed. Other considerations include the center-bed resolution problem, soft-field error during image reconstruction, and field fringe error.

3.6.2 Electrostatic Discharge

The presence of particulate matter can have a profound effect on electrostatic discharge especially were matters of safety and explosion hazard are concerned. The measurement of electrostatic discharges will not be covered here but can be found in the literature under the following categories (Glor, 1988; Jones and King, 1991): spark/lightning discharge, corona discharge, capacitance discharge, brush discharge (two surfaces), propagating brush discharge (three surfaces) - Lichtenberg, Maurer discharge (type of brush discharge).

3.6.3 Ignition and spark breakdown testing of powders

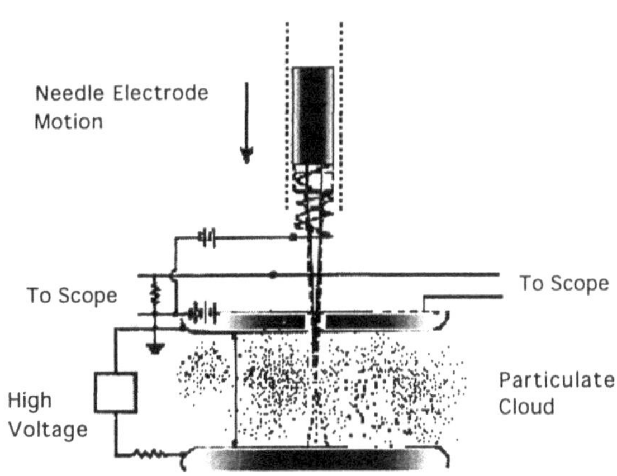

FIGURE 3-25 *EPS method used by Yu and Colver (1987, 1998) to suspend and measure sparking characteristics of copper suspensions.*

The electrostatic particulate method (EPS) of Fig. 3-25 has been used to measure various sparking and combustion characteristics of select powders (Colver et. al., 1996; Yu and Colver, 1987, 1998). Colver (1976) evaluated the breakdown distance for micro-discharge in an electric field between inductively charged small copper spheres and a conducting surface.

Other measurements of interest in multiphase systems include ignition energy, autoignition temperature, explosion pressure, and flammability limits in admixtures of oxygen with inert gases. Larger volume containers 15-20 liters are required to obtain ignition and flammability limits that are independent of wall effects (Cashdollar and Hertzberg, 1986). Once suspended, the cloud is ignited by any one of several means including spark energy, radiation (laser), chemical ignitors, and resistance heaters. The subject of powder combustion testing has been reviewed by Eckhoff (1991).

3.7 REFERENCES

Adamson, A. W, Physical Chemistry of Surfaces, 3rd. Ed., Wiley, NY, 1976, Chps. 4, 6.

Ally, M.R. and G. E. Klinzing, "Inter-relation of Electrostatic Charging and Pressure Drops in Pneumatic Transport," Powder Tech. (short Comm.) 44, 1985, pp. 85-88.

Asano, K. "Electrostatic Flow Measurement Techniques," in Handbook of Electrostatic Processes, Eds. J. S. Chang, A. J. Kelly, and J. M. Crowley, Marcelk Dekker, Inc., NY, 1995, pp. 265-269.

ASME/ANSI Power Test Code 28, "Determining the Properties of Fine Particulate Matter," 1973.

ASTM, Annual Book of ASTM Standards, ASTM Philadelphia, 1983 and 1986.

Bafrnec, M., and J. Bena, "Quantitative Data on Lowering of Electrostatic Charges in a Fluidized Bed," Chem. Eng. Sci., 27, 1972, pp. 1177-1181.

Bailey, A, W. L. Cheung, J. M. Smallwood, "A High-Resolution Probe for Space Potential Measurements," J. Electrostatics, 36, 1995, pp. 151-163.

Ban, H., J. L. Schaefer, and J. M. Stencel, "Particle Tribocharging Characteristics Relating to Electrostatic Dry Coal Cleaning," Fuel, 73, 7, 1994, pp. 1108-1113.

Ban, H., J. L. Schaefer, K. Saito, and J. M. Stencel, "Electrostatic Separation of Powdered Materials: Beneficiation of Coal and Fly Ash," Energeia, CAER, Univ. of Kentucky, 6, 4, 1995, pp. 1-3

Beck, M. and R. A. Williams, "Process Tomography: a European Innovation and its Application," Meas. Sci. Tech., 7, 1996, pp. 215-224.

Beck, M. S., T. Dyakowski, and, Wang, S. J., "Measurement of Fluidization Dynamics in a Fluidized Bed Using Capacitance Tomography," International Fine Particle Research Institute Annual Report, Nov. 30, 1995.

Bickelhaupt, R. E., "Surface Resistivity and the Chemical Composition of Fly Ash," J. Air Pol. Ctrl. Assoc., 25, 2, Feb. 1975, pp. 148-152.

Bleaney, B. I., and B. Bleaney, Electricity and Magnetism, Oxford, London, 1962.

Blitshteyn, M., "Measuring the Electric Field of Flat Surfaces with Electrostatic Fieldmeters," The Simco Co., Inc., 2257 N. Penn Rd., Hatfield, PA., 19440.

Boland, D., and D. Geldart, "Electrostatic Charging in Gas Fluidised Beds," Powder Tech., 5, 1971/72, pp. 289-297.

Bottomley, L. A., J. E. Coury, and P. N. First, "Scanning Probe Microscopy," Anal. Chem, 68, 1996, pp. 185R-230R.

Brodkey, R. S., The Phenomena of Fluid Motions, Addison-Wesley Series in Chemical Engineering, Robert S. Brodkey Pub., Columbus, OH., 1967, p. 114.
Brunton, J. D. and A. J. W. Rozelaar, "Factors Affecting The Force Between Electrical Contacts," Electronics Letters, 4, 26, Dec. 27, 1968, pp. 602-603.
Cashdollar K. L., and M. Hertzberg (Eds.), Industrial Dust Explosions, ASTM Special Technical Pub. 958, ASTM, Philadelphia, PA., June 10-13, 1986.
Castle, G. S. P. and M. R. Sewell, "An Ionization Device for Air Velocity and Mass Flow Measurements," IEEE Trans. Ind. Appl., IA-11, 1, Jan/Feb. 1975, pp. 119-124.
Castle, G. S. P. and M. R. Sewell, "General Model of Sphere-Sphere Insulator Contact Electrification," J. Electrostatics, 36, 1995, pp. 165-173.
Castle, G. S. P., I. I. Inculet, S. Lundquist, and J. B. Cumming, "Measurement of the Particle Space Charge in the Outlet of an Electrostatic Precipitator Using an Electric Field Mill," IEEE Trans. Ind. Appl., 24, 4, July/Aug. 1988, pp. 702-706.
Cheng, L., S, S. K. Tung and S. L. Soo, "Electrical Measurement of Flow Rate of Pulverized Coal Suspension," J. Eng. Power, Apr. 1970, pp. 135-149.
Cheng,L. and S. L. Soo, "Charging of Dust Particles by Impact," J. Appl. Phys., 41, 2, Feb 1970, pp. 585-591.
Ciborowski, J., and A. Wlodarski, "On Electrostatic Effects in Fluidized Bed," Chem. Eng. Sci., 17, 1962, pp. 23-32.
Clements, R. M. and P. R. Smay, "Collection of ions by Electric Probes in Combustion MHD Plasmas: an Overview," AIAA J. Energy, 2, 1978, pp. 53-58.
Clift, R., J. R. Grace, and M. E. Weber, Bubbles, Drops, and Particles, Academic Press, NY, 1978.
Cobine, James D. *Gaseous Conductors*, Dover, NY, 1958, p.254, 256,.
Colver, G. M. and A. M. Sarhan, "Pneumatic Transport of Solids with Electric Field," Second Internl. Particle Tech. Forum, 5th World Congress of Chemical Engineering, Vol. VI, AIChE, San Diego, CA., July 14-18, 1996, pp. 436-442.
Colver, G. M. and D. L. Howell, "Particle Diffusion in an Electric Suspension," Conference Record, IEEE Industry Applications Society Annual Meeting, Sept. 28-Oct. 3, 1980, pp. 1056-1062.
Colver, G. M. and J. S. Wang, "Bubble Stability Modeling in Fluidized Beds Utilizing Electric Fields," First Internl. Particle Technology Forum, Denver, CO, August 17-19, 1994, pp. 83-88.
Colver, G. M., "Bubble Control in Gas-Fluidized Beds with Applied Electric Fields," J. Powder Tech., 17, 1977, pp. 9-18.
Colver, G. M., "Dynamic and Stationary Charging of Heavy Metallic and Dielectric Particles against a Conducting Wall in the Presence of A DC Applied Electric Field," J. Appl. Phys., 47, 1976, pp. 4839-4849.
Colver, G. M., "Electric Suspensions Above Fixed, Fluidized and Acoustically Excited Beds", J. Powder Bulk Solids Tech., 4, 1980, pp. 21-31.
Colver, G. M., "Use of Electrical Resistivity in the Diagnostics of Powder Dynamics," International Powder and Bulk Solids Conference/Exhibition, Chicago, Illinois, May 12-14, 1981, pp. 89-96.
Colver, G. M., "Use of Powder Resistivity as a Diagnostic in Dense Phase Suspensions," AIChE Symposium Series 310, 92, 1996, pp. 168-173.
Colver, G. M., and G. S. Bosshart, "Heat and Charge Transfer in an AC Electrofluidized Bed," in Multiphase Transport: Fundamentals, Reactor Safety, Applications, 1-5, Hemisphere, Wash. DC, 1980, pp. 2215-2243.

Colver, G. M., and G. S. Bosshart, "Heat and Charge Transfer in an AC Electrofluidized Bed," in Multiphase Transport: Fundamentals, Reactor Safety, Applications, 1-5, Hemisphere, Wash. DC, 1980, pp. 2215-2243.
Colver, G. M., S. W. Kim, and Tae-U Yu, "An Electronic Method for Testing Spark Breakdown, Ignition, and Quenching of Powder," J. of Electrostatics, 37, 1996, pp. 151-172.
Corbett, R. P., "The Influence of Powder Resistivity and Particle Size on the Electrostatic Powder Coating Process," in Elektrostatische Aufladung, Dechema-Monographien Nr. 1370-1409, DECHEMA, Frankfourt, 1974, pp. 261-271.
Cotroneo, J. A. and G. M. Colver, "Electrically Augmented Pneumatic Transport of Copper Spheres at Low Particle and Duct Reynolds Numbers," J. Electrostatics, 5, 1978, pp. 205-223.
Cross, Jean, Electrostatics, Principles, Problems and Applications, Adam Hilger, Adam Hilger, Bristol, 1987, p. 30, Chp. 3.
Curtis, H. L., "Insulating Properties of Solid Dielectrics," , Bureau of Standards (Bulletin, Dept. of Commerce), 11, 1915, pp. 359-420.
Debeau, D. E., "The Effect of Adsorbed Gases on Contact Electrification," Physical Rev., 66, Nos. 1 & 2, July 1 and 14, 1944, pp. 9-16.
Dietz, P. W. and J. R. Melcher, "Momentum Transfer in Electrofluidized Beds". AIChE Symposium Series, 74, 175 , 1978A, pp.166-174.
Dietz, P. W. and Melcher, J. R., "Interparticle Electrical Forces in Packed and Fluidized Beds", Ind. Eng. Chem. Fundam., 17, 1, 1978B, pp. 28-32.
Donahoe, T. S. and G. M. Colver, "Bubble Rise Velocity in AC and DC Electrofluidized Beds," IEEE Trans. Ind. Appl., 1A-20, 2, March-April 1984, pp. 259-266.
Doremus, R. H., Glass Science, Wiley, NY, 1973, p 225.
Duckworth, H. E., Electricity and Magnetism, Holt, Rinehart and Winston, NY, NY, 1961.
Dutta, S. and C. Y. Wen, "A Simple Probe for Fluidized Bed Measurements," Can. J. of Chem. Eng., 57, Feb. 1979, pp. 115-119.
Eckhoff, R. K., Dust Explosions in the Process Industries, Butterworth-Heinemann, Oxford, England, 1991.
Fasso, L, B. T Chao and S. L. Soo, "Measurement of Electrostatic Charges and Concentration of Particles in the Freeboard of a Fluidized Bed," *Powder Tech.*, 33. 1982, pp. 211-221.
Fujino, M., Ogata, S., and H. Shinohara, "The Electric Potential Distribution Profile in a Naturally Charged Fluidized Bed and Its Effects," Intern. Chem. Eng., 25 (1), 1985, pp. 149-159.
Gajewski, J. B., "Mathematical Model of Non-Contact Measurements of Charges While Moving," J. Electrostatics, 15, 1984, p. 81.
Geldart, D. and J. R. Kelsey, "The Use of Capacitance Probes in Gas Fluidized Beds," Powder Technol., 6, 1972, pp. 45-60.
Gill, E. W. B. and G. F. Alfrey, Nature 163, 1949, p. 163.
Glor, M. Electrostatic Hazards in Powder Handling, Research Studies Press Ltd., Wiley, NY, 1988, pp. 27, 28.
Grace, J. R. and J. Baeyens, "Instrumentation and Experimental Techniques," in Gas Fluidization Technology, Ed. D. Geldart, Wiley, N. Y., 1986, Chp. 13.
Graham, W. and E. A. Harvey, "The Electrical Resistance of Fluidized Beds of Coke and Graphite," Canadian J. of Chem. Engr., 43, 3, June, 1965, pp. 146-149.

Greaves, J. R. and B. Makin, "Measurement of the Electrostatic Charge from Aerosol Cans," Conference Record, IEEE Industry Applications Society Annual Meeting, Sept. 28-Oct. 3, 1980, pp. 1075-1080.
Haenen, H. T. M., "The characteristic decay with time of surface charges on dielectrics," J. Electrostatics, 1, 1975, pp. 173-185.
Harper, W. R., "How do solid surfaces become charged?," in Static Electrification, Proceedings of Conference, Conf. Series No. 4, The Institute of Physics and the Physical Society Static Electrification Group, London, May 1967, pp. 3-10.
Harper, W. R., Contact and Frictional Electrification, Oxford university Press, 1967.
Hassler, H. E. Birgitta, "A New Method for Dust Separation Using Autogenous Electrically Charged Fog," J. Powder & Bulk Solids Tech., 2, 1, Internl. Powder Inst., Chicago, IL., Spring 1978, pp. 10-14.
Haus, H. A. and J. R. Melcher, Electromagnetic Fields and Energy, Prentice Hall, Englewood Cliffs, NJ, 1989, art. 1.6.
Hayakawa, T., W. Graham and G. L. Osberg, "A Resistance Probe Method for Determining Local Solid Particle Mixing Rates in A Batch Fluidized Bed," Can. J. of Chem. Eng., June 1961, pp. 99-103.
Hill, K. M. and J. Kakalios, "Magnetic Resonance Imaging of Granular Media Segregated in a Rotating Drum," Second Internl. Particle Tech. Forum, , 5th World Congress of Chemical Engineering, VI, AIChE, San Diego, CA., July 14-18, 1996, pp. 216-219.
Holland, L., The Properties of Glass Surfaces, Chapman and Hall, London, 1966., p. 480.
Holm, R., Electric Contacts, Springer-Verlag NY Inc., Berlin, 1967, pp. 10, 14.
Horenstein, M. N., "Measurement of Electrostatic Fields, Voltages, and Charges," in Handbook of Electrostatic Processes, Eds. J. S. Chang, A. J. Kelly, and J. M. Crowley, Marcelk Dekker, Inc., NY, 1995, pp. 225-246.
Ikazaki, F. and M. Kamamura, "Electric Adhesion Force of a Single Particle and of a Powder at Room Temperature and Above Ambient Temperature and its Application to a Fluidized Bed," Particle Sci. Tech., 2, 3, 1984, pp. 271-283.
Inculet, I. I., N. H. Malak, and J. A. Young, "Corona Charging of Immobilized Spherical Particles," in Electrostatics 1983, Ed. S. Singh, Conf. Ser., 66, Inst. Phys, 1983. pp. 98-105.
Israelachvili, J. N. and P. M. McGuiggan, "Forces Between Surfaces in Liquids," Science, 241, Aug. 12, 1988, pp. 795-800.
Jackson, J. D., Classical Electrodynamics, Wiley, NY, 1962.
Jimbo, G. and R. Yamazaki, "The Development and Evaluation of New Measuring Methods of the Adhesion Force of Single Particles," Kona, Powder Science and Technology in Japan, 1, 1983, pp. 40-47.
Johnson, T. W. and, J. R. Melcher, "Electromechanics of Electrofluidized Beds," Ind. Eng. Chem., Fundam, 14, 3, 1975, pp. 148-153.
Jones, T. B. "Diploe moments of conducting particle chains," J. Appl. Phys. 60, 7, 1 Oct. 1986, pp. 2226-2417.
Jones, T. B. and J. L. King, Powder Handling and Electrostatics, Understanding and Preventing Hazards, Lewis Pub., Chelsea Michigan, 1991.
Jones, T. B., "Dielectric Measurements on Packed Beds," GE Report No. 79CRD131, Tech. Info. Series, June 1979.
Jones, T. B., "Dielectric Measurements on Packed Beds," Report No. 79CRD131, General Electric Co., Schenectady, NY, June, 1979, p. 26.

Jones, T. B., "Effective dipole moment of intersecting conducting spheres," J. Appl. Phys. 62, 2, 15, July 1987, pp. 362-365.
Jones, T. B., Electromechanics of Particles, Cambridge Univ. Press, NY., 1995, p. 26.
Keithley, Test and Measurement Catalog, 1995/96, p. 237.
Kennedy, J. B. and A. M. Neville, Basic Engineering Statistics, Harper and Row, N. Y., 1976, pp. 239-243.
Kingery, W. D., H. K. Bowen, and D. R. Uhlmann, Introduction to Ceramics, 2nd Ed., Wiley, NY, 1976, pp. 875, 876, art. 17.5.
Kisel'nikov, V. N., Vyalkov, V. V., and V. M. Filatov, "On the Problem of Electrostatic Phenomena in a Fluidized Bed," Intern. Chem. Eng., 7, 3, 1967, pp. 428-431.
Klinzing, G. E., A. Zaltash, and C. A. Myler, "Particle Velocity Measurements Through Electrostatic Field Fluctuations Using External Probes," Particle Sci. Tech., 5, 1, 1987, pp. 95-104.
Kobashi, M., "Particle Agglomeration Induced by Alternating Electric Fields," HTGL (High Temp. Gasdynamic Lab.; Mech. Eng. Dept.), Report No. 111, Stanford Univ., Dec. 1978.
Krupp, H., "Particle Adhesion, Theory and Experiment," In Adv. in Colloid and Interface Sci., 1967, p. 170.
Kuhn, F.T., J. C.Schouten, C. M. Van den Bleek, and B. Scarlett, "Electrical Capacitance Tomography Applied to Fluidization Dynamics for Chaos Analysis," Second Internl. Particle Tech. Forum, 5th World Congress of Chemical Engineering, VI, AIChE, San Diego, CA., July 14-18, 1996, pp. 216-219.
Kuhn, F.T., J. C.Schouten, R. F. Mudde, J. C. M. Marijnissen, C. M. Van den Bleek, H. E. A. van den Akker, and B. Scarlett, "Electrical Capacitance Tomography for Flow Imaging in Fluidised Beds," First Internl. Particle Tech. Forum, Part I, AIChE, Denver CO., Aug. 17-19, 1994, pp. 512-517.
Kunii, Daizo, and Octave Levenspiel, Fluidization Engineering, 2nd Ed., Butterworth-Heinemann, Boston, 1991, pp. 80,81.
Kunkel, W. B., "The Static Electrification of Dust Particles on Dispersion into a Cloud," J. Appl. Phys., 21, Aug. 1950, pp. 820-832.
Lacharme, J. P., "Ionic jump processes and high field conduction in glasses," J. Non-Crystalline Solids 27, 1978, pp. 381-397.
Lamarre, E. and W. K. Melville, "Instrumentation for the Measurement of Void-Fraction in Breaking Waves: Laboratory and Field Results," IEEE J. of Oceanic Eng., 17, 2, April 1992, p. 204-.
Lampert, M., and P. Mark, Current Injection in Solids, Academic Press, N. Y.,1970, Chps. 2, 5.
Lapple, C. E., "Electrostatic Phenomena with Particulates," Advances in Chemical Engineering, Academic Press, NY, 1970, pp. 2-96.
Law, S. E, "Electrostatic Atomization and Spraying," in Handbook of Electrostatic Processes, Eds. J. S. Chang, A. J. Kelly, and J. M. Crowley, Marcelk Dekker, Inc., NY, 1995, pp. 413-439.
Lawton, J. and Weinberg, F. J., Electrical Aspects of Combustion, Clarendon Press, Oxford., 1969, Chps.2, 5.
Liu, X., and G. M. Colver, "Capture of Fine Particles on Charged Moving Spheres: A New Electrostatic Precipitator," IEEE Trans. Ind. Appl., 27, 5, Sept/Oct 1991, pp. 807-815.
Loeb, L. B., Static Electrification, Springer-Verlag, Berlin, 1958, Chp. IV d.

Louge, M., "Experimental Techniques, in Circulating Fluidized Beds," Eds. J. R. Grace, A. A. Avidan, and T. M. Knowlton, Blackie Academic & Professional (Chapman Hall), NY, 1997, Chp. 9.
Martin, C. M., M. Ghadiri, U. Tuzun and B. Formisani, "Effect of the Electrical Clamping Forces on the Mechanisms of Particulate Solids," Powder Tech., 64, 1991, pp. 37-49.
Martin, C. M., P. A. Arteaga, M. Ghadiri, and U. Tuzun, "Characterization of the Single Contact Electrical Clamping Force," (pt. II), AICHE First Internl. Particle Technology Forum, Denver, CO, Preprints, Aug. 17-19, 1994, pp. 77-82.
Masuda, H., T. Itakura, k. Gotoh, T. Takahashi, and T. Teshima, "The Measurement and Evaluation of the Contact Potential Difference Between Various Powders and a Metal," Adv. Powder Tech., 6, 4, 1995, pp. 295-303.
Mathur, M. P. and G. E. Klinzing, "Measurement of Particle Velocity in Pneumatic Transport of Coal Using Cross-Correlation Technique," Particulate Sci. and Tech., 2, 1984, pp. 223-235.
Matsuyama, T. and H. Yamamoto, "Electrification of Single Polymer Particles by Successive Impacts with Metal Targets," IEEE Trans. Ind. Appl., 31, , 1995, pp. 1441-1445.
Maxwell, J.C., A Treatise on Electricity & Magnetism, 3 rd. ed., Dover, NY, 1, 1954, pp. 276, p. 440.
Mazumder, M. K., "E-SPART Analyzer: Its Performance and Application to Powder and Particle Technology Processes," KONA, No. 11, 1993, pp. 105-118.
Mazumder, M. K., R. E. Ware, T. Yokoyama, B. J. Rubin, and D. Kamp, "Measurement of Particle Size and Electrostatic Charge Distributions on Toners Using E-SPART Analyzer," IEEE Trans. Ind. Appl., 27, 1991, pp. 611-619.
McLean, K. J., "Cohesion of Precipitated Dust Layer in Electrostatic Precipitators," J. Air Pol. Contrl. Assoc., 27, 11, Nov. 1977, pp. 1100-1103.
Misev, T. A., Powder Coatings, Chemistry and Technology, Wiley, NY, 1991, p. 292.
Miyanami, K., and Terashita, K., "Direct Shear Test of Powder Beds," Kona, Powder Science and Technology in Japan, 1, 1983, pp. 28-39.
Mizuno, A. and M. Otsuka, "Development of Charge-to-Radius Ratio Measuring Apparatus for Submicron Particles," IEEE Trans. Ind. Appl., IA-20, 3, May/June 1984, pp. 703-708.
Monroe, Electrostatic Instrumentation Catalog, 1997, p. 9.
Montgomery, D. J., Solid State Physics, *Advances in Research and Applications*, Eds. F. Seitz and D. Turnbull, 9 Academic Press, NY, 1959.
Morey, G. W., The Properties of Glass, Reinhold, NY, 1954, p. 476.
Mort, J., The Anatomy of Xerography, Its Invention and Evolution, McFarland & Co. Inc, Jefferson, NC., 1989.
Moslehi, B., "Electromechanics and Electrical Breakdown of Particulate Layers," High Temperature Gasdynamics Laboratory Report No. T-236, Stanford Univ., Dec. 1983.
Moslehi, G. B., and, S. A. Self, "Current Flow Across a Sphere with Volume and Surface Conduction," J. Electrostatics, 14, Mar. 1983, pp. 7-17.
Nakajima, Y., Y. Komuro, and T. Sato, "Electrostatic Scavenging of Submicron Particles Aided by the Hydrodynamic Effect of Particle Vibration," Adv. Powder Tech., 7, 4., 1996, pp. 255-270.

Nieh, S., B. T. Chao, and S. L. Soo, "An Electrostatic Induction Probe for Measuring Particle Velocity in Suspension Flow," Part. Sci. and Tech., 4, 1986, pp. 113-130.
Northrop, R. B., Introduction to Instrumentation and Measurement, CRC Press, New York, 1997, sec. 8.2.
Plaskowski, A, M. S. Beck, R. Thorn, and T. Dyakowski, Imaging Industrial Flows, (Applications of Electrical Process Tomography), Inst. of Physics, Philadelphia, 1995.
Pratt, T. H., Electric Ignitions of Fires and Explosions, Burgoyne Inc., Marietta, GA, 1997.
Rhim, W. K. and A. J. Rulison, "Measuring Surface Tension and Viscosity of a Levitated Drop," NASA Tech. Briefs, July 1996, pp. 64-65.
Rietema, K., The Dynamics of Fine Powders, Elsevier, New York, 1991.
Robinson, K. S. and T. B. Jones, "Electromechanics of Electropacked Beds," Conference Record, IEEE Industry Applications Society Annual Meeting, Sept. 5-9, 1980, pp. 1068-1074.
Robinson, K. S. and T. B. Jones, "Particle-Wall Adhesion in Electropacked Beds," Conference Record, IEEE Industry Applications Society Annual Meeting, Oct. 4-7, 1982, pp. 1013-1017.
Robinson, K. S. and T. B. Jones, "Slope Stability of Electropacked Beds," Conference Record, IEEE Industry Applications Society Annual Meeting, Oct. 5-9, 1981. 1036-1042
Sarhan, Ahmed., "Effect of Electrically Driven Particles on Air Flow in a Rectangular Duct," Ph.D. thesis, Dept. of Mechanical Eng., Iowa State Univ., Ames, IA.1989.
Schaefer, J. L., H. Ban, and J. M. Stencel, "Triboelectrostatic Dry Coal Cleaning," Proceedings of the Eleventh Annual Coal Conf., Sept. 12-16, Pittsburgh, pp. 624-629.
Schein, L. B., Electrophotography and Development Physics, 2 nd Ed., Springer-Verlag, Berlin, 1992, pp. 77, 78, 82, art. 4.4.4.
Self, S. A., R. Assaad, E. Kushner, P. Paul and E. Pejack, "A Charge-Analyzer Probe for Aerosol Flows," Conference Record, IEEE Industry Applications Society Annual Meeting, 1979, pp. 218-225.
Smith C. P., S. R. Snyder, and H. S. White, "Measurement of Surface Forces," in Electrochemical Interfaces: Modern Techniques for In-Situ Interface Characterization, Ed. H. Abruna, VCH Pub., NY, 1991, pp. 157-191.
Smythe, W. R., Static and Dynamic Electricity, McGraw-Hill, NY, 1968, arts. 6.04, 6.10.
Soo, S. L., "State of Multiphase Instrumentation," Vol. XI, Developments in Theoretical and Applied Mechanics, University of Alabama, Huntsville, AL., 1982, pp. 563-567.
Soo, S. L., D. A. Baker, T. R. Lucht, and C. Ahu, "A Corona Discharge Probe System for Measuring Phase Velocities in a Dense Suspension," Rev. Sci. Instrum. 60, 11, Nov. 1989, pp. 3475-3477.
Soo, S. L., Fluid Dynamics of Multiphase Flow, Blaisdell, Waltham Mass, 1967, art. 10.4.
Stoy, R. D., "Force on Two Touching Dielectric Spheres in a Parallel Field," J. Electrostatics, 35, 1995, pp. 297-308.

Takahashi, H., K. Sato, S. Sakata, and T. Okada, "Charge Leakage Characteristic of Glass Substrate for Liquid Crystal Display," J. Electrostatics, 35, 1995, pp. 309-322.

Tardos, G. I., R. W. L. Snaddon, and P. W. Dietz, "Electrical Charge Measurements on Fine Airborne Particles," IEEE Trans. Ind. Appl., IA-20, 6, Nov/Dec 1984, pp. 1578-1583.

Tardos, G., and R. Pfeffer, "A Method to Measure Electrostatic Charge on a Granule in a Fluidized Bed," Chem. Eng. Commun., 4, 1980, pp. 665-671.

Taylor, D. and P. Secker, Industrial Electrostatics, Wiley, NY, 1994.

Teyssedou, A., A. Tapucu, and M. Lotie, "Impedance Probe to Measure Local Void Fraction Profiles," Rev. Sci. Instrum., 59, 4, April 1988.

Timoshenko, S. and J. N. Goodier, Theory of Elasticity, 2nd Ed., Mcgraw Hill, NY., 1951, art. 125.

Tombs, N. T. and T. B. Jones, "Digital Dielectrophoretic Levitation," Rev. Sci. Instrum, 62, 4, Apr 1991, pp. 1072-1077.

Tombs, N. T. and T. B. Jones, "Effect of Moisture on the Dielectrophoretic Spectral of Glass Spheres," IEEE Trans. Ind. Appl., 29, 2, Mar/Apr 1993, pp. 281-285.

Tombs, N. T., "Electrostatic Force on a Moist Particle Near a Ground Plane," J. Adhesion, 51, 1995, pp. 15-25.

Touchard, G., "Flow Electrification of Liquids," in Handbook of Electrostatic Processes, Eds. J. S. Chang, A. J. Kelly, and J. M. Crowley, Marcelk Dekker, Inc., NY, 1995, pp. 83-87.

Trek, Product and Systems Catalog, 1998, p. 18.

Tucholski, D. and G. M. Colver, "Electrostatic Separation of Coal Pyrite in a Circulating Fluidized Bed," Proceedings of the 2nd International Conference on Appl. Electrostatics, Beijing China, Nov. 4-7, 1993B, pp. 412-416.

Tucholski, D. and G. M. Colver, "Triboelectric Charging in a Circulating Fluidized Bed," Proceedings of the 2nd International Conf. on Appl. Electrostatics, Beijing China, Nov. 4-7, 1993A,. pp. 287-296.

Tucholski, D., "Coal Beneficiation and Triboelectric Effects in a Circulating Fluidized Bed," MS Thesis, Dept. of Mechanical Engineering, Iowa State University, 1992.

Turner, G. A., and M. Balasubramanian, "The Frequency Distribution of Electrical Charges on Glass Beads," J. of Electrostatics, 2, 1976, pp. 85-89.

Vercoulen, P. H. W., Electrostatic Processing of Particles, Doctoral Thesis, Technical University of Delft, 1995, p. 81.

Vercoulen, P. H. W., J. C. M. Marijnissen, B. Scarlett and R. A. Roos, "The Development of an Instrument for Measuring Electric Charge on Individual Particles," Proc. of the PARTEC Conf. Nurnberg, 2, 1992, p. 593.

Wang, S. J., T. Dyakowski, and M. S. Beck, "An Application of Electrical Capacitance Tomography to Measure Gas-Solid Motion in a Fluidized Bed," AIChE Symposium Series, National Heat Transfer Conf., Houston, Ed, M. S. El-Genk, 92, 310, Aug. 3-6, 1996, pp. 155-160.

Weast, R. C., (Ed.), Handbook of Chemistry and Physics, The Chemical Rubber Co., Cleveland, OH., 1970.

Weissler, G. A., "Resistance Measurements on Copper Powder Using High DC Currents," J. Powders & Bulk Solids Tech., 2, 1, 1978, pp. 38-40.

White, H. J., Industrial Electrostatic Precipitation, Addison-Wesley, Reading MS, 1963, p.135.

Williams, R. A. and M. S. Beck (Eds), Process Tomography, Butterworth-Heinemann, Oxford, 1995.
Yamamoto, H. and B. Scarlett, "Triboelectric Charging of Polymer Particles by Impact," Part. Charac. 3, 3, Oct. 1986, pp. 117-121.
Yan, Y., "Spatial Filtering Method of Particle Velocity Measurement in a Pneumatic Suspension Using a Single Capacitance Sensor," Second Internl. Particle Tech. Forum, 5th World Congress of Chemical Engineering, Vol. VI, AIChE, San Diego, CA., July 14-18, 1996, pp. 581-586.
Yu, Tae-U and G. M. Colver, "Spark Breakdown in Clouds of Conducting Particles and Related Particle Dynamics," World Congress on Particle Technology 3, IChE, Brighton, UK, July 6-9, 1998, paper #77 (CD-ROM), Abstracts p. 42.
Yu, Tae-U and G. M. Colver, "Spark Breakdown of Particulate Clouds: A New Testing Device," IEEE Trans. Ind. Appl., 1A-23, 1, Jan./Feb. 1987, pp. 127-133.
Zacher, D. M. and B. T. Willimam, "An AC Feedback Electrostatic Voltmeter," ESA 1995 Annual Meeting Proceedings, Eds., J. M. Crowley, M. Zaretsky, and D. Rimai, June 20-23, 1995, Univ. of Rochester, Laplacian Press, Morgan Hill, CA., pp. 159 - 164.
Zaltash A., C. Myler and G. E. Klinzing, "Stability Analysis of Gas-Solid Transport with Electrostatics," J. Pipelines, 7, 1988, pp. 85-100.

Notation

a = cap (contact) radius of spherical particle (m)

A_b = bed cross-section (m^2)

A = sample or capacitor cross-section (m^2)

C = capacitance (F)

d = diameter of particle (m)

D = diameter of ring electrode or bed

D_w = diameter of ring electrode or bed

E = electric field strength (V/m)

E_o = far-field electric field strength (V/m)

f = frequency (s^{-1})

F_{dc} = dc field particle force (N)

$F_{ac/dc}$ = ac or dc field particle force (N)

F_g = force on particle by gravity (N)

g = gravity (m^2/s)

h = height of powder (m)

I = current (A)

J = current density (A/m^2)

K = equation constants, Cunningham correction

L = sample length (m)

m = mass of particle (kg)

$\dot{m}_d$ = mass flux of dispersed phase (kg/m^2-s)

M_d = mass of powder in the bed

N_p = number of particles

n_d = dispersed phase concentration ($number/m^3$)

Q = (particle) charge (C)

Q_{el} = (particle) triboelectric charge (C)

q_s = surface charge density (C/m^2)

q_v = charge-to-mass ratio (C/kg)

r = radius of particle (m), of cylinder (m)

R = resistance (Ω), outer radius of cylinder (m)

R.H. = relative humidity (%)

R_s = surface resistance (Ω)

R_c = particle contact resistance (Ω)

R_p = resistance of single particle (Ω)

S = surface area to mass ratio of particle (m^2/kg)

t = time (s)

T = period of cycle (s)

u_d= dispersed phase velocity (m/s)

V = voltage (V) (= electric potential difference)

V_o = applied voltage (V)

V_c = contact potential (V)

$\dot{Vol}$ = volume flow rate of gas

W = sample width (m)

Z = impedance (Ω)

Greek Notation

Subscripts: b = bulk (bed), c = continuous phase (gas or liquid), d = disperse phase (solid particle or droplets), i = phase i, p = particle, s = surface, v = volume, bold = vector quantity

α = void fraction (dimensionless)

γ = constant = 0.5772; material surface resistivity (Ω/sq)

ε= material permittivity (phase i) (F/m)

ε_o = permittivity of free space (F/m)

ε_b = effective bulk permittivity of bed (F/m)

Ω = electric potential or contact potential (V)

κ = dielectric constant (= $\varepsilon/\varepsilon_o$)

λ = mean free path of gas (m)

μ = viscosity (N-s/m^2)

ρ = material density (kg/m^3)

ρ_q = volume charge density (C/m^3)

σ = material conductivity (phase i) (S/m); surface conductivity (N/m)

σ_q = surface charge density (C/m^2)

τ = time constant (s)

ω = radial frequency for *ac* field (radian/s)

$\Re$ = material resistivity (Ω-m)

4

Fiber Optics

Shaozhong Qin, Mooson Kwauk

4.1 INTRODUCTION

Since fiber optics was introduced (Kapany, 1957, for instance), this new sensing technique has developed rapidly. The application of optic fibers as sensors in measurements has the advantages of high sensitivity, fast response, large dynamic range, small volume and light weight, fire- and shock resistance and corrosion proof, freedom from disturbance by electric and magnetic fields, insulation against high voltage, and suitability for remote transmission and multi-channel detection. As light may produce reflection, refraction, interference, polarization and diffraction, combination of these phenomena with the fiber optic technique results in a variety of optic fiber sensors.

With development in application of optic fibers in communication and imaging, special demands have called forth different types of optic fibers each with its unique performance. According to materials, optic fibers can be divided into three categories: glass optic fibers, plastic optic fibers and liquid core fibers. Based on refractivity distribution, optic fibers can be classified into two types: step-index optic fibers and gradient-index optic fibers. As to transmission mode, there are single-mode and multi-mode optic fibers. Figure 4-1 shows the range of core diameters and the refractivity distribution for three principal types of optic fiber.

The optic fiber shown in Figure 4-1a consists of a core with refractivity of n_1 and a coating with refractivity of n_2, for which the condition of $n_2 < n_1$ should be satisfied. Light waves are transmitted in total reflection. This kind of optic fiber has been widely used due to its large core diameter and good performance. The optic fiber shown in Figure 4-1b, is a special example in which the refractivity n_1 of the core material varies radially from the central axis outward. When light is transmitted in this kind of optic fiber, its transfer

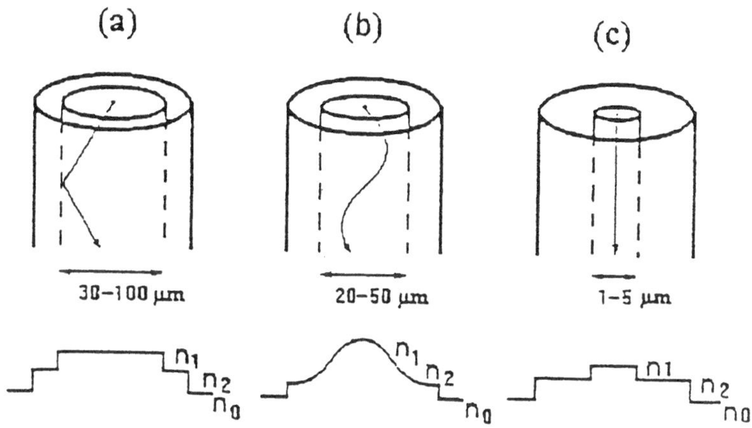

FIGURE 4-1 *Refractivity and core diameter ranges for three types of optic fibers. (a) Multi-mode step-index optic fiber, (b) Multi-mode gradient-index optic fiber, (c) Single model step-index optic fiber.*

trajectory is a sine curve. This optic fiber has the ability of focusing and transferring images in a single fiber. With this kind of optic fiber, lens fibers of several tens of μm can be constructed for direct transmission of images. The optic fiber shown in Figure 4-1c is called single mode fiber because its core diameter is so small (usually several μm) that only a single beam of axial light could be transmitted in it. This kind of optic fiber has a very narrow frequency band and little signal distortion, and is thus suitable for remote communication.

Based on the basic performance of optic fiber sensors, Krohn (1986) divided optic fiber sensors into two basic classes. In the first class, the transmission of the fiber is directly affected by the physical phenomena being sensed and is referred to as an intrinsic optic fiber sensor. The second class is for optic fiber position sensors which detect position changes and are sensitive to changes in physical property. There are usually five types of sensors according to their different working principles: intensity modulated, transmitting, reflective, micro bending and intrinsic.

Optic fiber sensors used in multiphase flows are based on the reflectivity of particles against incident light. This kind of optic fiber sensor, the so-called intensity modulated optic fiber sensor, is simple in structure, easy in operation and high in sensitivity.

The application of optic fiber probes to the measurement of local concentration of solids and particle velocity, will be described below separately.

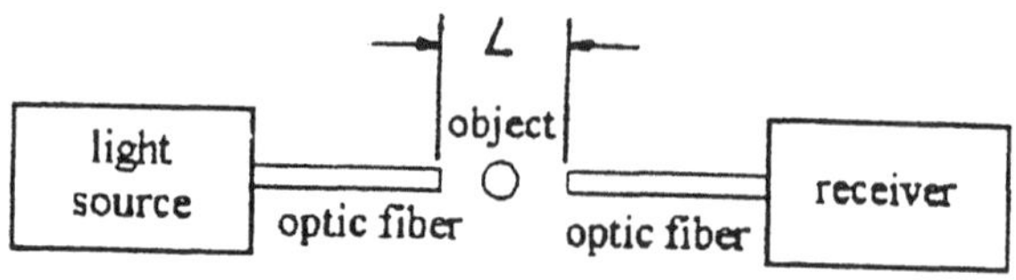

(a) transmission – type

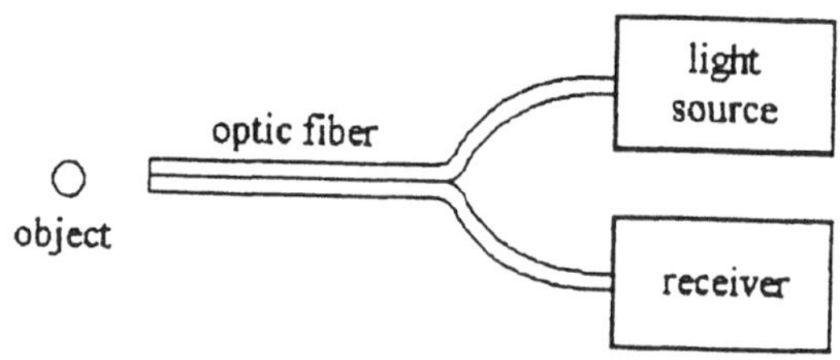

(b) reflection – type

FIGURE 4-2 *Two different arrangements of optic fiber probes (Matsuno et al., 1983).*

4.2 MEASUREMENT OF LOCAL CONCENTRATION OF SOLIDS

The application of optic fiber probes to the measurement of local concentration of solids is based on the principle that the particles in the fluid produce scattering of incident light. Thus structurally this kind of probes consists of two parts: light input and light output. Usually, for light input, a fiber is connected to a light source, while for light output, it is connected to a photoelectric converter. However, the difference in location of light input and light output for optic fiber probes may result in different modes of signals.

Figure 4-2 shows two different arrangements of optic fiber probes. Type (a) is called the transmission-type probe, in which light input and light output are coaxial, i.e., the object to be measured is located between the two probe tips, and the effective volume of measurement is dependent on the distance *L*, between the two probe tips, diameter and the numerical aperture of the probe. The output signals of the probe may attenuate due to the forward scattering of particles against the incident light, and the signal is

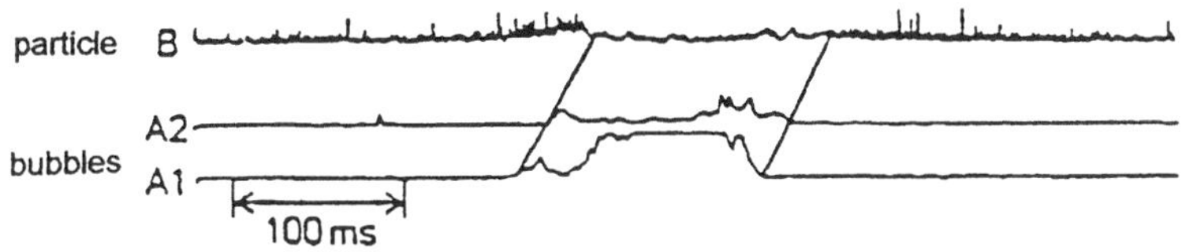

FIGURE 4-3 *Comparison of particles (B) and bubbles (A1, A2) signals Hatano and Ishida, 1983).*

thus independent of the chromaticness of particles, i.e., a single white or black particle may produce the same output signals.

Type (b) is called the reflection-type probe. The probe has only one tip and the effective volume of measurement depends on the diameter, numerical aperture, overlap region of the capture angles and the optic sensitivity of the photoelectric converter. The output signals of the probe are produced by the back scattering of incident light by particles, and are thus dependent on the chromaticness and reflectivity of the particles.

4.2.1 The TransmissionType Probes

The measurement of local particle concentration with the transmission type probes was developed on the basis of opto-electric turbidimetry. This simplest probe structure is based on the forward scattering of particles against incident light. Hatano and Ishida (1983) used this type of probe in the measurement of bubbles in fluidized beds and compared the resulting signals with those obtained by the reflection-type probe. Figure 4-3 illustrates the signals for bubbles and particles velocity obtained at different probe locations. It can be seen from Figure 4-3 that the bubble signals (A_1, A_2) obtained by the transmission-type probe are more prominent than those by the reflection-type probe for velocity measurement, but when the fluidized bed is in a bubbleless highly concentrated state, there would be no obvious difference between particle velocity signals (B). Therefore, the measuring range of volumetric concentration of the transmission-type probe could not be too large, and the distance, L, between the incident light and the receiving light could not be too long, or otherwise the waveforms of signals would be flat and without obvious response due to the overlap of particles in the dispersed phase. It is especially important that the structure of the transmission-type probe should ensure the light input and output exactly coaxial, since any displacement may produce variation in the effective

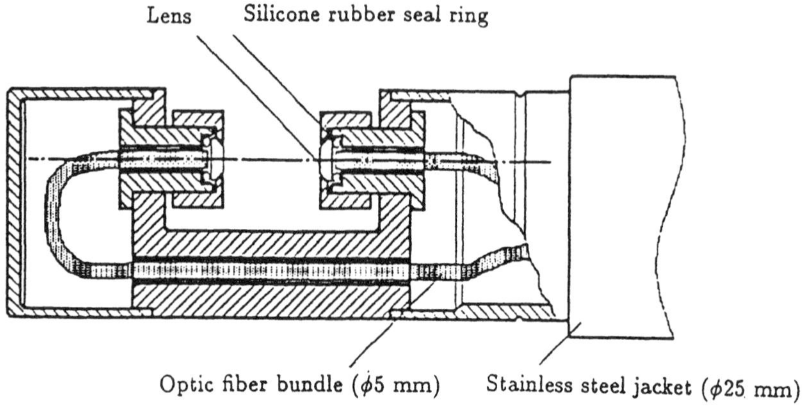

FIGURE 4-4 *Industrial optic fiber probe for cell concentration measurement.*

fluid medium or vibration of the device.

Nakajima et al. (1990) improved the structure of the transmission-type probe by parallel arrangement of the two fibers, thus avoiding the difficulty of keeping the two tips of probes coaxial, and reducing the influence of vibration. Moreover, in order to maintain a constant distance *L*, the incident light and receiving light of the fiber could be bent by 90°, i.e., the tips of the two large fibers were ground into 45° reflecting surfaces.

Cutolo et al. (1990) measured high time-averaged solids volumetric concentration (up to 0.16) in gas--solid suspensions. These authors' method was based on a properly modified version of the forward scattering of laser light. When a parallel beam of light was incident on a collection of dielectric particles uniform in size and in spatial distribution, the energy fraction that emerges without experiencing any deflection from the original direction of propagation was given by Dobbins and Jizmagian (1966):

$$T = \exp\left[-3E(\alpha,m)C_v L/2d_p\right] \tag{4.1}$$

where T is the optical transmittance, L the optical path length, E the extinction coefficient, m the refractive index of the particles relative to the surrounding medium, α the size number equal to $\pi\ d_p/\lambda$, and λ is the wavelength of the incident light in the surrounding medium. The experimental apparatus basically consists of a solids feed hopper and a 41 mm i.d. by 1 m high plexiglass pipe. The solids were narrow cut glass beads, with a 90 μm average diameter. A smaller tube (33 mm i.d.) was coaxically placed at the bottom of the 41 mm pipe. This allowed the separation of solids falling along the walls of the larger tube which were not subject to measurements.

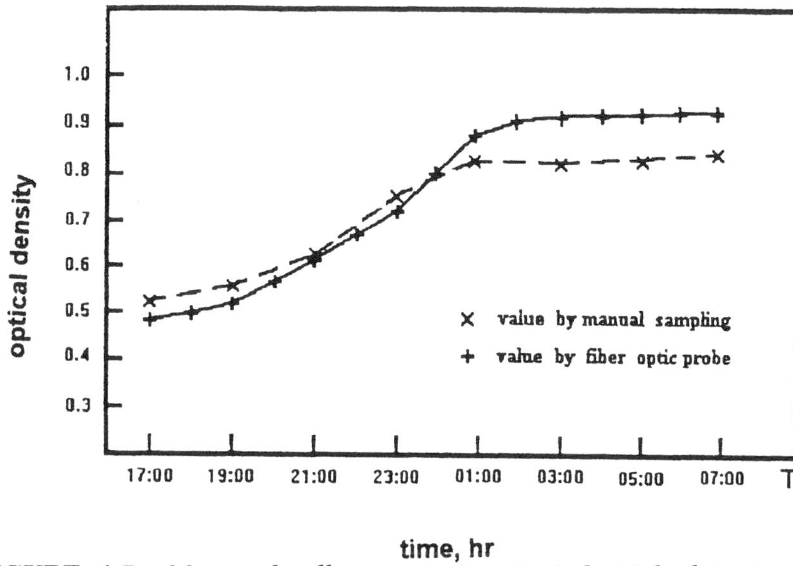

FIGURE 4-5 *Measured cell concentration in industrial glutamic acid fermentation.*

Qin and Liu (1982) measured concentration of cells with the transmission-type optic fiber probe in a 5-ton reactor for glutamic acid fermentation. The structural features of the probe used for measurement in industrial reactors are shown in Figure 4-4. This probe stood up long periods of steam disinfection at 130°C and vibration of stirring etc., and large amounts of data on concentration of cells during fermentation were obtained. The measuring range of this type of probe was from zero to 10^8 cells/ml. In order to avoid the influence of change in fermentation liquor on the wavelength of incident light, an interference filter with fixed wavelength of 650 nm was used in the apparatus, and flexible optic fiber bundles were used instead of the traditional optic parts such as reflector, focusing lens, cassette etc., so that solid state light circuitry could be employed, resulting in improvement in vibration resistance of the optic probe that is required in industrial application. In electrical circuitry the automatic compensation method was adopted by using a dual-light path modulation system. By means of a light chopper, the photo-multiplier picked up alternatively measuring signals, light source reference signals and zero standard signals in the measuring cycle, and the light source reference signals and zero standard signals were compensated automatically by two feedback circuits, eliminating errors caused by unavoidable instability of light source and zero drift of the system. Figure 4-5 shows the measurements made by the transmission-type probe for cell concentration, as compared to those obtained by a spectrophotometer at a wavelength of 650 nm after manual

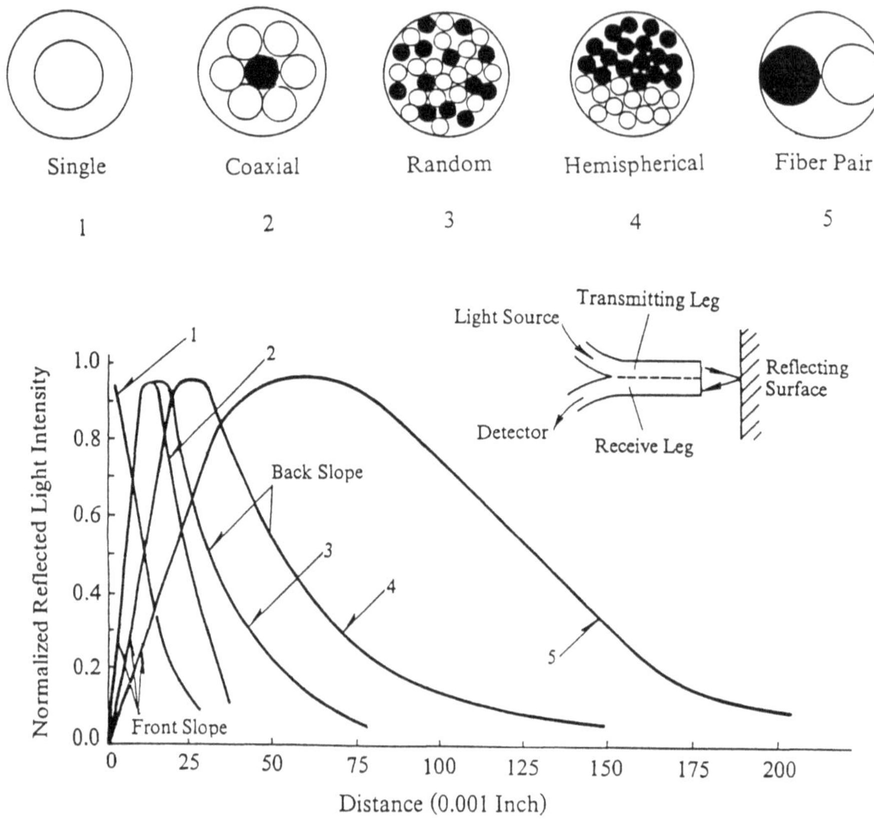

FIGURE 4-6 *Representative configurations of reflection type optic fiber probes (Krohn, 1986).*

sampling every two hours. It can be seen from this figure that under normal conditions fermentation for producing glutamic acid needs about 14 hours. This probe and the related instrument are important for monitoring fermentation and other biochemical processes and their on-line computerized control.

4.2.2 The Reflection-Type Probes

The application of the reflection-type optic fiber probes to measurement of local concentration of solids was developed on the basis of the optic fiber displacement sensor. Salins (1975) and Krohn (1986) summarized the arrangement of different optic fiber probes and their response curves, as illustrated in Figure 4-6. It shows that when the displacement sensor detects

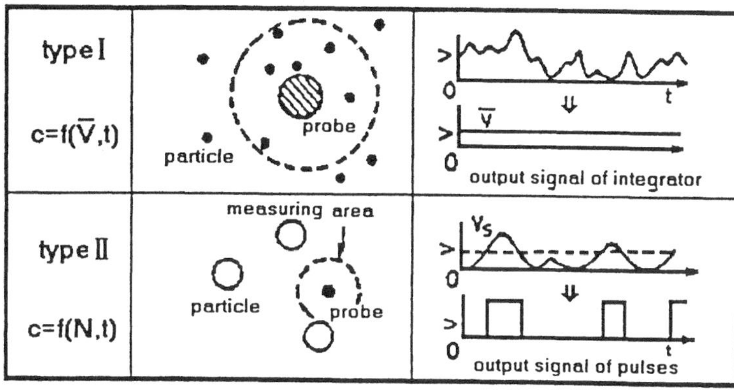

FIGURE 4-7 *Comparison of probes having different particle-to-probe diameter ratios (Matsuno, 1983).*

a plate reflecting surface, its output signals always consist of an upward part and a downward part as displacement increases. The same intensity of reflected light may arise from two different corresponding displacement points. Hence, this type of displacement sensor can usually be applicable to only the rising slope with best linearity when measuring displacement. Contrary to the objective of the displacement sensor, the measurement of particles concentration requires the location of particles in an effective volume to be independent of the intensity of the reflected light, an impossibility. Therefore, the output signals of the reflection-type probe include errors caused by the variable distance of the particles, and the greater the gradient of the responding curves of the probe, the greater the stochastic error. However, the output signals of the reflection-type probe depend, to a large extent, on the aforementioned chromaticness of the particles, and the intensity of reflected light of white and black particles of the same diameter may differ by several times, and the roughness of particle surfaces may also have a similar influence. Hence the reflection-type probes need to be calibrated and adjusted according to the optical sensitivity of the materials to be measured.

Matsuno (1983) classified the reflection-type probes into two categories according to the ratio of particle diameter to fiber diameter, as shown in Figure 4-7. Type II, for which the particle is larger than the probe is derived from the particle velocity probe, the output signals from the light receiver being converted into pulses at some threshold level V_s, and the pulse count corresponding to the number of particles. The reflected light of particles at long distances is eliminated if it falls below the threshold level V_s. Therefore the effective volume of measurement is rather limited.

Type I probe has a diameter larger than that of the particle, the output signals are all generated by back scattered light from the particles, and the integrated values of the output signals can be correlated with the concentration of particles by any calibration method, from which the instantaneous concentration can be obtained by output signals analysis. These probes have been widely used in concentration measurement of particles.

Qin and Liu (1982) used a probe in which 0.015 mm fibers were arranged by alternate-layers to form a 2x2 mm sectional area. This probe structure allowed large receiving signals of reflection light under the same intensity of incident light. Probes of the similar structure were used by Tung et al. (1988), Zhang et al. (1991), Wang et al. (1992) and Zhou et al. (1994).

Matsuno et al. (1983) used a pair of parallel plastic fiber to form a simple optic probe of rather low sensitivity. The surface of the tip was easily roughened by abrasion by particles. Improvement by Reh and Li (1990) includes bending the front part of probe so that the axes of light crossed each other. Figure 4-8 shows a comparison of measured volume and response curves of parallel and cross beams. As shown in the figure, for the defined measurement volume, β has to be greater than θ. Otherwise, particularly for $\beta = 0°$, the measuring volume will be infinite, and the optic fiber arrangement becomes the parallel probes, and for $\beta = 90°$, the optic fiber arrangement becomes the transmission type. The maximum length of the measuring volume as shown in Figure 4-8 is:

$$l_{max} = d_f \sin(\beta/2) + d_f \cos(\beta + \delta/2)/\tan(\beta - \theta) \qquad (4.2)$$

The cross beam probe can limit the measured volume, which is very important for gas-solid two-phase flow with bubbles.

Hartge et al. (1988) further simplified the parallel optic fiber probe. Fig. 4-9 illustrates such probe structure, in which the incident light and the reflected light are transferred in a common single fiber. This was realized by a beam splitter that separates light signals before receiving the reflected light, thus making the probe much finer, but the beam splitter may cause a loss of more than a half of the light signals, and may need compensation by increasing the intensity of the incident light or the sensitivity of the light detector. Another feature of this type of probe is that both concentration and velocity measurement of particles can be made with the same probe, thus realizing the measurement of solid mass flow rate in two-phase flow.

Two different designs of the multi-fiber probe were described by Hartge et al. (1986). Probe design I consisted of one light emitting fiber surrounded by 6 fibers for the transmittal of the reflected light, each fiber having a diameter of 0.5 mm. Probe design II consisted of a bundle of 700 fibers, each fiber with a diameter of 0.05 mm. About half of the fibers were emitting light

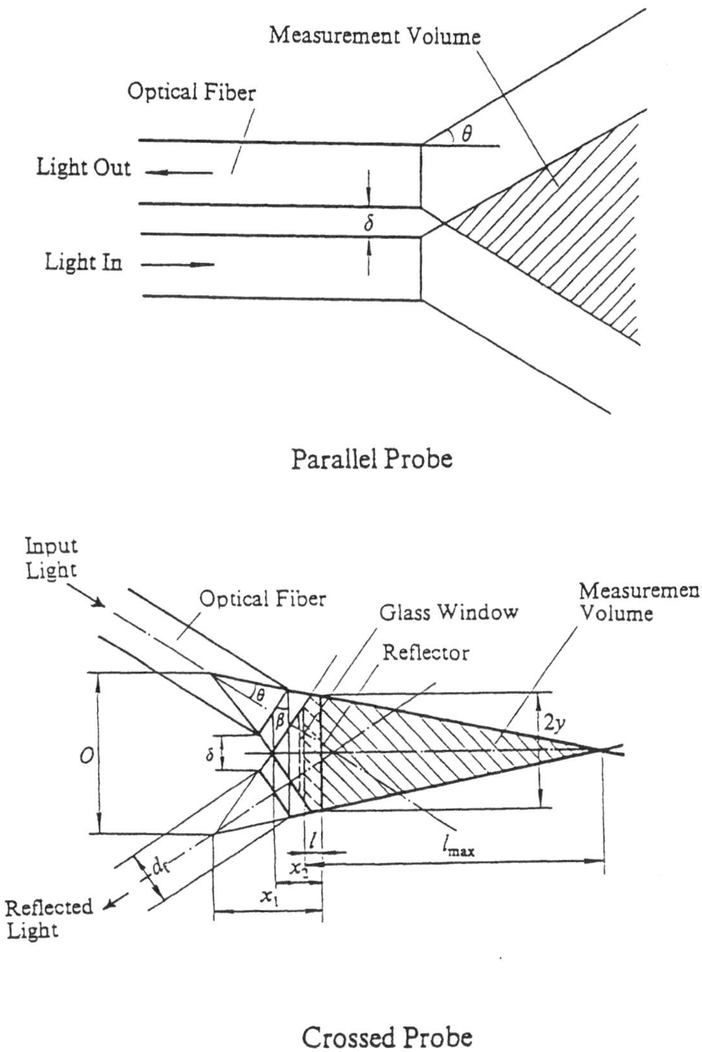

FIGURE 4-8 *Comparison of measurement volume between parallel and crossed optic fiber probe (Reh and Li, 1990).*

and the other half receiving light. Light emitting and light receiving fibers were randomly mixed at the probe tip. Figure 4-10 shows the results of calibration of these two types of probes in liquid-solid fluidized beds. It can be seen from the figure that for the same 2mm i.d. optic fiber bundle operating on a 25 mW HeNe light source, the sensitivity of the randomly arranged fine fibers is higher than that of the larger diameter coaxially arranged fibers.

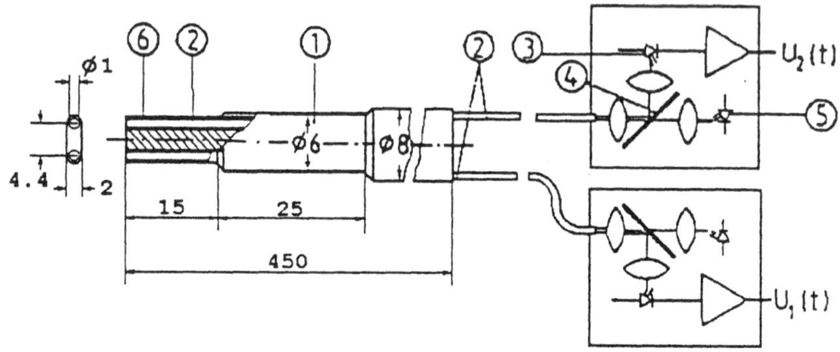

FIGURE 4-9 *Opto-electronic measuring system of Hartge et al. (1988).* 1-Probe, 2-Optical fiber, 3-Photo-diode, 4-Beam splitter, 5-Laser, 6-Steel capillary, 2 mm o.d.

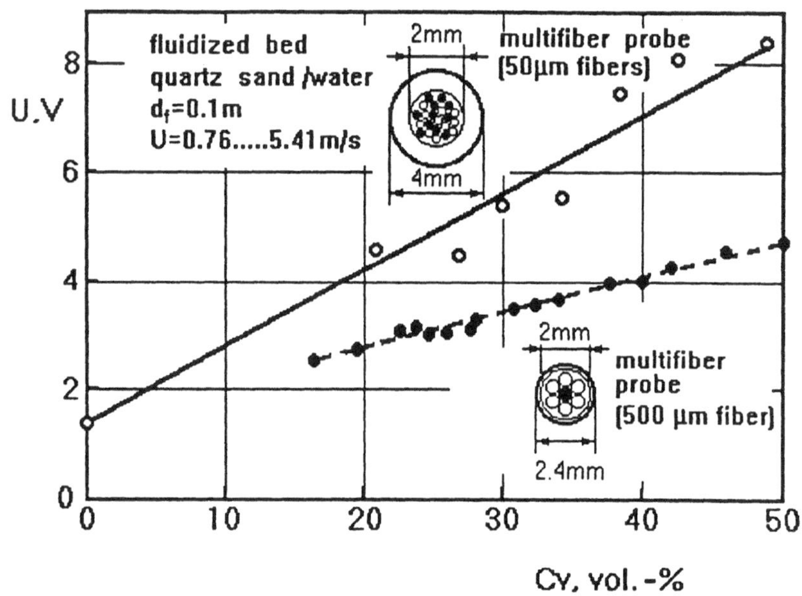

FIGURE 4-10 *Two types of multi-fiber probes and their calibration results (Hartge et al., 1986).*

scattering of light from particles were given. Bemer (1978) found that the intensity of the back scattered light was a function of solids concentration and mean particle size. Qin and Liu (1982) described the relationship between output signals and voidage ε and particle diameter d_p based on regular arrangement of particles. Nevertheless, the expressions were based on many assumptions, for instance, particles are all spherical with a common diameter of d_p, reflectivity of incident light for all particles is the same, and there is no overlap between particles in the foreground and background. Furthermore, all coefficients obtained were based on the particular materials under the given experimental conditions. Therefore, for the various types of probes, calibration method and data processing become important problems in the measurement of solids concentration.

4.2.3 Calibration Method

Both the transmission-type probe and the reflection-type probe, need be calibrated for their measuring range in local solids concentration. The calibration of optic fiber probes is known to be a difficult problem. Calibration methods fall into two categories: the first is to calibrate a probe against agitated or fluidized liquid—solid systems; the second is to use particle free-fall in gas--solid systems or the traditional pressure drop method for fluidized solids; the third is in a flow system with particle density deduced from mass flux of particles and measurement where phase velocities were nearly equal.

Qin and Liu (1982) calibrated their probes with two kinds of river sands with diameters of 0.3 mm and 0.9 mm, in a liquid--solid fluidized bed 40 mm in diameter. Voidage can be obtained from height L of the expanded bed as:

$$\varepsilon = 1 - (L_o/L)(1-\varepsilon_o) \qquad (4.3)$$

where L_o is the height of a packed bed and ε_o is the voidage of the packed bed. The results of the calibration is shown in Fig. 4-11. The integral time of the measured signals was 60 s.

The calibration curves (Figure 4-10) of Hartge et al. (1986) were obtained by a similar method. They found that with quartz sand with a size distribution around an average diameter of 56 μm, calibration was difficult in the high voidage region. For voidages above 80% the bed surface was obscure.

Zhou et al. (1994) divided the calibration methods for concentration probes used in liquid--solid systems into two parts. For voidages less than 0.8, calibration was carried out in a fluidized bed because particles are quite uniformly distributed in such a system. Calibrations at high voidages was carried out in a beaker: a certain volume of solid particles is put into a beaker and mixed with a known volume of water, and the liquid--solid mixture is then stirred until the particles are uniformly distributed in the water. Different voidages were attained by mixing different volumes of particles

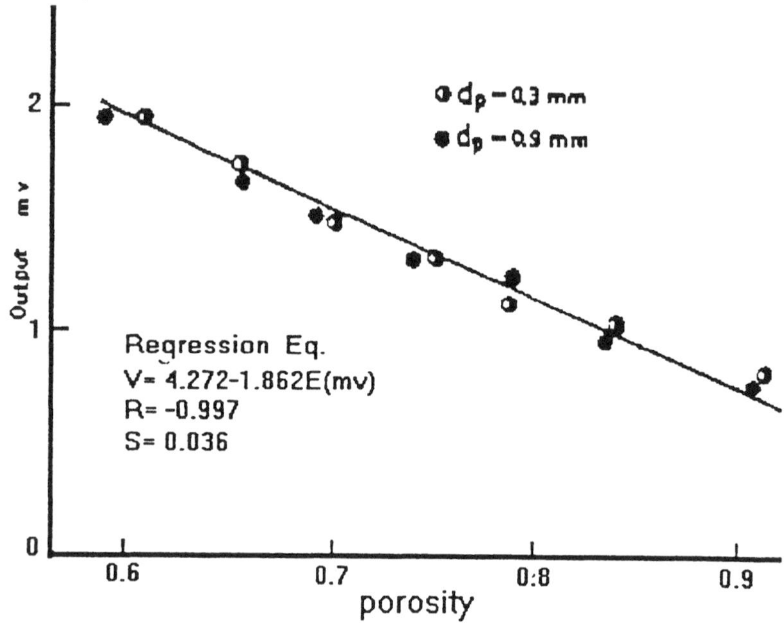

FIGURE 4-11 *Results of calibrating optic fiber probe in liquid-solid fluidized bed (voidage between 0.6 to 0.9) (Qin and Lin, 1982).*

into water. The calibration curve was very nearly linear over the entire voidage range of interest. This calibration method is applicable to all multi-fiber probes with the probe tips in direct contact with the fluid and the particles.

Yamazaki et al. (1992) calibrated his probe (Figure 4-12) and in a liquid-solid stirred mixer. Figure 4-13 gives the calibration curves for different particles. It shows that particles of different sizes and materials have different sensitivities and linear regions. Saturation may occur because of light scattering by particles of different surfaces and concentrations.

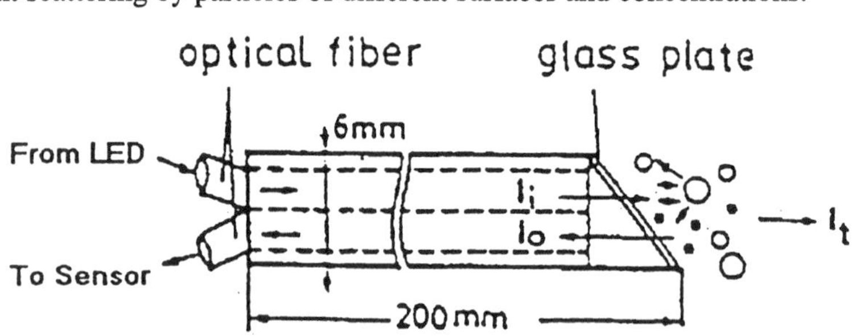

FIGURE 4-12 *Optic fiber probe used by Yamazaki et al.(1982).*

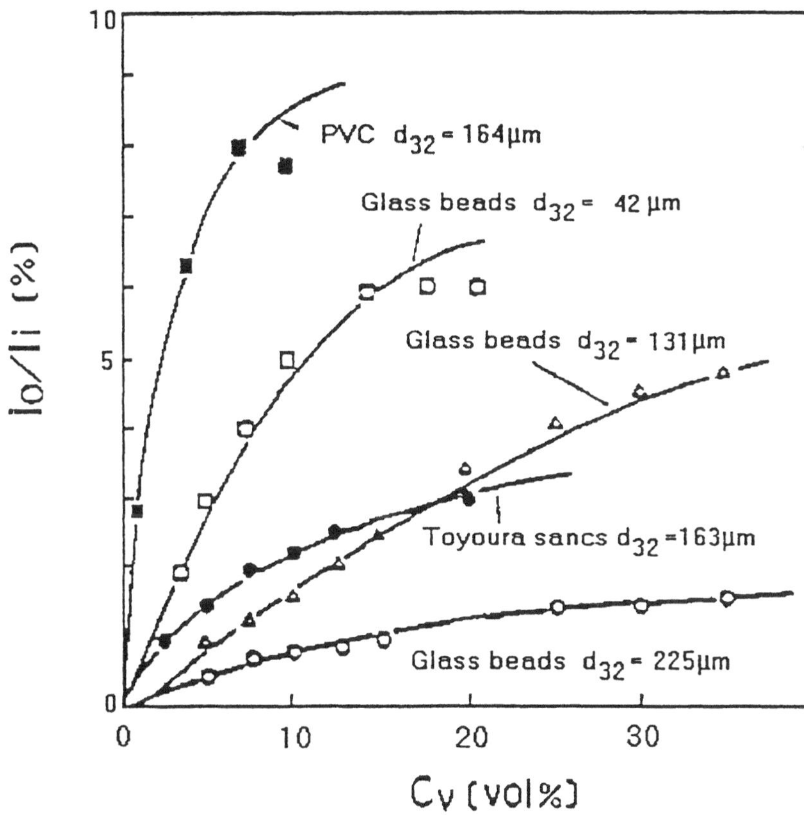

FIGURE 4-13 *Calibration cureves for various solid particles (Yamazaki et al., 1983).*

The purpose of probe calibration in liquid--solid systems is to ascertain if the responding curves of the probe and the measuring system are linear for the test materials. Since in liquid-solid systems particles are distributed uniformly, the reproducibility would be perfect. If the calibration results of the whole system are linear, for gas-solid systems the only parameter which needs to be changed is the index of refraction of the continuous phase. Experiments showed that in the case of a gas--solid system, under the same experimental conditions the output signals of concentration were larger that those for the liquid--solid systems.

One of the calibration methods for probes used in gas-solid systems employs the free-falling particles at their terminal velocities after having traveled a certain distance. The calibration method of Matsuno et al. (1983) utilized (Figure 4-14) particles falling at uniform velocity from a vibrating sieve at a sufficient height. The particle concentration was varied by

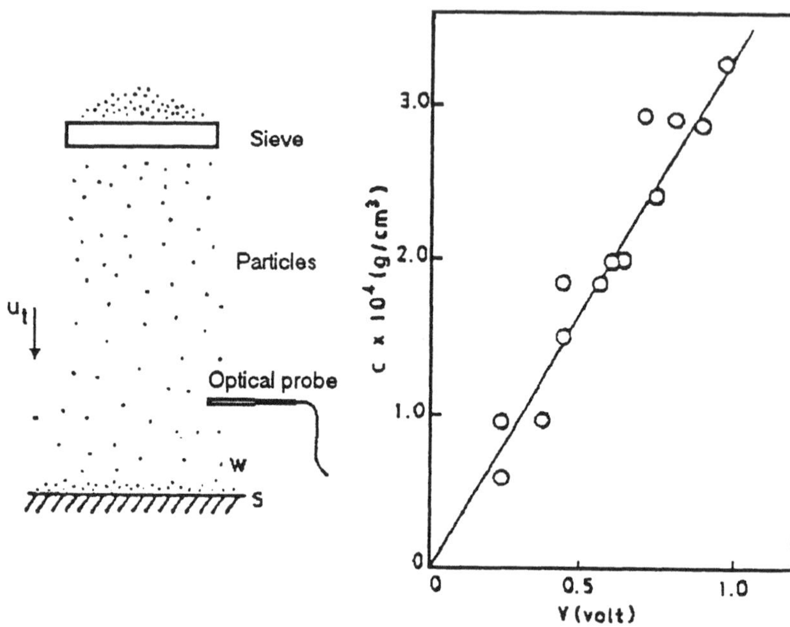

FIGURE 4-14 *Calibration with free-falling particles, and relation between integrator voltage and particle concentration (Matsuno et al., 1983).*

changing the weight of particles on the sieve and also by using sieves of different apertures. Glass beads of average diameter 65.5 μm and density 2.52 g/cm^3 ($C_v \sim 10^{-4}$) were used, and the particle density is calibrated by:

$$\rho_s = \Delta W / S V_t \Delta t \tag{4.4}$$

where ΔW is the cumulative weight of particles sampled on the cross-sectional area S within time Δt, and V_t is the terminal velocity of particles. The density range of the calibration curves in the figure has to be limited to a relatively dilute state. However, this method proves that in the calibrated concentration range the output signals of the probe for gas--solid systems are linear too, although a comparatively uniform distribution of solid particles should be satisfied.

Cutolo et al. (1990) calibrated his probe with a collecting vessel placed on a platform balance which is mounted at the bottom of the device. The solids rate is directly evaluated by weight and time measurements. The average volume fraction of solid $C_{v,w}$ in the pipe of cross section S is:

$$C_{v,w} = W / \rho S U(C_{v,w}) t_m \tag{4.5}$$

where W is the weight of the particles, having material density ρ collected during the time interval t_m, and U is the average falling velocity of the particles. In general, $U(C_{v,w})$ is a rather involved function of pressure and composition of the gas and of the size, concentration, and weight of the particles. This is a calibration method for the measurement of concentration in gas-solid two-phase flow with the transmission-type probes. Experimental results show that it has good linearity when the volume fraction is below 0.1.

Herbert et al. (1994) calibrated probes with a similar method, with the difference that the FCC catalyst particles used flow from a fluidized bed, through an orifice in the center of a porous metal grid, into a square tube (8x8 mm) where they fall 2.5 m into a collection pot. The mass flow rate was determined by particle collection and weighing over a known time period, and the volume fraction range calibrated was only 0.01to 0.1.

Tung et al. (1988) calibrated their probe directly in a 90mm diameter fast fluidized bed with FCC catalyst. A multiple regression method was developed for calibrating the probe by traversing it through the bed and

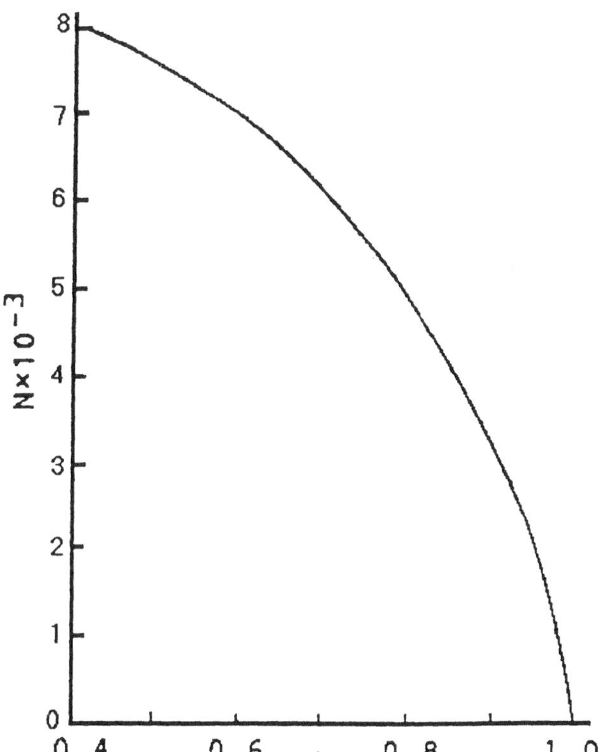

FIGURE 4-15 *Calibration in a fast fluidized bed and typical multiple regression curve (Tung et al., 1988).*

comparing the resulting measurements with average volume fractions inferred from static-pressure gradients. The average voidages were obtained from pressure drops across the two nearest taps located above and below the probe. If m runs under different operating conditions were made, the readings with respect to average voidage $\bar{\varepsilon}$ and corresponding values of output signals N_i at n radial coordinates, would form a system of m simultaneous equations:

$$\bar{\varepsilon} = a + \frac{b_1}{n}\sum_{i=1}^{n} N_{ij} + \frac{b_2}{n}\sum_{i=1}^{n} N_{ij}^2 + \frac{b_3}{n}\sum_{i=1}^{n} N_{ij}^3 + \frac{b_4}{n}\sum_{i=1}^{n} N_{ij}^4 \tag{4.6}$$

$j = 1, 2, 3, .. m$

The coefficients a, b_1, b_2, b_3 and b_4 can therefore be obtained by solving the above equations. By using this calibration method, a plot of voidage versus probe output signals N is shown typically in Figure 4-15. These are readings,

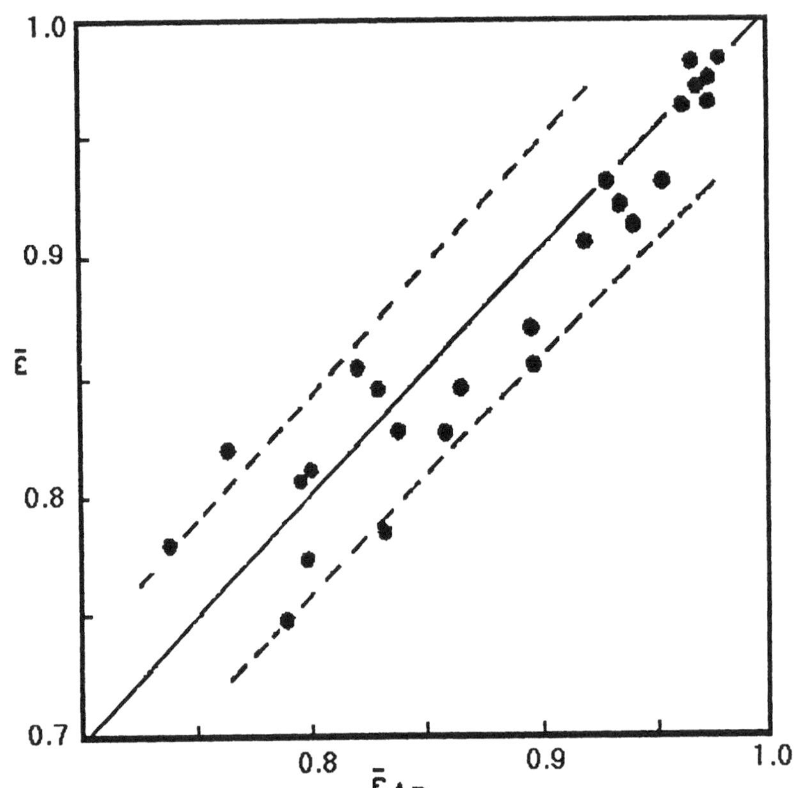

FIGURE 4-16 *Comparison of average voidage given by optic fiber* $\bar{\varepsilon}$ *and that given by pressure drop* $\bar{\varepsilon}_{\Delta p}$ *(Tung et al., 1988)*

with $m = 9$ and $n = 8$, which cover conditions from incipient fluidization to pneumatic transport. Figure 4-16 shows a comparison of the average voidage ($\bar{\varepsilon}$) computed from radial voidage profiles calibrated above, with values determined by pressure difference measurements ($\bar{\varepsilon}_{\Delta p}$).

Tung's calibaration method provides a comparison between the measured results obtained with a optic fiber probe used in a practical apparatus under different operating conditions and those using the traditional pressure-drop method. The deviation of this calibration method may depend upon the amount of pressure taps and circular rings of equal areas, because the probe measures only the concentration at some specified point.

Lischer and Louge (1992) described an optic fiber probe that measured the particle volume fraction in a dense suspension using light from a 5 mW He-Ne laser. Particles illuminated by the central fiber scattered light to the surrounding fibers connected to a photodiode. The output was calibrated against a capacitance probe. Figure 4-17 is a schematic of the calibration set

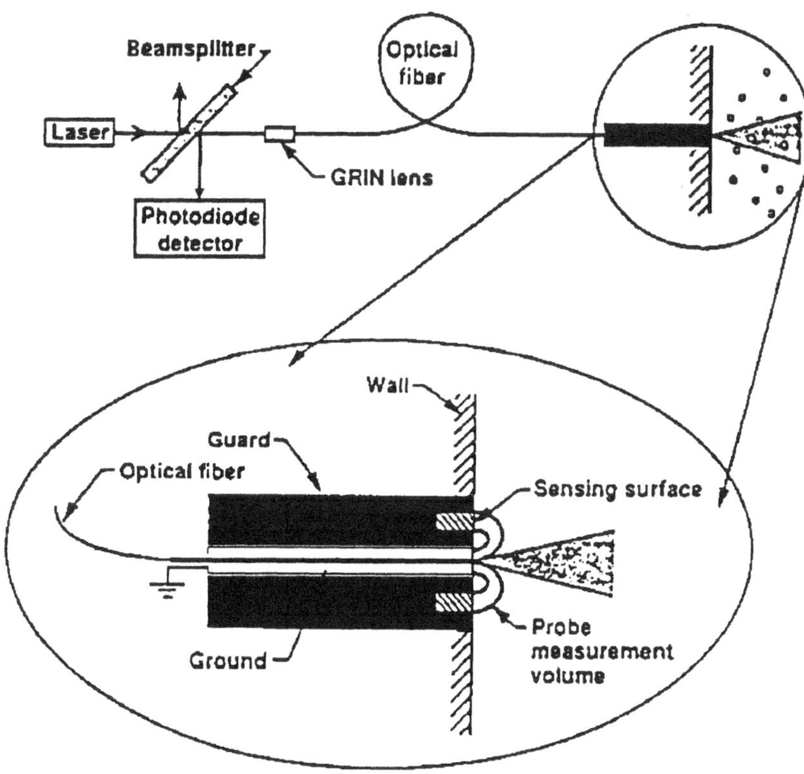

FIGURE 4-17 *Setup for calibration of optic fiber probe against a capacitance probe (Lischer and Louge, 1992).*

up. The experimental results were obtained by pouring particles randomly along the probe assembly. They showed that the accuracy of the probe measurement increased with a decreasing ratio of particle diameter to the probe diameter, d_p/d_f. They also found that measurement of transparent materials such as glass beads in water was problematic.

Reh and Li (1990) calibrated the crossed and parallel probes in a 100 mm i.d. gas-solid fluidized bed. The average bed voidage was obtained by pressure drop measurements. These two probes were also calibrated in a liquid--solid fluidization system. It can be seen from the comparison of two calibration curves in Figure 4-18 that both these two kinds of probes had linear response when calibrated in a liquid--solid system, but the nonlinearity of the parallel probe became more evident in the calibration in a gas--solid system. It is obvious that whether the continuous phase is water or air, the average intensity of light output indicates the difference in refractivity. The reason for the difference is that in the gas-solid calibration system the particles were not as uniformly distributed as in the liquid-solid system. Random bubbles and agglomerates exist in the former. Signals of local voidage from these two probes are shown in Figure 4-19. Bubbles, to which the responses of the crossed probe should be zero, cannot be correctly detected by the parallel probe, because the reflection from the bubble boundaries causes a certain level of output. Thus for the modified cross probe, the near-linear response curve is attributed to its localed measurement. These authors concluded from their optic voidage measurements that if the measuring volume of a probe is reasonably small, its response to the bed density would approach linearity. Therefore, it is important to limit the measuring volume and to use appropriate data processing method for random signals in determining the local particle concentration in gas-solid flow using fiber optic probes.

4.2.4 Analysis of Signals

Since dispersion of solids in a fluid-particle system is in a random state of movement, and the signals of light output from both the transmission-type and the reflection-type probes are dependent on diameter, morphology, chromaticness, distance and refractivity of the particles, the signals produced by particles in random movement can only be described by random data analysis.

One of the simplest way is to use the mean value, i.e.,

$$M_x = (1/T) \int_0^T x(t)dt \qquad (4.7)$$

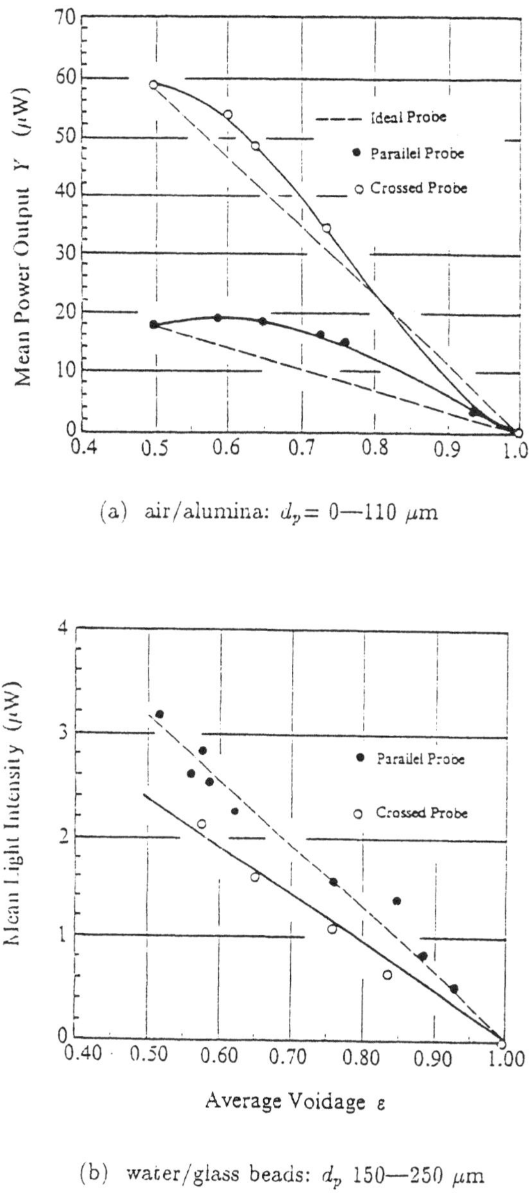

FIGURE 4-18 *Calibration curves for parallel and crossed optic fiber probes (Reh and Li, 1990).*
(a) air-alumina: d_p = 0-110 μm, (b) water-glass beads, d_p = 150-250 μm.

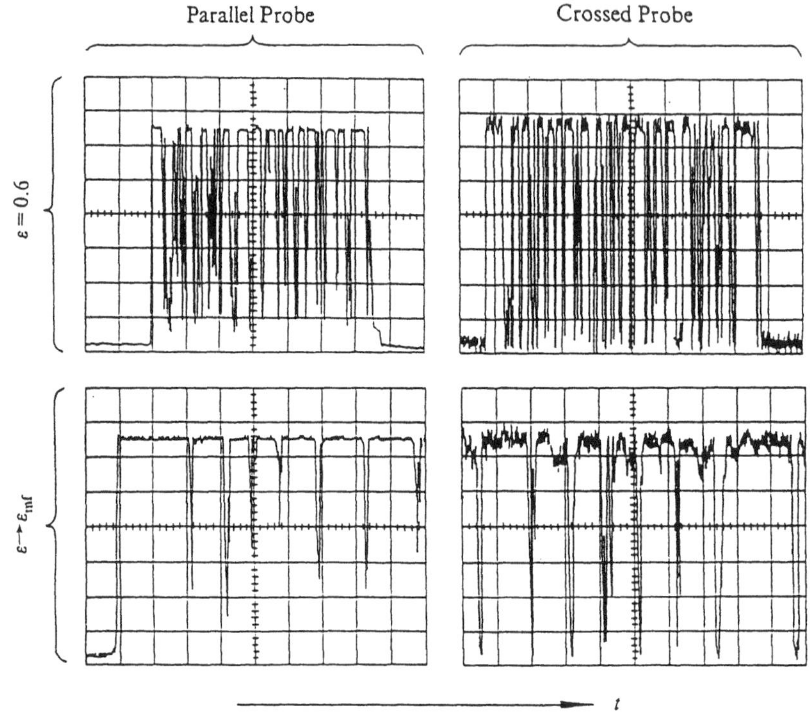

FIGURE 4-19 *Local voidage signals for parallel and crossed optic fiber probes (Reh and Li, 1990). (0-110 μm alumina in a 90 mm i.d. bed)*

Usually the signals of the intensity of light are converted into analog signals through the light detector. Through an A/D converter the time-averaged values of signals can be determined. In addition, the waveforms of the original signals in the sampling process can be observed. The instantaneous readings of each point enables calculation of the variance. In order to determine the sampling time, a simple way is to increase the sampling time after stable operation is achieved until the reproducibility is controlled within some preset range of deviation. The time-averaged value obtained by whatever method was not complete for the measurement of local concentration of particles, which may have the random fluctuations. For instance, and the existence of bubbles in gas--solid two-phase flow would not be accounted for in the time-averaging process. Therefore the probability density function and the cumulative probability distribution function have been widely used in the analysis of concentration signals.

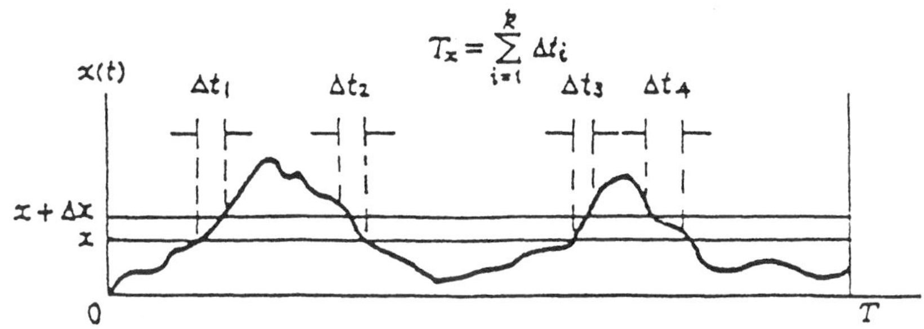

FIGURE 4-20 *The nature of probability density measurement of signals.*

The probability density function of random data describes the probability that the data will assume a value within some defined range at any instant of time. Consider the sample record of $x(t)$ illustrated in Figure 4-20. The probability that $x(t)$ assumes a value within the range between x and $x + \Delta x$ may be obtained by taking the ratio of T_x/T, where T_x is the total time that $x(t)$ falls within the range $(x, x+\Delta x)$ during an observation lasting up to T. This ratio will approach an exact probability description as T approaches infinity, or in equation form,

$$Prob\ [x < x(t) \leq (x + \Delta x)] = \lim_{T \to \infty}(T_x / T) \tag{4.8}$$

For small x, the probability density function $Prob(x)$ can be defined as:

$$Prob\ [x < x(t) \leq (x + \Delta x)] \approx P(x)\Delta x \tag{4.9}$$

According to the above principle, a probability density function measurement of data is to establish a probabilistic description for the instantaneous values of solids concentration.

Qin and Liu (1982) showed in Figure 4-21 the probability density function at the center, 1/3 and 2/3-way from the center and at the wall of a fast fluidized column of 90 mm i.d. and 8 m high, and operating with a solids mass flow rate of 19.6 kg/m^{2}s. It can be seen from the figure that when $\varepsilon = 0$ and $\varepsilon = 1$ are defined, the probability of voidage signals of the probe at different locations can be clearly presented. $P(x)$ is defined as the probability at the instant for value $x(t)$ larger or smaller than certain x value, which is equal to the integration of the probability density function. $P(x)$ is called the cumulative probability distribution function and should be between 0 to 1. The probability for $x(t)$ falling within any domain (x_1, x_2) is

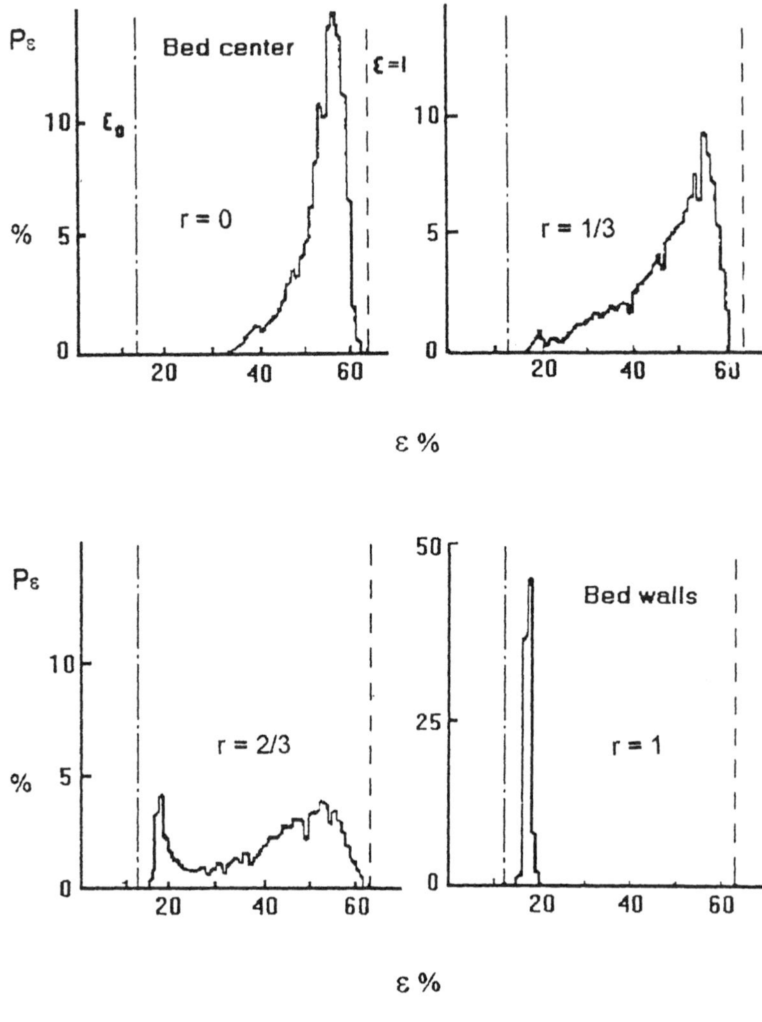

FIGURE 4-21 *Probability density function for voidage ε (Qin and Lin, 1982).*

$$P(x_2) - P(x_1) = \int_{x_1}^{x_2} P(x)dx \qquad (4.10)$$

Hartge et al. (1988) measured radial solids concentrations and probability density and cumulative probability distribution of ash particle in a 400 mm i.d. and 8 m high circulating fluidized bed, the results of which are shown in Figure 4-22. It can be found from comparison of data for the same

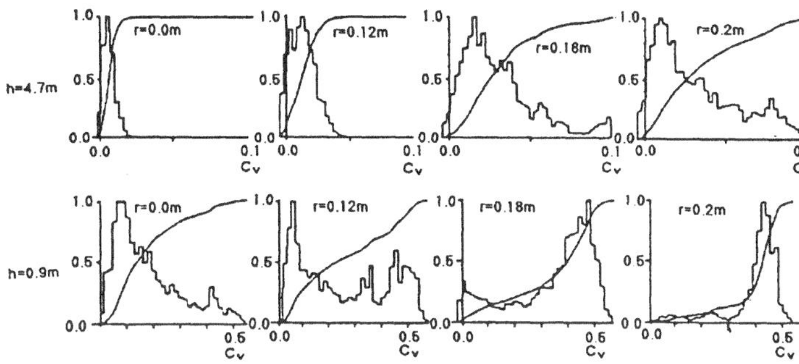

(a) Local solids concentration distributions

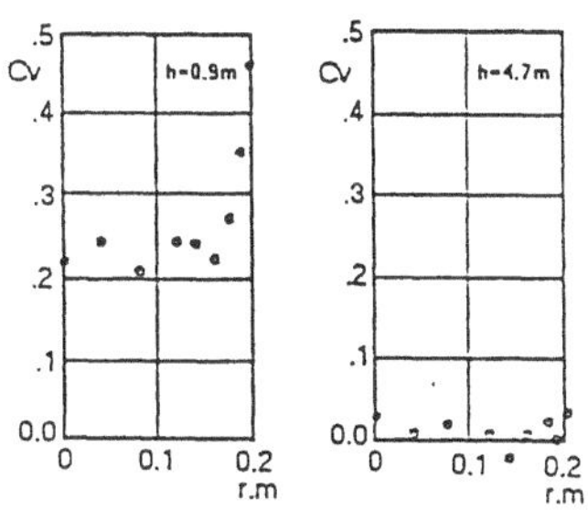

(b) Radial solids concentration profiles

FIGURE 4-22 *Measurement of ash concentraion in a circulating fluidized bed (400 mm i.d., 8 m height) u = 3.7 m/s, G_s = 30 kg/m²s (Hartge et al., 1988). (a) Local concentration distributions, (b) Radial concentration profiles.*

height and operating conditions that when $h = 0.9$ m and probe in the central position, the time-averaged value of volume fraction $C_v = 0.22$, but the maximum probability in the figure occurs at $C_v \approx 0.08$. It is obvious that the time-averaged value ignores the local instantaneous concentration profile, because during the process of time averaging of concentration there were bubbles.

Since the optic fiber probe has time response characteristic of the nano-second level, limited by the response of the light detector, the response of signals will have essentially no time lag. Thus another method for describing optic output signals for particle concentration is to analyze the power spectrum. The power spectral density function of random signals can express the frequency structure of signals by the mean square value. The power spectral density function $S_x(f)$ can be defined as:

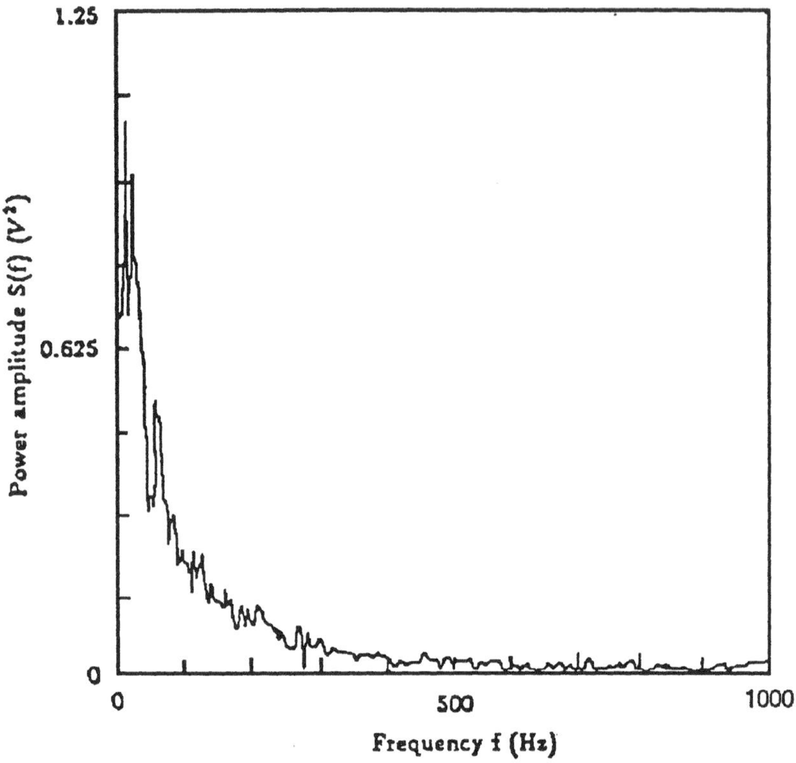

FIGURE 4-23 *Typical power spectrum density for fast fluidized bed (Xia et al., 1992).*

$$S_x(f) = \lim_{\Delta f \to 0} \frac{M_x^2(f, \Delta f)}{\Delta f} \tag{4.11}$$

where $M_x^2(f, \Delta f)$ is the mean value.

Xia et al. (1992) applied this signal analysis method to study the oscillatory behavior of light output signals in a fast fluidized bed. Figure 4-23 shows the typical power spectral density of optic output signals in the fast fluidized bed. The oscillatory behavior of the optic output signals has no characteristic time scale, or a deterministic frequency response, but forms fractal time characteristics.

The major methods for local solids concentration measurement with fiber optic probes are summarized in Table 4-1.

Table 4–1 Major methods for local solids concentration measurement with optic fiber probes

Workers	Apparatus	Materials	Probe type	Measuring system	Calibration
Qin and Liu (1982)	fast fludized bed 90 mm i.d. 8 m height	FCC d_p 0.054 ρ_s 9295	alternate-layer arrangement f_d 0.015 mm, p_d 5mm	photomultiplier timing digital integraph probability density analyser	liquid-solid bed 40 mm i.d., 3.5 height Sand:d_p 0.300, 0.900
Hatano et al. (1983)	fluidized bed 150 mm i.d., 2.25 m height	FCC d_p 0.075	transission type l 5 mm, f_d 0.5 mm	8-bit personal computer	
Matsuno et al. (1983)	freeboard of the fluidized bed 81 mm i.d.	glass beads d_p 0.056, ρ_s 2520	pair of bundles P.F.O. p_d 5mm	halogen lamp J-FET type operational amplifier	calculated by $C = \Delta W/SU_t\Delta t$
Hartge et al. (1986)	fast fluidized bed ϕ 0.4×7.8 m ϕ 0.05×3.3m	quartz sand d_p 0.056	probe I probe II coaxial random f_d0.5, p_d 2.4 f_d0.05, p_d 4	25 mW HeNe laser computer (HP 1000)	liquid-solid bed quartz sand d_p 0.056
Horio et al. (1988)	CFB 50 mm i.d., 2.79 m height	FCC d_p 0.060 p_s 1000	three fiber of vertical arrangement P.O.F., f_d 0.5 mm p_d 2.5 mm	photomultiplier computer probability distribution	
Hartge et al. (1988)	CFB 400 mm i.d. 7.8 m height	FCC d_p 0.085, ρ_s 1500 CFBC ash d_p 0.12, ρ_s 2600	single fiber for C_v and V_p by a same probe f_d 1mm, p_d 2mm	laser diode, computer probability density and cumulative density function	
Nakajima et al. (1990)	turbulent fluidized bed 200mm i.d. 6 m height	FCC: d_p 0.064, ρ_s 880 silica sand: d_p 0.087 ρ_s 1200	transission type l=3 mm, p_d 6 mm	laser and photo transistor A/D converter to computer	

(To be continued)

Table 4-1 Major methods for local solids concentration measurement with optic fiber probes

(Continued)

Workers	Apparatus	Materials	Probe type	Measuring system	Calibration
Reh and Li (1990)	fluidized bed 100 mm i.d.	glass beads: d_p 0.04~0.08 alumina powders: d_p 0~0.11	pair of P.O.F. crossed, parallel probe f_d 1mm, p_d 6 mm window thickness: 0.4 mm	mean powder output (μ W)	fluidized bed with air and liquid pressure drop
Nowak et al. (1990)	CFB: 205 mm i.d.	mixture of FCC d_p 0.046, ρ_s 2300 silica alumina d_p 3.0	coaxial, f_d 1 mm, p_d 6 mm for C_v and V_p by a same probe	multichannel data recorder personal computer	
Yamazaki et al. (1992)	flat-bottom cylindrical tank 60 mm i.d.	glass beads d_p 0.042, ρ_s 2350 sand d_p 0.163, ρ_s 2650 PVC powder d_p 0.164, ρ_s 1500	pair of P.O.F. f_d 2 mm, p_d 6 mm with glass window	LED modulated at 460 Hz personal computer	in liquid-solid system
Zhou et al. (1994)	CFB square cross-section 146 x 146 mm	sand: d_p 0.213, ρ_s 2640	alternate-layer f_d 0.015 mm, p_d 5 mm	sampling frequency 483 Hz probability distribution personal computer	in liquid-solid system
He et al. (1994)	spouted bed 152 mm i.d., 1.4 m height	glass beads: d_p 1.41, ρ_s 2503	random arrangement fiber bundle 0.50 mm, p_d 5 mm	A/D converter to personal computer	
Herbert et al. (1994)	CFB 50 mm i.d., 8 m height	FCC d_p 0.08, ρ_s 1400	single fiber for C_v and V_p by a same probe f_d 2.0 mm, p_d 8 mm	one packet of 512 points sampled frequency 10 kHz	in fluidized bed with particle collection and weighing

Note: P.O.F.—plastic optic fiber, LED—light emitting diode.

4.3 MEASUREMENT OF LOCAL PARTICLE VELOCITY

Particle velocity is one of the fundamental parameters in the study of fluid-particle flow systems. The initial application of optic fibers was devoted to the measurement of particle velocity in fluidization. Of interest to researchers are the instantaneous value, average value and profile of particle velocity, and at the same time, it is desirable to measure solids concentration as well as the distribution of particle diameters. The range of measurement should be as wide as possible, while the probe diameter should be as small as possible in order to minimize its influence on the flow field. Since the optic fiber probe can satisfy these requirements, it was widely applied and developed.

Measurement of particle velocity with optic fibers is based on traversing a distance l by a particle between two known points over a transit time (or the time lag) t or the velocity $V = l/t$. However, the instantaneous velocity can be obtained only when the measuring distance between the two detector points is short enough to avoid interference. The diameters of the optic fiber and of the particle can both fit in the same range, and the response of probe against light signals is almost without time lag. Thus this kind of probes is very suitable for the measurement of instantaneous particle velocity.

The probe for local particle velocity measurement is generally comprised of two separate sets of fibers, one for the incident light and one for the reflected light from the particles. One or more fibers are generally used to project the light emitted by a light source on the particles, and two or more fibers to detect the light reflected by the solid particles. The structure of fiber optic probes varies little, but they may be provided with different arrangements for satisfying different requirements. They include different velocity ranges and design characteristics of the probe, e.g., for measurement of bubble velocity, moving direction of particles, local concentration and solid flow rate, etc. Figure 4-24 illustrates typical arrangements of various optic fiber probes. The maximum of cross-correlation function for average velocity and the local discrimination for instantaneous velocity will be described below.

4.3.1 Cross-Correlation Method

In order to obtain the time lag between two sets of reflected light signals of known distance apart, points a and b, for particles in random movement, the cross-correlation function method is generally used in treating the signals. The cross-correlation function of two sets of random signals $a(t)$ and $b(t)$ express the independence of the two sets of sampled data, i.e.,

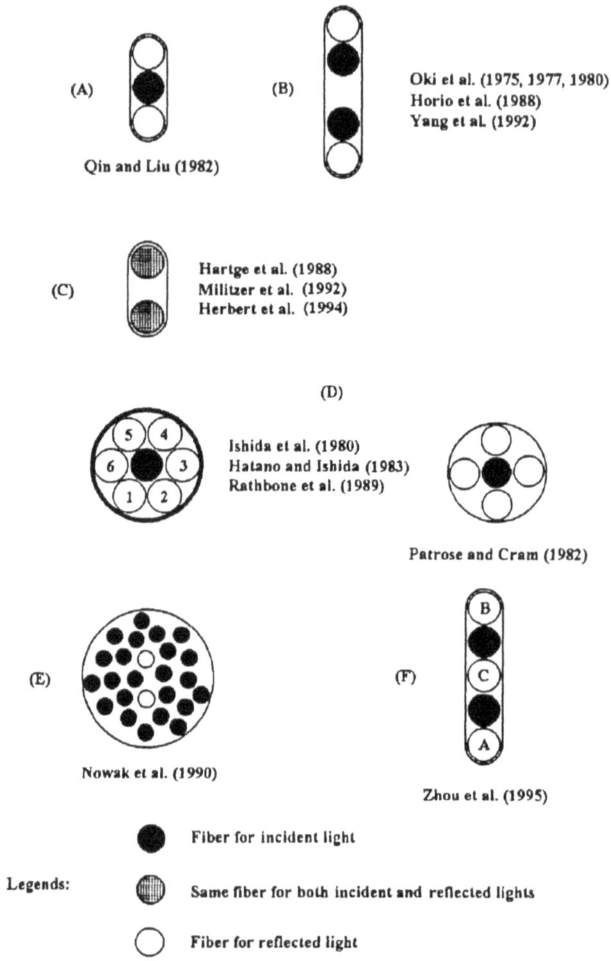

FIGURE 4-24 *Different types of fiber optic probes for particle velocity measurement.*

$$R_{ab}(t) = \lim_{T\to\infty}(1/T)\int_0^T a(t)b(t+\tau)dt \tag{4.12}$$

When $T \to \infty$, the mean product would tend to a correct cross-correlation function. In addition, when a and b are interchanged, $R_{ab}(\tau)$ in the diagram of cross-correlation is symmetric, i.e.,

$$R_{ab}(-\tau) = R_{ba}(\tau) \tag{4.13}$$

Sometimes the cross-correlation coefficient is used to express the maximum time lag, $C^*(\tau_m)$, i.e.,

$$C^*(\tau_m) = \lim_{T\to\infty}(1/T)\int_0^T[a(t-\tau_m)-\overline{a(t)}][b(t)-\overline{b(t)}]dt/\delta_a\delta_b \qquad (4.14)$$

where

$$\delta_a = \{\lim_{T\to\infty}(1/T)\int_0^T[a(t)-\overline{a(t)}]^2dt\}^{1/2},\ \overline{a(t)} = \lim_{T\to\infty}(1/T)\int_0^T a(t)dt$$

and δ_b is given by an analogous relation by replacing symbol a with b in the same relation. The time required for particles to travel between these two points is the transit time τ_m where the cross-correlation function is a maximum. The particle velocity V_p, therefore, can be calculated according to:

$$V_p = l/\tau_m \qquad (4.15)$$

where l is the effective distance between the two detecting fibers.

Improving on the system used by Oki et al. (1977), Horio et al. (1988) used a probe of Type B in Figure 4-24 for measuring velocities of FCC particles and clusters in a circulating fluidized bed. The fiber diameter was 0.5 mm, and the distance between two fibers was 3.1 mm. A special FFT analyzer was used for analyzing the cross-correlation or power spectrum to give τ_m. Figure 4-25(a) shows an example of the signals of the probe. It indicates that groups of particles were passing the probe tips at certain time intervals. The time lag between two signals can be easily known by comparing the time of corresponding peaks and valleys. Such a procedure can be automatically carried out by computing a cross-correlation function.

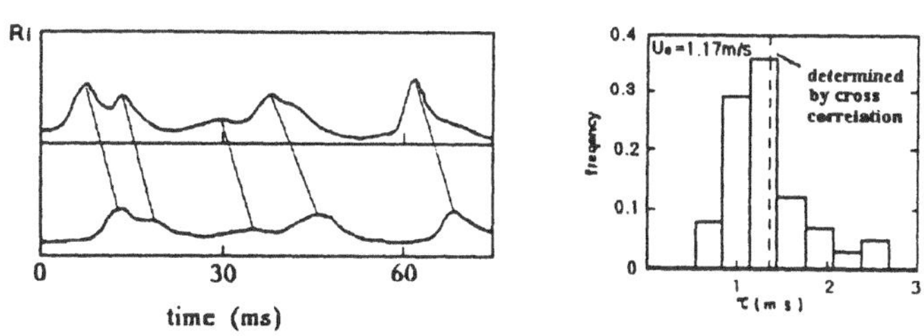

(a) Typical probe signals

(b) Distribution of measured time lag and mean value from cross correlation

FIGURE 4-25 *Measurement of FCC particle velocity (Horio et al., 1988).*

In Figure 4-25(b) the real velocity of each group of particles was obtained from the time lag between corresponding peaks of the signals from the two probe tips, and it is compared with the average time delay obtained from the FFT analyzer. The average time delay from the correlation agrees fairly well with that calculated from the original signals.

Yang et al. (1992) also used the above mentioned probe in the measurement of particle velocity in a circulating fluidized bed. Experiments were carried out in a 140 mm i.d. bed with FCC particles at operating gas velocities ranging from 1.5 to 6.5m/s. As shown in Figure 4-26, it is worth mentioning that, in the dilute zone at 6.6 m axial position, the measured particle velocity agreed well with those measured by laser Doppler velocimetry (LDV).

Hartge et al. (1988) developed a optic fiber probe of the minimum size for velocity and concentration measurements. This kind of probe which allows the transmission of incident and reflected lights in a single fiber, as already shown in Figure 4-9, falls into Type C in Figure 4-24. The authors detailed their measuring method and experimental results, and pointed out

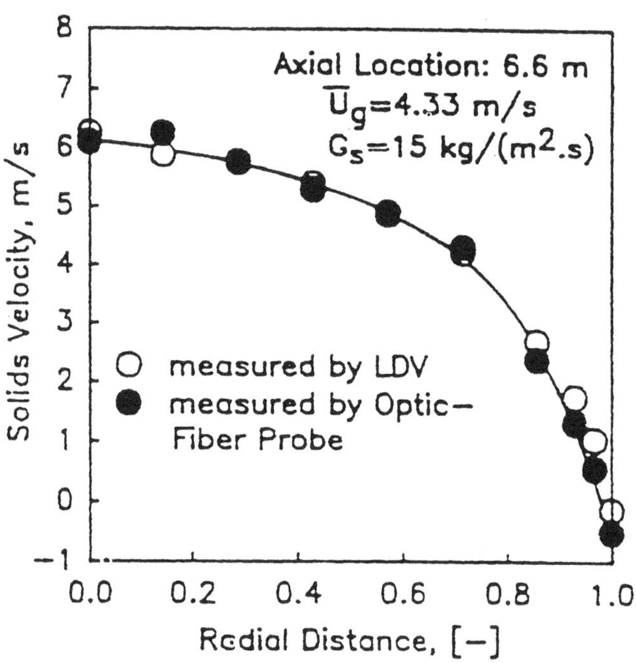

FIGURE 4-26 *Comparison of particle velocities determined by LDV and cross-correlation of optic fiber measurement (Yang et al.,* 1992)

that the effective optic distance between the two fibers for receiving light and the distance of fibers having different geometric arrangements should be calibrated by using a rotating metal plate with color marks, since *l* was one of the main factors influencing data error. Besides, in order to calculate cross-correlation function from Eq. (4.14) , it means that in practical cases the integration time *T* must be sufficiently long. In a circulating fluidized bed, they have to take into account the structure elements of the flow pattern, e.g., clusters or strands of solids which may have significantly different velocities. In order to detect instantaneous velocities, it was therefore decided to use very short integration time (*T*). Values of *T* between 10 and 30 ms were found to be sufficient to yield reproducible values of instantaneous velocities in the present case, where the sampling frequency was 25 kHz and 8.3 kHz, respectively. An integration period *T* of 20 ms corresponds to a vertical length of 100 mm for a typical gas velocity in the fluidized bed operating at 5 m/s, and this seems sufficiently short to detect individual velocities of the structure elements.

Figure 4-27(a) gives representative of velocity measurements for FCC particles taken at two different heights above the distributor and at different distances, *r*, from the vessel center line. The results were plotted as probability densities and cumulative probability densities of the velocities calculated from the time delay τ_m of the cross-correlation functions. It is interesting that in all measurements, positive as well as negative velocity values, were registered. It is only the proportion of upward and downward velocities, respectively, which varies with the locus of measurement. As a general tendency, it is the upward velocity which dominates in the vessel center whereas the downward velocity is more pronounced in the wall region. Such a display of data with probability statistic distribution is visual and detailed. If compared with the profiles of the local average solids velocity, more information can be obtained. For example, in Figure 4-27(b), at the wall position of r =0.2 m at h =0.9 m, , the average solids velocity is zero. Figure 4-27(a) shows basically symmetric probability distributions of both positive and negative velocities, but it can be seen from the cumulative density function curves that the positive velocity larger than 5 m/s still constitutes about 8\%.

Militzer et al.(1992) used probes of similar structure (Type C in Figure 4-24) for measuring particle velocities and improved the computer software for processing cross-correlation signals (as will be mentioned later in this chapter). The probe contains two parallel plastic fibers at a distance of approximately 3 mm from each other. Each of these fibers is connected to a light emitting diode (LED) and to a photocell. The same fiber is used both to send and to receive the light signals. The intensity of the signal reflected by the solids passing in front of the fibers depends on the composition of the

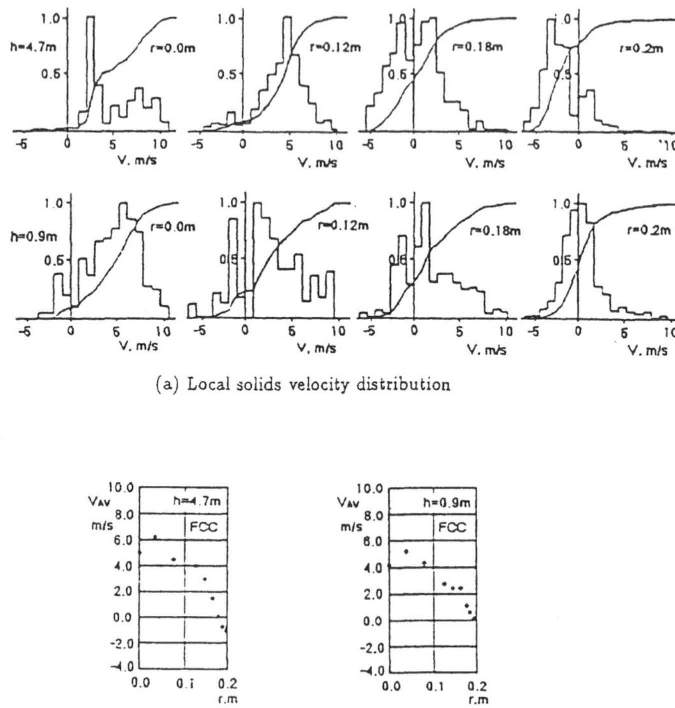

(a) Local solids velocity distribution

(b) Radial profiles of local average solids velocity

FIGURE 4-27 *Measurement of FCC velocity in a circulating fluidized bed of 400 mm i.d., 8 m height, u = 2.9 m/s, G_s = 49 kg/m²s (Hartge et al., 1988).*

particles, their shape, size distribution and concentration. The signals are transferred to a file on a PC with an A/D data acquisition card. Usually, 512 samples are taken from each channel with a frequency of between 10 and 50 kHz per channel. For a sample with 512 values, this gives a sampling period of between 20 and 100 ms. Figure 4-28 shows the velocity of 150 μm average diameter sand particles falling under gravity out of a hopper. Samples were taken at 21 kHz per channel. According to Guigon (1987) the velocity of a stream of particles falling in air for height of up to 15 cm is very close to the free-fall velocity in vacuum. The free-fall velocity in vacuum is given by $(2gh)^{1/2}$ where g is the acceleration of gravity and h is the vertical distance from the opening of the hopper. This is a simple and convenient calibration method.

Herbert et al. (1994) measured both velocity and concentration of particles with Type C probe in Figure 4-24. At each position, five files were

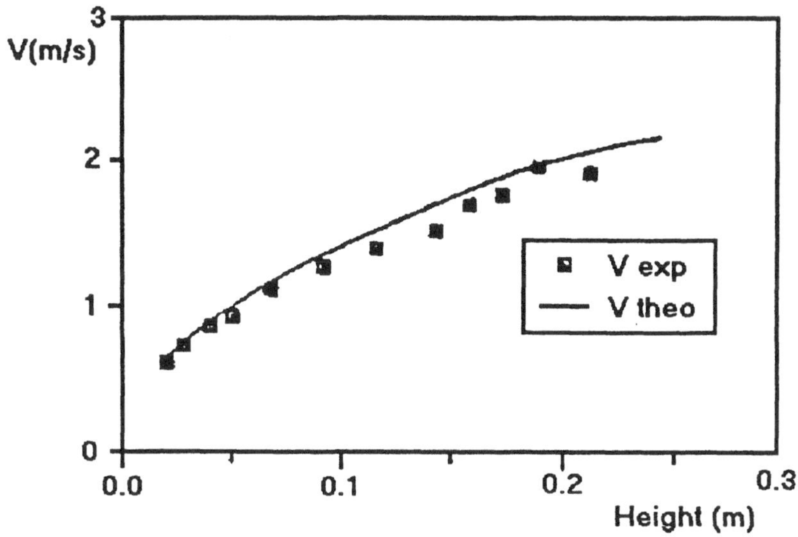

FIGURE 4-28 *Comparison of experimental and theoretical velocities for particles falling from a hopper (Militzer et al., 1992).*

recorded, each containing 50 individual velocity values. Each velocity was calculated from the cross correlation of two signals of 512 points sampled at the frequency of 50 kHz. This frequency was the maximum attainable with present DMA data acquisition card. The velocity measurements were shown to be reproducible with errors of between 10 and 15\% for particle velocities of up to 8 m/s.

The two kinds of probe structures shown under Type D in Figure 4-24 were employed for measuring both velocity and direction of moving particles in fluidized beds by Ishida and Shirai (1980). The probe consisted of 7 fibers 0.2 mm in diameter. Light emitted by a lamp was guided through the central fiber to project onto the particles around the tip of the probe. The light reflected on the surface of each particle was received by the peripheral six fibers. Particle velocity was obtained by measuring the time required for the particles to travel from the position of Fiber No.1 to that of Fibers No. 2, 3, 4, 5 or 6, and vice versa, depending on the direction of the particle flow, where No.1 plays the role of the reference signal. The 7 channel analog data recorded by the data recorder were converted to a sequence of 8-bit digital signals in the following order: No.1, 2, No. 1, 3, No. 1,4, No. 1, 5 and No. 1, 6. In most cases, those data were sampled in an interval of 21.75 μs. Hence, the sampling interval for the reference signal No. 1 becomes 43.5 μs and that for the master signals No.2 through No. 6 was 348 μs. The range of the velocity scale was set between 2 cm/s and 2 m/s. The development of high speed multi-channel A/D card and processing capacity of computer as well as statistical softwares in recent years have facilitated the use of this type of probe and the method.

Rathbone et al. (1989) used a similar probe to determine particle velocity parallel to the surface. The difference is that the central fiber is the sensor of the Fiber Optic Doppler Anemometry (FODA), and it transmits light from a 5 mW He-Ne laser. Particles illuminated by the central fiber scatter light to the surrounding fibers as they passed the probe. When the particles passed over two fibers in succession, the transit time at cross correlation maximum can be obtained. In principle, by correlating the strongest signals from the fibers, it should be possible to determine the direction or different components of particle velocity. However, the measurement was carried out in a two-dimensional fluidized bed, and relatively few data were obtained.

Nowak et al. (1990) reported a probe shown in Figure 4-24 as type E for measuring velocity of mixed particles of different sizes. The bed particles were FCC particles averaging 46 μm in diameter, mixed with porous silica alumina particles with a mean diameter of 3 mm as the large particles. The large particles were coated with a water-soluble fluorescent dye which has a strong visible light color under UV illumination. The total concentration of large particles was 2%, and the inventory of the bed was 200 kg. The optic fiber probe consists of several plastic fibers; silica fibers were used as the light source while plastic fibers acted as the light receiver. Ultraviolet light from a low-pressure mercury lamp, filtered, leaving the silica fibers, illuminates the fluorescent particles. A portion of the visible light reflected at the surface of the particle passed through the plastic fibers and UV cut filter to a photo multiplier, the electric signals of which were collected in a data recorder. Figure 4-29 shows the measured velocities of the particles of different diameters in a 205 mm i.d. circulating fluidized bed. It can be seen from the experimental results that in the region from the center to the position of $r/R = 0.8$, the change in velocity of large particles is not obvious, while the maximum velocity of fine particles in the center decreases along the bed wall. These data were obtained with the FFT method by deriving the transit time from the maximum value of the cross-correlation function. The fluorescent tracing measuring technique can detect and display given objects in motion, e.g., particles. Since there is apparent difference between the emitting light and the fluorescent light, the use of a filter can easily distinguish the signals of a given object from others.

Little has been discussed on error about measurement of particle velocity with the cross-correlation method. Usually, the data presented are the average velocity profiles of each measuring point. It is hard to know the exact sampling time at any position, the time interval of data processing, and at

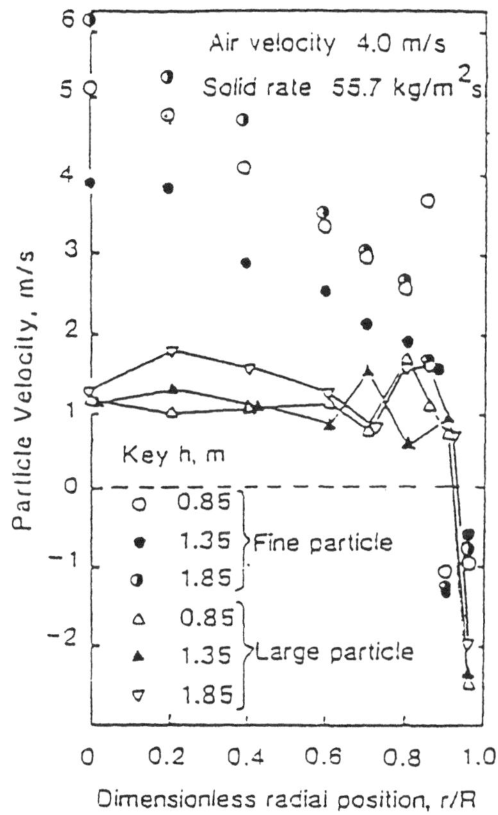

FIGURE 4-29 *Profiles of measured large and small particle velocities (Nowak et al., 1990).*

what cross-correlation functions the maximum value τ_m of the average transit time is obtained, and therefore, the scope of errors cannot be clearly explained. And if the measuring system was calibrated, it was, in most cases, calibrated in rotating beds. Obviously, under these circumstances, the cross-correlation functions are always large. The errors in the measurement of particle velocity may result from the distance between two measuring fibers, sampling time and frequency, particle size distribution and the sharpness of cross-correlation curves.

In designing a probe, the distance between the two detecting points, *l*, is essential. Ideally, a particle passing through the upstream detecting point should also pass through the downstream detecting point. If this can be satisfied, the cross-correlation function would approach to 1. In practice, the cross-correlation function may become small due to collision and friction between particles and between particles and the probes. Oki et al. (1977) qualitatively described the relationship between l and d_p. When $l < d_p$, the coherence is high, the transit time is small and the directional characteristics is dull. If $l > d_p$ the above relationship is reversed. When the distance l is greater than five times the size of particle d_p, the maximum value of cross-correlation coefficient $C(\tau_m)$ is smaller than 0.2, and thus it becomes difficult to locate the peak of $C(\tau)$.

Generally, for signals with known frequency f_s the sampling frequency of 2 ~ 5 times of f_s can be easily satisfied, because in most cases the velocity at a measuring point is approximately known, and the signal frequency produced at a given fiber diameter can be estimated. Since the cross-correlation function is a discrete function of the time lag of multiples of the sampling interval, that is, inverse of the sampling frequency, a significant error is introduced in the velocity calculation when only a few points are sampled during a particle's passage between the two fibers. Herbert et al. (1994) suggested that the error, E, can be defined as the difference in velocity as calculated from two neighboring points, or:

$$E = fl\,[M^{-1} - (M+1)^{-1}] \tag{4.16}$$

where f is the sampling frequency, l is the distance between the fibers, and M is the position of the cross-correlation maximum which is an integer multiple of the sampling interval. The distribution of particle diameter directly affects the velocity distribution of particles. The free-fall velocity of a particle is proportional to its diameter. Figure 4-30 shows that the normalized standard deviation of the measured free-fall particle velocity increases exponentially with the normalized deviation of the particle diameter. It is therefore necessary to conduct many more measurements when the particles exhibit a large distribution in order to determine a reliable mean value. The sharpness of the cross-correlation function is the main factor of deviation of particle velocity measurement, because τ_m based on the maximum of the cross-correlation function is considered as the mean value of particle velocity during the sampling period, with its accuracy indicated by the sharpness of its curve.

Militzer et al. (1992) took notice of this feature. Figure 4-31 shows three examples for correlation and time lag curves. Curve (a) is for a low velocity flow with well correlated signals, (b) is for a high velocity flow with well correlated signals and (c) is for a low velocity flow with poorly correlated signals. In order to distinguish the data of different correlation coefficients, they used two sets of criteria in their program to reject or accept calculated velocities. The first one used the value of the normalized correlation coefficient, the default value used in the program being 0.5,

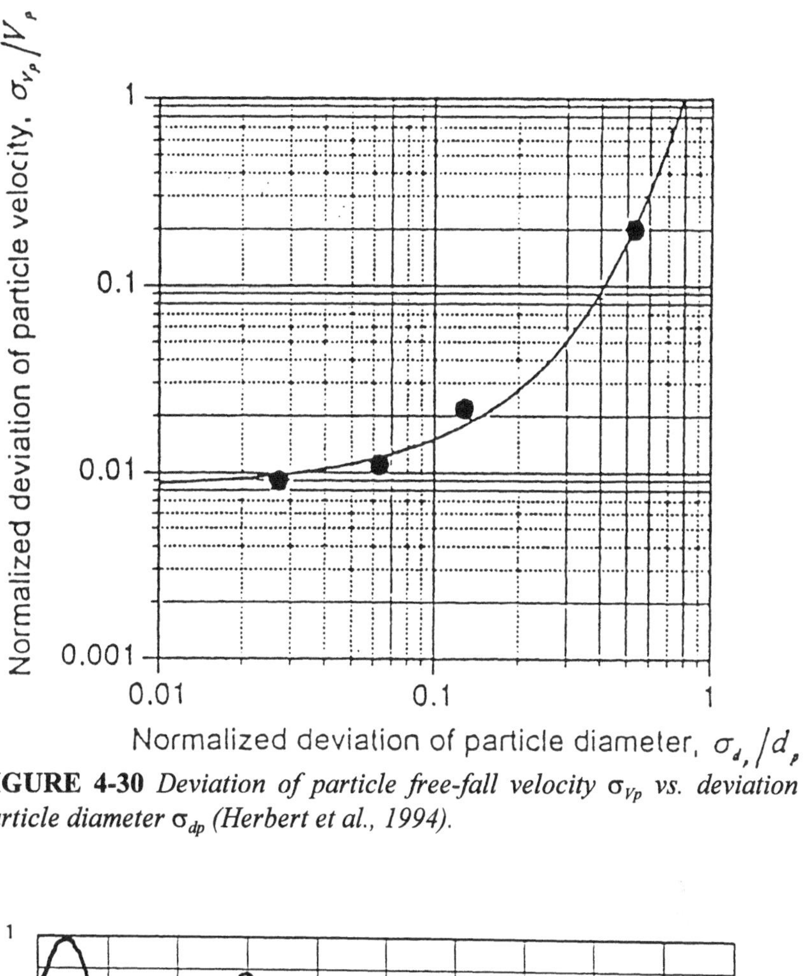

FIGURE 4-30 *Deviation of particle free-fall velocity* σ_{Vp} *vs. deviation of particle diameter* σ_{dp} *(Herbert et al., 1994).*

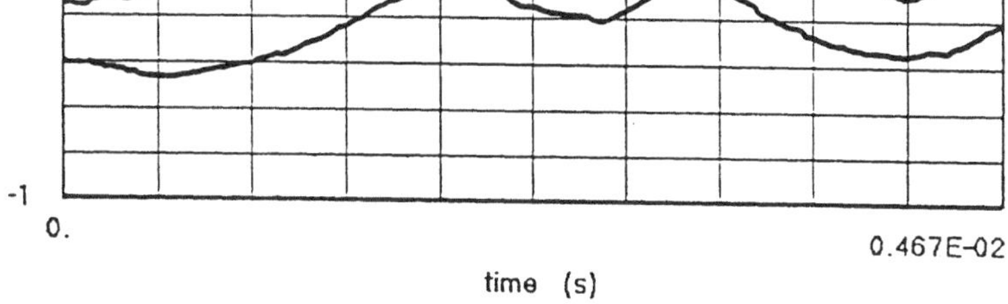

FIGURE 4-31 *Example of correlation versus time delay curves (Militzer et al., 1992).*

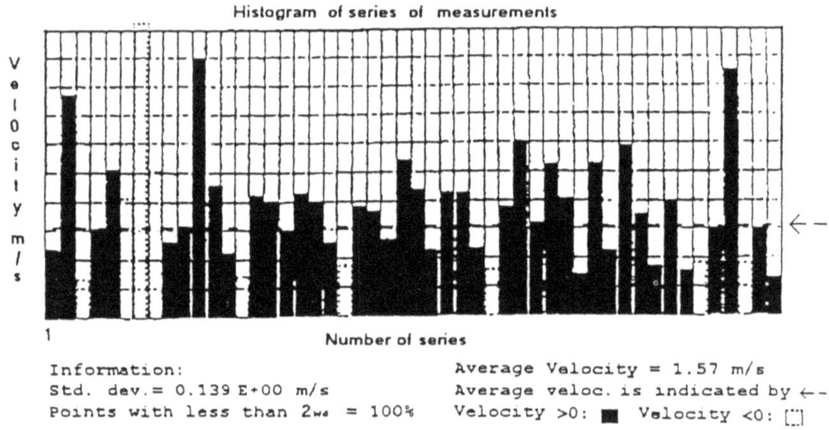

FIGURE 4-32 *Histogram of solid particle velocity (Militzer et al., 1992).*

though it is dependent on flow. The second criterion that can be invoked eliminates from the average and standard deviation calculation all those velocities outside of two standard deviation interval around the average of the original sample. The program takes each individual velocity and compares it with the average of the remaining velocities. Those fell outside of two standard deviation interval around the average were eliminated from the calculation. This procedure was repeated for each of the velocities from the sample. The results of this treatment is shown in Figure 4-32, the histogram of a series of 50 measurements at the wall of an industrial pilot CFB. The solid particles were 200 micron mean diameter sand particles at a fluidizing gas velocity of approximately 3 m/s. It was found from the histogram that even the two criteria were added in the program, some data for particle gas velocities larger than 3 m/s were not meaningful, although their cross-correlation function is larger than 0.5 and between the two standard deviations. The maximum of the cross-correlation function represents the dependence between the two sets of signals of varying quality. Since the τ_m obtained from different cross-correlation functions will be used directly in the calculation of average particle velocity, it would be difficult to relate various sets of data or to determine their deviation. The following logic discrimination method is thus developed in consideration of the above factors.

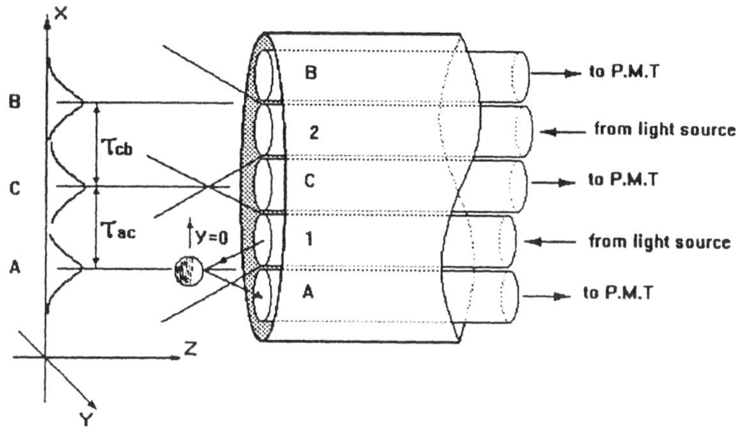

FIGURE 4-33 *Schematic of the five-fiber optic probe (Particles move along the x-axis)*

4.3.2 A logical discrimination method

In order to obtain the instantaneous particle velocity directly from the signals of the reflected light of the probe for particle velocity measurement, Qin and Liu (1982) proposed a method to acquire the instantaneous particle velocity distribution with a logic discrimination method.

As can be seen in Figure 4-33, that for a five-fiber probe (Type F in Figure 4-24), it can be found from the behavior of particles at the tip of a probe and the signals of reflected light that if the particles move along the x axis in a three-dimensional space, the intensity of the signals mainly depends on the location of particles on the z axis. And if a threshold level is set, the lateral range of x axis can be limited, thus eliminating the influence of remote particles and background light, etc. For the x and y axes, if the particles deviate from the x axis, the intensity and width of signals would become small, which can be eliminated by the threshold level; and if $y = 0$, it would be a standard qualified particle signal and the transit time of the particle, which is also the instantaneous velocity of the particle, at the measuring point can be obtained directly from the peak-value signals A and B. However, when the particles move in turbulence or the moving direction of particles changes frequently due to collision at high concentration, the construction of the three-fiber probe would produce relatively desirable pseudo-signals. Since the optic fiber probe can only resolve the transit time of two sets of signals and could not judge the continuity of a signal source, i.e., signals A and B may be produced by different particles, phenomena such as broader distribution of particle velocity and some unreasonable data may occur. In order to discriminate and limit the moving deviation of particles in the x and y directions, a reference fiber was added in the probe, as shown by

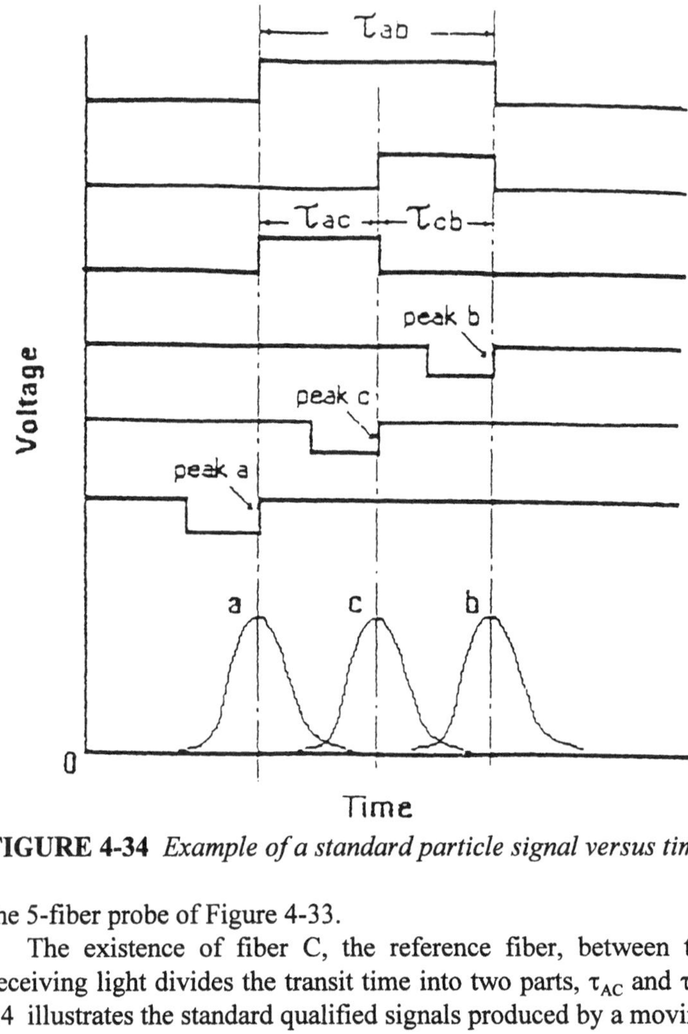

FIGURE 4-34 *Example of a standard particle signal versus time.*

the 5-fiber probe of Figure 4-33.

The existence of fiber C, the reference fiber, between the fibers for receiving light divides the transit time into two parts, τ_{AC} and τ_{CB}. Figure 4-34 illustrates the standard qualified signals produced by a moving particles at the tip of the 5-fibers probe. When the particles move along the x axis, three sets of electrical signals are produced in channels A, C and B with a delay in time, the sequence of which is τ_A - τ_C - τ_B or τ_B - τ_C- τ_A, depending on the moving direction of the particle. Therefore, the measuring system can obtain the results of velocity distribution and number of particles in the positive and the negative *x* direction. If any one of the three signals is lacking in the above mentioned sequence, it can be ascertained that the particle did not move completely along the *x* axis or the signal is not produced by the same particle. More precisely, if particles move along the *x* axis, τ_{AC}, τ_{CB}, or τ_{BC}, τ_{CA} will be equal, because the movement of particles within the distance of less than 1 mm should be equal in velocity without the disturbance of external force, and all signals with unequal τ_{AC}, τ_{CB} or τ_{BC}, τ_{CA} transit time will be considered as signals not produced by the same particle, and such pseudo-correlation signals, once produced, will be canceled, so that the system can meet the requirement of velocity measurement of particles even under the condition of high concentration and turbulent flow. Nevertheless, such strict condition of $\tau_{AC} = \tau_{CB}$ would reduce the number of qualified signal data and sampling time would be correspondingly longer. In practical application, a computer would easily adjust the acceptable range of the above conditions to control the deviation of instantaneous velocity of particles within the known range. As a matter of fact, a velocity could be made acceptable if these two measured velocities are within a certain tolerance, say, within 1% of each other, i.e.,

$$V_p = (V_{AC} + V_{CB})/2 \tag{4.17}$$

for

$$2\left|\frac{V_{AC} - V_{CB}}{V_{AC} + V_{CB}}\right| \leq 0.01$$

where $V_{AC} = l_{AC}/\tau_{AC}$ and $V_{CB} = l_{CB}/t_{CB}$. For the purpose of accurately determining the effective separation distance between fiber A and fiber C and between fiber C and fiber B, not to obtain l value simply from the length of fiber arrangement, a simple and effective way is to calibrate the probe and measuring system with a D.C. speed adjustable motor and a oscilloscope. Fix a fine rod with a diameter equal to that of the particle to be measured onto the shaft of the motor, adjust the distance between the fine rod and probe tip (i.e., the z axis). Cycle time and transit time can be taken from the oscilloscope, i.e., calculating l from cycle velocity and τ_{AB}. Since the calibration signal was produced cyclically and is reproducible, the signal at each channel was observable on the oscilloscope. In order to inspect the effects of particle diameter change on the peak signals, experiments were carried out with 6 fine rods of diameters varying from 0.05 to 0.4 mm, with a probe made from 0.2 mm fibers. Experimental results show that the velocity tolerance produced by peak value detection was not larger than ±2%.

Figure 4-35 shows the schematic diagram of the measuring system of the 5-fiber probe. As shown in Figure 4-33, fibers 1 and 2 of the probe are used for illuminating the moving particles and the other three fibers, A, B and C, are used to catch the light reflected on the surfaces of those particles. The variations of the reflected light are converted to electric signals by photomultipliers and processed by amplifiers, filters and peak detectors respectively. After the peak detectors, the signals are converted to pulse trains and then the pulse trains are sent to the logic circuit and the computer to analyze and calculate the delay times, τ_{AC} and τ_{CB}. The computer was used

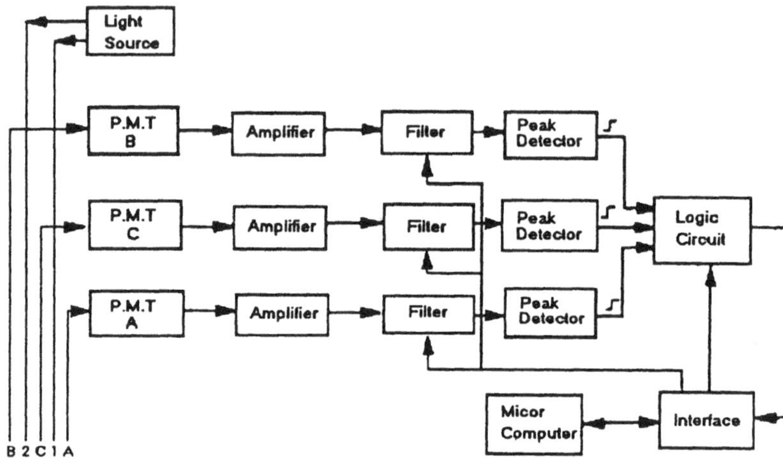

FIGURE 4-35 *Block diagram of five-fiber optic probe measuring system.*

to control the interface to acquire, calculate and analyze the data by a specially written computer program.

Figure 4-35 shows the block diagram of the logic discrimination functions of the measuring system. These functions are implemented by circuits or programs, with main consideration placed on the realization of fast processing of the acquired instantaneous velocity distribution of particles. Figure 4-36 shows a typical particle velocity distribution determined by a five-fiber optic probe. Through the analytical and statistical software in the personal computer, data such as the statistical distribution of particle velocity, total particle number, effective particles data, sampling time, maximum, minimum and average velocity, peak velocity and standard deviation of samples, and so on, can be acquired by the monitor display.

Zhou et al. (1995) used a five-fiber optic probe and measuring system in a 9 m high circulating fluidized-bed riser of 146×146 mm square cross-section. The particle velocity distribution for a typical case determined by the optic fiber particle velocity probe, is shown in Figure 4-38 (mean particle velocities were 5.85 m/s upward (1991 particles), and -0.97 downward (9 particles). From the measuring system, the velocities of both ascending and descending particles can be obtained, and the relative percentage of sampled particles which are being carried upwards and downwards. Lateral profiles showing the local percentage of sampled particles which were being carried upwards (rather than descending) are shown in Figure 4-39. The percentage of ascending particles increased from the wall to the center. Almost all particles moved upwards in the central region of the riser, while most particles moved downwards near the wall. Lateral profiles of particle

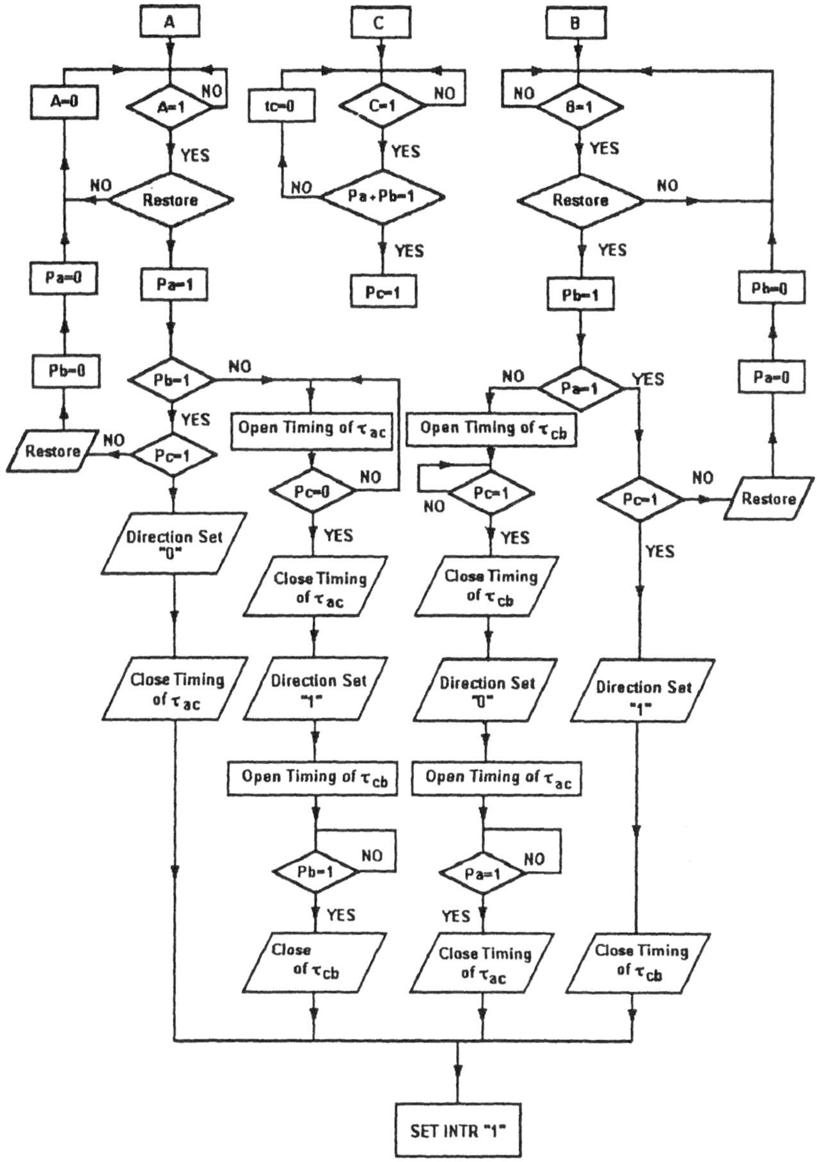

FIGURE 4-36 *Schematic of the logic distribution functions of the five-fiber measuring system.*

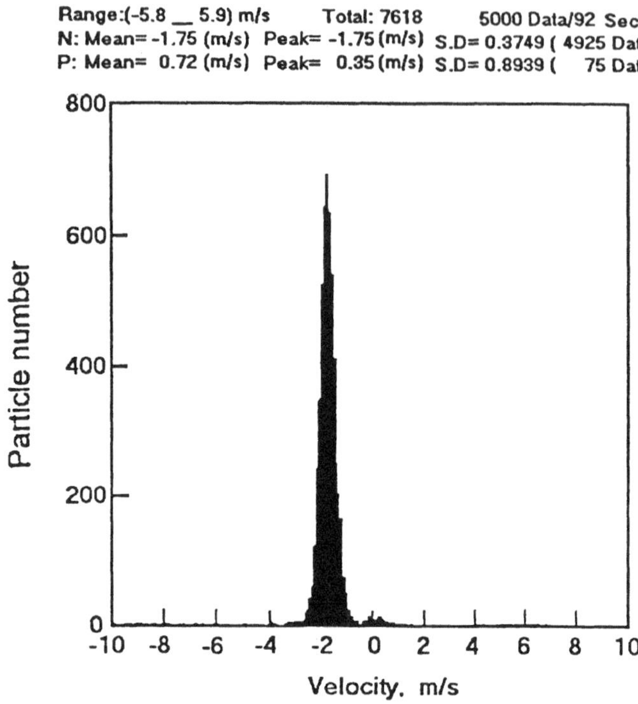

FIGURE 4-37 *Typical particle velocity distribution as measured by a five-fiber optic probe (Sand, d_p of 150 μm, by free fall system).*

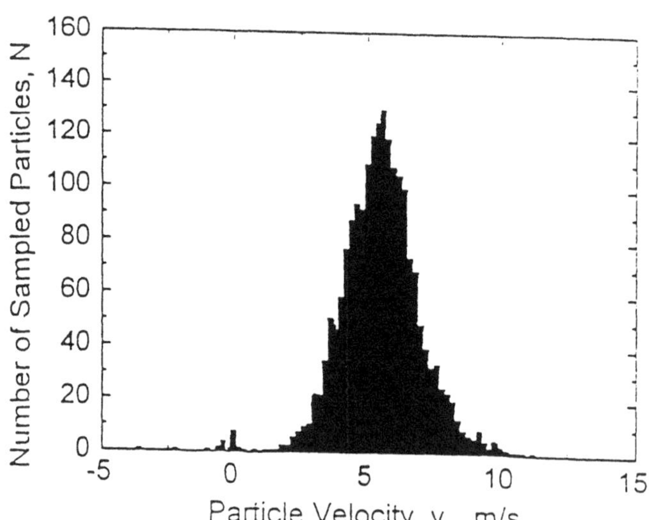

FIGURE 4-38 *Particle veocity distribution in the center of a circulating fluidized bed for $U_g = 5.5$ m/s, $G_s = 20$ kg/m^2 (Zhou et al., 1995).*

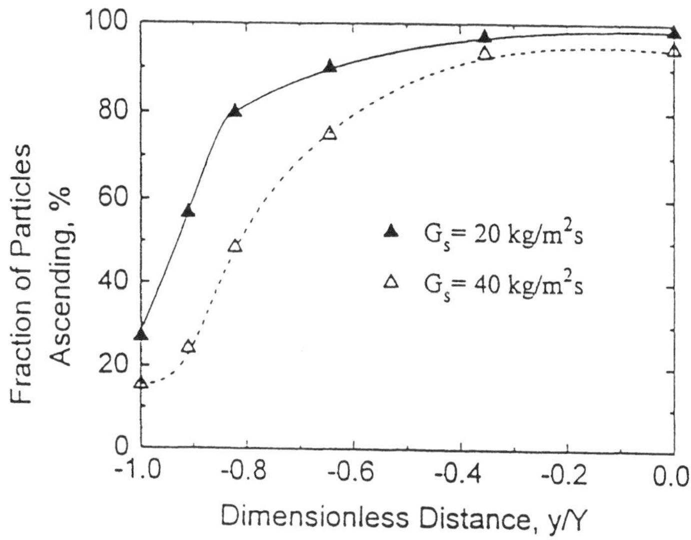

FIGURE 4-39 *Lateral profiles of fractions of particles travelling upwards for different solids fluixes: Ug = 5.5 m/s, z = 6.2 m, x/X =0(Zhou et al, 1995).*

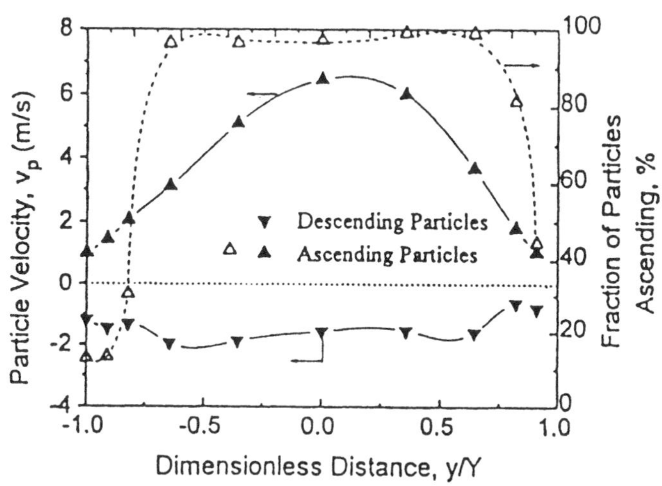

FIGURE 4-40 *Lateral profiles of particle velocities and fractio of particles ascending at top of column: Ug = 5.5 m/s, Gs = 40 kg/m2s, x = 8.98 m, x/X = 0 (Zhou et al., 1995).*

velocities and fraction of sampled particles ascending near the top of the riser are presented in Figure 4-40.

Since the average velocities of the ascending and descending particles can be obtained simultaneously during the sampling period, the number of the ascending and descending particles can thus be calculated, and the data for each measuring point can be easily taken from the monitor of the computer and the dynamic changes of velocity profiles during the sampling period be displayed on the screen by a statistic software until the whole sampling process is ended. This discrimination method for particle velocity measurement, which uses the five-fiber optic probe, well defines reflected light signals of particles to judge whether or not the signals are produced by the same particle. If the qualified particle signals $\tau_{AC} \approx \tau_{CB}$ obtained through logic discrimination are treated for τ_m with the cross-correlation method, then cross-correlations for these signals approach unity. Therefore, the error of the logic discrimination method can be controlled within a very narrow range, and the time taken for data processing is much shorter than that required by calculating the cross-correlation function with the FFT method.

Notations

C	concentrationor volume fraction, e.g., C_v = solids volume fraction
d_p	diameter of a particle, mm
f	frequency, e.g., f_s = sampling frequency
f_d	fiber diameter, mm
G_s	solids circulation rate, kg/m^{2}s
L	distance between the two probe tips, mm
l	distance between incident light and receiving light, mm
N	number of sampled particles
n_o	air refractivity
n_1	core refractivity
n_2	coat refractivity
P	probability density function
p_d	probe diameter, mm
R	cross-correlation function
r	distance from center-line
t	time, e.g., t_m = measuring time, s
U_g	superficial gas velocity, m/s
V_{AC}, V_{CB}	particle velocity obtained from fibers A and C, C and B, m/s
x, y	horizontal coordinates, m
X, y	half-width of column cross-section, m
Z	vertical coordinate measured from the primary air distributor, m

Greek symbols

ε	voidage, e.g., ε_o = voidage of the packed bed
λ	wavelength, nm
ρ	density, e.g., ρ_s = particle density, kg/m^3
τ	transit time, e.g., τ_m = maximum transit time, s

REFERENCES

Bemer, G., "A Simple Light Backscatter Technique to Determine Average Particle Size and Concentration in a Suspension", *Powder Tech., 20*, 133 (1978).

Cutolo, A., Rendina, L., Arena, U., Marzocchella, A., and Massimilla, L., "Optoelectronic Technique for the Characterization of High Concentration Gas--Solid Suspension", *Applied Optics, 29*, 1317 (1990).

Dobbins, R.A. and Jizmagian, G.S., ``Optical Scattering Cross Sections for Polydispersions of Dielectric Particles", *J. Opt. Soc. Am., 56*, 1345 (1966).

Hartge, E., Li, Y., and Werther, J., ``Analysis of the Local Structure of the Two Phase Flow in a Fast Fluidized Bed", *in Circulating Fluidized Bed Technology* (Edited by Basu, P.), Toronto, p.153, 1986.

Hartge, E., Rensner, D., and Werther, J., ``Solids Concentration and Velocity Patterns in Circulating Fluidized Beds", *in Circulating Fluidized Bed Technology II* (Basu, P., and Large, J.F., eds.), France, p.165, 1988.

Hatano, H., and Ishida, M., ``Study on the Entrainment of FCC Particles from a Fluidized Bed", *Powder Tech., 35*, 201 (1983).

He, Y., Lim, C.J., Grace, J.R., Zhu, J., and Qin, S., "Measurements of Voidage Profiles in Spouted Beds", *Can. J. Chem. Eng., 72*, 229 (1994).

Herbert, P.M., Gauthier, T.A., Briens, C.L., and Bergougnou, M.A., ``Application of Fiber Optic Reflection Probes to the Measurement of Local Particle Velocity and Concentration in Gas--Solid Flow", *Powder Tech., 80*, 243 (1994).

Horio, M., Morishita, K., Techibana, O., and Murata, N., ``Solid Distribution and Movement in Circulating Fluidized Beds", in *Circulating Fluidized Bed Technology II* (Basu, P., and Large J.F., eds.), France, p.147, 1988.

Ishida, M., Shirai, T., and Nishiwaki, A., ``Mesurement of the Velocity and Direction of Flow of Solid Particles in a Fluidized Bed", *Powder Tech., 27*, 1 (1980).

Kapany, N. S., "Fibre Optics," Parts 1-6, *J. Opotical Soc. Of America, 47*, 413, 494 (1957); *49*, 770, 1109 (1959).

Krohn, D. A., SPIE Vol.718, *Fiber Optic and Laser Sensors IV*, p.2 (1986).

Lischer, D.J., and Louge, M.Y.,``Optical Fiber Measurements of Particle Concentration in Dense Suspensions: Calibration and Simulation", *Applied Optics, 31*, 5106 (1992).

Matsuno, Y., Yamaguchi, H., Oka, T., Kage H., and Higashitani, K., ``The Use of Optic Probes for the Measurement of Dilute Particle Concentration: Calibration and Application to Gas-Fluidized Bed Carryover", *Powder Tech., 36*, 215 (1983).

Militzer, J., Hebb, J.P., Jollimore, G., and Shakowrzadch, K., ``Solid Particle Velocity Measurements", in *Fluidization VII* (Potter, O.E., and Nicklin, D.J., eds.), Engineering Foundation, New York, 1992.

Nakajima, M., Harada, M., Asai, M., Yamazaki, R., and Jimbo, G., ``Bubble Fraction and Voidage in an Emulsion Phase in the Transition to a Turbulent Fluidized Bed", in *Circulating Fluidized Bed Technology III* (Basu, P., Horio, M., Hasatani, M., eds.), Japan, p.79, 1990.

Nowak, W., Mineo, H., Yamazaki, R., and Yoshida, K., ``Behaviour of Particles in a Circulating Fluidized Bed of a Mixture of Two Different Sized Particles", in *Circulating Fluidized BedTechnology III* (Basu, P., Horio, M., and Hasatani, M., eds.), Japan, p.219, 1990.

Oki, K., Walawender, W.P., and Fan, L.T., ``The Measurement of Local Velocity and Solid Particles", *Powder Tech., 18*, 171 (1977).

Qin, S., and Liu, G., ``Application of Optical Fibers to Measurement and Display of Fluidized Systems", in *Fluidization Science and Technology* (Kwauk, M., and Kunii, D., eds.), China, p.258, 1982.

Rathbone, R.R., Ghadiri, M., and Clift, R., ``Measurement of Particle Velocities and Associated Stresses on Immersed Surface in Fluidized Beds", in *Fluidization VI* (Grace, J.R., Shemilt, L.W., and Bergougnou, M.A., eds.) Endineering Foundation, New York, p.629, 1989.

Reh, L., and Li, J., ``Measurement of Voidage in Fluidized Beds by Optical Probes", in *Circulating Fluidized Bed Technology III* (Basu, P., Horio, M., and Hasatani, M., eds.), Japan, p.163, 1990.

Salins, R.B., ``Plastic Optical Fiber Displacement Sensor for Study of the Dynamic Response of a Solid Exposed to an Intense Pulsed Electron Beam", *Review of Scientific Instruments, 46*, 879 (1975).

Tung, Y., Li, J., and Kwauk, M., ``Radial Voidage Profiles in a Fast Fluidized Bed", in *Fluidization Science and Technology* (Kwauk, M., and Kunii, D., eds.), China, p.139, 1988.

Wang, Z., Bai, D., and Jin, Y., ``Hydrodynamics of Cocurrent Downflow Circulating Fluidized Bed", *Powder Tech., 70*, 271 (1992).

Xia, Y., Zheng, C., and Li, H. ``Characterizing Fast Fluidization by Optic Output Signals", *Powder Tech., 72*, 1 (1992).

Yamazaki, H., Tojo, K., and Miyanami, K., ``Measurement of Local Solids Concentration in a Suspension by an Optical Method", *Powder Tech., 70*, 93 (1992).

Yang, Y., Jin, Y., Yu, Z., and Wang, Z., ``Investigation on Slip Velocity Distribution in the Riser of Dilute Circulating Fluidized Bed", *Powder Tech., 73*, 67 (1992).

Zhang, W., Tung, Y., and Johnsson, F., ``Radial Voidage Profiles in Fast Fluidized Beds of Different Diameters", *Chem. Engng. Sci., 46*, 3045 (1991).

Zhou, J., Grace, J.R., Qin, S., Brereton, C.M.H., Lim, C.J., and Zhu, J., ``Voidage Profiles in a Circulating Fluidized Bed of Square Cross-Section", *Chem. Engng. Sci.*, 49, 3217 (1994).

Zhou, J., Grace, J.R., Lim. C.J., and Brereton, C.M.H., ``Particle Velocity Profiles in a Circulating Fluidized Bed Riser of Square Cross-Section", *Chem. Engng. Sci, 50*, 237 (1995).

5

Instrumentation for Fluid-Particle Flow: Acoustics

Shu-Haw Sheen, Hual-Te Chien, and Apostolos C. Paul Raptis

5.1 INTRODUCTION

In Chapters 5 and 6, we describe some practical instrumentation for measuring fluid/particle flow. The emphasis on practical instrumentation is inspired by the need for flow instruments in industrial processes. Because the environment of most industrial processes is adverse, flow instruments must meet several stringent requirements. Typically, the instruments must be able to perform under high pressure and temperature while withstanding erosion, corrosion, and vibration. In addition, they must be rugged, low cost, and easy to operate. In many processes, flow diversion and constriction are undesirable because the additional flow perturbation may cause plugging and measurement errors. Thus, conventional single-phase flowmeters, such as orifice and venturi meters, may not be applicable to solid/fluid flows, especially those with high solid loading. Nonintrusive flowmeters that do not flow are the top choice of industry. Commercially available nonintrusive flowmeters can be categorized by technique into electromagnetic (EM) (magmeter), thermal, Coriolis, and acoustic/ultrasonic.

Electromagnetic flowmeters can be applied only to solid/liquid slurries that contain electrical conducting liquids. A common problem of EM flowmeters is their temperature dependence for magnet stability. Thermal flowmeters, which measure the heating power that is required to maintain a constant temperature difference between two points along the stream, have not been successful in multiphase streams because varying composition changes the thermal conductivity of the flow rate.

The Coriolis flowmeter consists of a U-shaped tube through which the material whose flow is to be measured passes. The tube is given an oscillatory angular motion by a driver and counterbalance assembly. The material flows in opposite directions in the two legs of the U and experiences oppositely directed oscillatory Coriolis forces, which are transmitted to the tube. The resulting torque produces a twisting motion of the tube about the x-axis. Examination of the equations of motion shows that the amplitude of this oscillatory twisting motion is directly related to the total mass flow of the flowing material when the flow and density profiles are symmetrical with respect to the xy-plane. Flowmeters based on this principle have been accepted by the industry as a standard for solid/liquid flows, but their application is still limited because of their high cost and difficult installation requirements. Field tests show that the Coriolis flowmeter was successful in steady homogeneous flows but somewhat erratic in unsteady flows.

Acoustic/ultrasonic techniques that have been developed into flow-monitoring instruments are Doppler, cross-correlation, and transit-time methods. An ultrasonic Doppler flowmeter has been applied to single-phase turbulent flows and mixed-phase (solid/liquid or gas/liquid) flows. The cross-correlation technique is mainly for mixed-phase flows, whereas the transit-time method has been applied to single-phase flows, either liquid or gas, in large pipes.

In this chapter, we will review the principles of acoustic measurement techniques and describe, in detail, acoustic flowmeters for solid/liquid and solid/gas pipe flows. Because field engineers who work with solid/liquid flows find it important to monitor liquid viscosity, we also describe the use of ultrasound to monitor. To conclude the chapter, we enumerate future sensor research needs.

5.2 PRINCIPLES OF ACOUSTIC FLOW-MEASUREMENT TECHNIQUES

Acoustic techniques applicable to solid suspensions are essentially based on measurements of three parameters: velocity, attenuation and scattering, and their dependence of these parameters on frequency (and angle of the scattering). Experimental variables that are important for optimization of the measurements are transducer geometry, frequency range, and signal processing. The choice of transducer geometry depends on the process environment. For example, in a high-temperature process, one must use waveguides to isolate transducers from the process stream. The operating frequency determines the resolution; the higher the frequency, the better the resolution in measuring particle size. However, the tradeoff is the measurement distance, because, typically, attenuation is inversely proportional to the frequency squared. Signal processing is the fastest growing area of measurement technology. Because of advances in microprocessor technology, more complex signal processing can be performed faster; thus, real-time and intelligent instruments can be made.

5.2.1 Signal-to-Noise Criteria

The figure of merit for a sonic system of measurement that utilizes the generation, transmission, and reception of sound is defined as the maximum allowable transmission loss. This loss is derived from the sonar equation, which simply states that the received signal is equal to the background noise plus a measurement requirement, where the measurement requirement is set by the particular measurement under consideration. This method is not unique to sonics and may be applied to radar, astronomy, and communication theory. The advantage of the sonar equation that underlies the figure-of-merit computation is the fact that it permits quantitative comparison of various measurement schemes

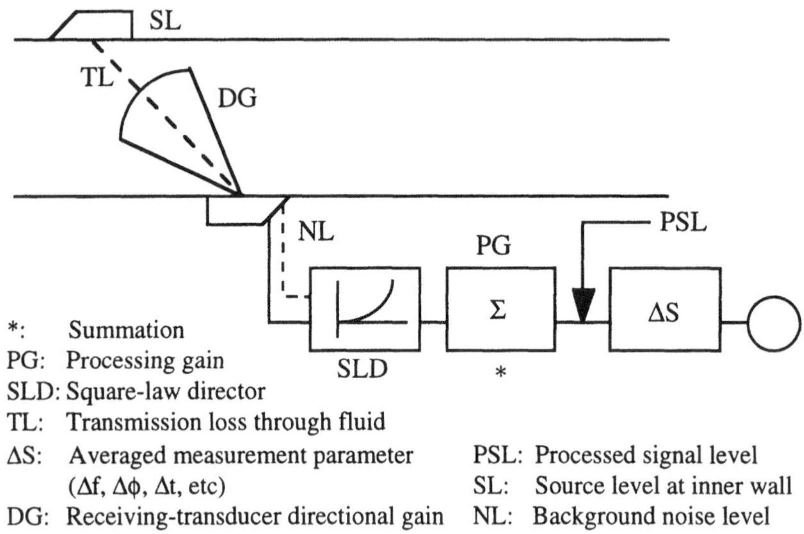

FIGURE 5-1 *Elementary sonic-transmission scheme for measuring flow-velocity.*

and selection of operating parameters on a common basis. A simple form of the sonar equation is

Received Signal Level = Background Noise + Measurement Requirement. (5.1)

The sonar equation for a sonic flowmeter can be stated as

$$SE = SL - TL + PG + DG - NL + MR, \qquad (5.2)$$

where SE represents the detectable flow signal; SL, the source level; TL, the transmission loss; PG, the processing gain; DG the directivity gain; NL, the noise level; and MR, the measurement requirement that is set by the developer of the instrument and dictates the amount of signal above the background noise that a sonic instrument must have for satisfactory performance.

Typically, each term in the sonar equation is given in decibels per unit bandwidth (1 Hz). A representation of a sonic flowmeter is presented in Fig. 5-1. The processed signal level (PSL) of a one-way transmission system, as shown in Fig. 5-1, may be defined by the equations

$$PSL = SL - TL + PG + DG, \qquad (5.3)$$

which states that PSL is the received signal level increased by a factor that accounts for the gain in the signal-to-noise ratio S/N due to signal processing. Equation 5.3 is commonly used to establish the feasibility of a sonic-flow metering technique.

To illustrate an approach based on Eq. 5.3, we describe a feasibility study for developing an acoustic/ultrasonic technique to measure mass flow in coal

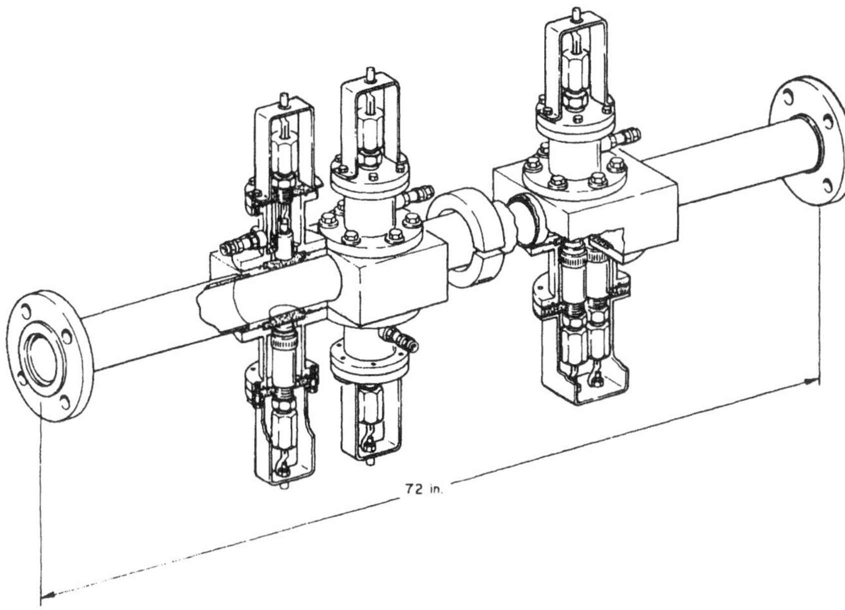

FIGURE 5-2 *HYGAS transducer spool.*

conversion process streams. The study (Raptis et al., 1978) was performed at the Illinois Gas Technology (IGT) HYGAS facility in 1978. The HYGAS facility was running 18-33% coal/toluene/benzene slurries with coal particle size ranging from 1 to severalhundred microns. The first step in the feasibility study was the determination of the background noise in the frequency range of interest (0.1-1 MHz for the HYGAS case) and the sound attenuation in the coal/toluene/benzene slurries. To measure the background noise and attenuation, a transducer spool (a 2-in. schedule 40 pipe) was designed to hold the wide-band (0.1-2 MHz) acoustic emission (AE) transducers (Acoustic Emission Model 500) that were used; see Fig. 5.2. Teflon wave guides were used to isolate the transducers from the slurry. The waveguide insertion loss and transducer sensitivity were initially calibrated. Figure 5-3 shows the spectra of the HYGAS background noise of slurry flows with 0, 18, and 33% coal. It was found that most of the acoustic energy of the background noise is below 500 kHz. Some peaks were observed near 1.5 MHz, but they are attributed to stray RF pickup. It was also observed that the background noise level decreased with increased coal concentration. This decrease in background noise is due mainly to noise attenuation in the coal slurry, specifically, the scattering loss caused by the increased number of coal particles in the slurry. This observation implies that the concentration of coal in a slurry line may be obtained by measuring the RMS value of the background noise. However, it is known that flow noise depends on flow rate; thus, using the noise level to indicate the coal concentration is not reliable. This lack of reliability was confirmed in the HYGAS tests, especially for slurries with high concentrations of coal.

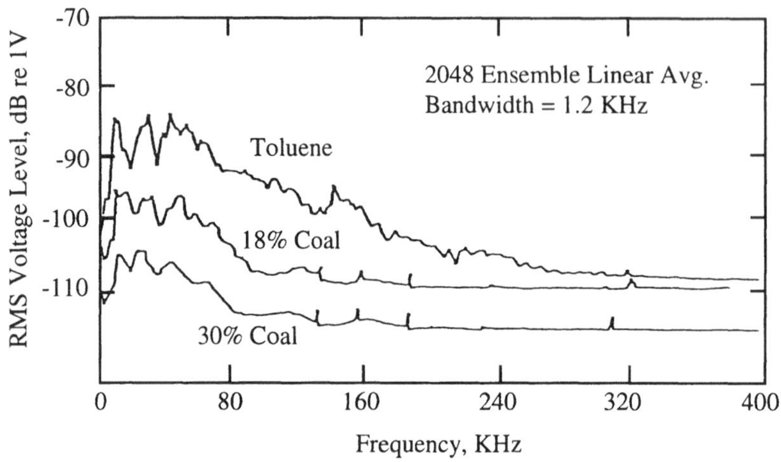

FIGURE 5-3 *Background Noise spectra obtained with acoustic-emission transducers.*

The sound transmission through the coal/toluene/benzene slurry was examined in the frequency range of 0.1-1 MHz. Measurements were made by pulsing an AE transducer with a sine tone burst. The tone burst travels through the medium and is received by another AE transducer directly across the pipe. Acoustic emission transducers were chosen for this study because of their flat frequency response. A sine tone burst with a pulse width of 75 μsec and a repetition rate of 0.2 msec was used to drive the transmitter. Because the AE transducer is not designed for pulsing, the output pressure was not constant over the frequency range. Therefore, a calibration curve of the driving voltage vs. the received voltage with only toluene/benzene in the line was used to normalize the results.

To obtain the absolute sound attenuation in the coal slurry, the diffraction loss, the acoustic mismatch loss, the attenuation due to the Teflon window, and the oil coupling must be calculated. Thus, it is difficult to accurately determine the absolute attenuation. In practice, one measures the relative attenuation with respect to a standard. The attenuation of ultrasonic waves in a solid suspension is attributed to three major factors, namely, scattering, viscosity, and thermal effects. Although the presence of particles affects the fluid viscosity and thermal conductivity, the primary source of attenuation may be due to particle scattering. Hence, one may define the relative attenuation of the HYGAS coal slurry by comparing the slurry attenuation with that of the carrier fluid, i.e., the toluene/benzene mixture. This can be expressed by the equation

$$\alpha_r = [20 \log (V_1/Vo)]/D, \tag{5.4}$$

where α_r is the relative attenuation in dB/cm, D is the separation between two Teflon windows (5.08 cm), V_0 is the ratio of the received voltage to the driving voltage in toluene/benzene, and V_1 is the ratio of the received voltage to the driving voltage in a coal/toluene/benzene slurry.

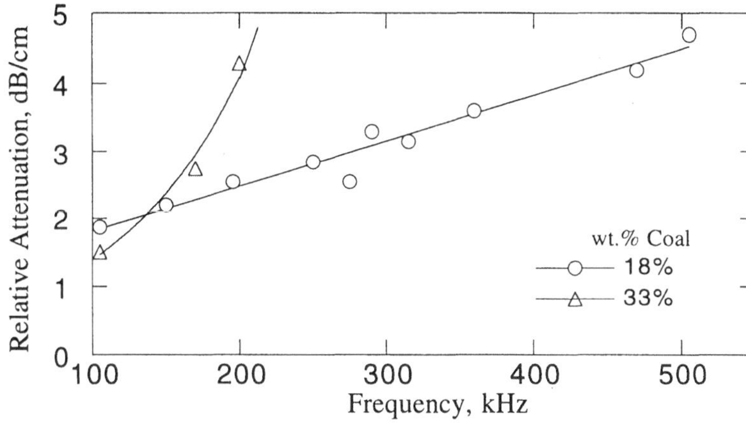

FIGURE 5-4 *Relative attenuation vs. frequency for 18 and 33% coal in toluene/benzene.*

Sound attenuation in slurries with 18 and 33% coal is shown in Fig. 5-4, which shows that sound attenuation increases linearly at a rate of 0.7 dB/cm/100 kHz for the 18%-coal slurry, whereas, for the 33%-coal slurry, the attenuation increases quadratically. The maximum input power used in this measurement was ~10 W across 50 Ω.

A least-squares linear fit was applied to the attenuation data of Fig. 5-3. An empirical equation,

$$\alpha = B_1 + B_2(f - 100) + B_3 f^2, \tag{5.5}$$

represents the attenuation data. In this equation, α is expressed in dB/cm; f is the sound frequency in kHz; for 18% coal, B_1 = 1.86 and B_2 = 0.0067; for 33% coal, B_1 = 1.14 and B_2 = 0.0287; and $B_3 = 1.04 \times 10^{-8}$. The last term in Eq. 5.5 is the attenuation in a toluene/benzene mixture with an ~0.2 mole fraction of benzene. The overall transmission loss at r centimeters from the transmitter is given as

$$TL = \alpha r + DL + 20 \log r, \tag{5.6}$$

where $20 \log r$ is the beam spreading loss, and DL is the transducer diffraction loss, which is neglected in the present work because the transducer separation is within the near field (d^2/λ, where d is the transducer diameter and λ is the wavelength).

The noise level shown in Fig. 5-3 was analyzed with a 50-kHz bandwidth and was found to have a frequency dependence of $f^{-2.8}$. Based on this dependence, the noise level (NL) in dB re Pa (Pascal) can be represented by

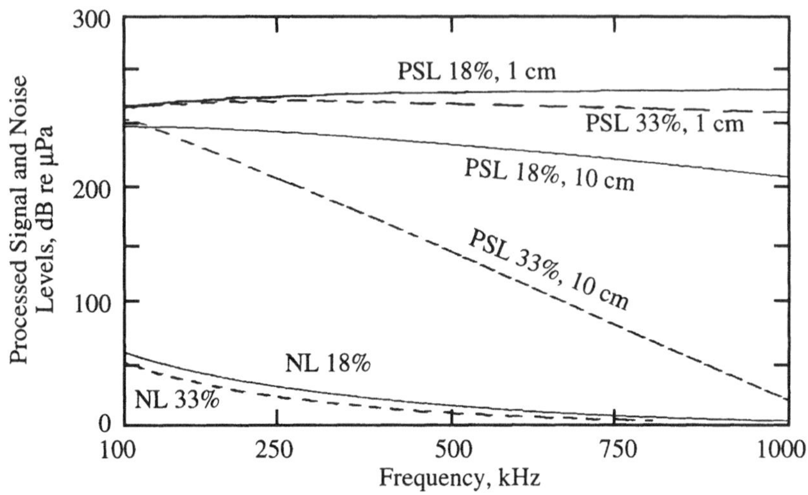

FIGURE 5-5 *Processed signal level (PSL) and noise level (NL) vs. frequency.*

$$NL = A - 56 \log f, \tag{5.7}$$

where f equals the frequency in kHz; and a constant A equals 170.9 and 163.9 for 18 and 33% coal, respectively.

For comparison, the values used in this study for SL and PG are SL = 180 dB at 1 cm at 100 kHz and PG = 5 dB. Substituting Eqs. 5.5 and 5.6, and the values for SL and PG in Eq. 5.3, we obtain

$$PSL = 185 - 20 \log r - [B_1 + B_2 x (f - 100) + B_3 f^2]\, r. \tag{5.8}$$

The PSL and NL for the HYGAS test in the frequency range of 100-500 kHz are plotted in Fig. 5-5, which shows that the PSL is higher than the NL for the entire frequency range of interest. The results also indicate that the transmission loss is the primary factor that will limit the utility of any acoustic/ultrasonic device. The background noise is relatively small, especially in the higher frequency range.

5.2.2 Transit-Time Technique

The transit-time technique measures the difference in the sound velocity between two wave propagation paths across the flow. The simplest geometry of a transit-time flowmeter is a two-transducer contrapropagation flowmeter. The transducers are separated by a known distance along the flow stream., If the speed of sound in the flow is known, the flow velocity V is determined from upstream and downstream transit times. Figure 5-6 is a schematic representation of a single-path contrapropagation flowmeter. Assume that the flow is parallel to the pipe axis and intersects the acoustic path at an angle θ. The upstream and downstream transit times t_u and t_d can be calculated from

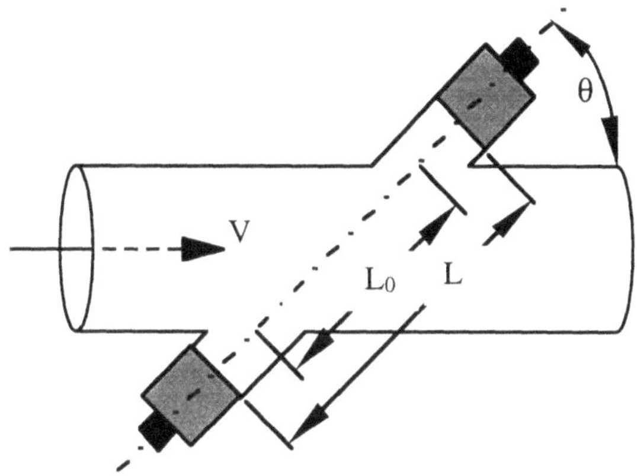

FIGURE 5-6 *Schematic representation of a single-path contrapropagation flowmeter.*

$$t_u - t_m = L_0 / (C - V\cos\theta) \tag{5.9}$$

and

$$t_d - t_m = L_0 / (C + V\cos\theta)\,, \tag{5.10}$$

where t_m is the travel time in dead space $(L - L_0)$ and C is the speed of the sound. Combining Eqs. 5.9 and 5.10, and neglecting the term involving V^2/C^2, we obtain

$$V = C^2 (t_u - t_d)/(2\, L_0 \cos\theta)\,. \tag{5.11}$$

The speed of the sound can also be measured from the transit times. Let $T_u = t_u - t_m$ and $T_d = t_d - t_m$. Then

$$C = L_0/2\, (T_u + T_d)/\, T_u\, T_d. \tag{5.12}$$

From Eqs. 5.10 and 5.11, it is clear that the accuracy in measuring V is determined by the error in measuring transit times, more specifically, the transit-time difference. In practice, to achieve an accuracy of 1% in flow velocity measurement, the resolution in transit-time measurement must be 1 nsec. Thus, this technique is often applied to large pipes that give longer transit times. It has also been demonstrated (Foster et al., 1985) that use of a multipath, such as a zigzag path, can improve the accuracy of the measurement. Unfortunately, an increase in path length causes additional signal attenuation; therefore, the technique is seldom applied to mixed-phase flows that exhibit high attenuation.

5.2.3 Doppler Technique

The Doppler technique measures the frequency shift of scattered waves with respect to incident sound waves. The technique, therefore, requires the presence of scatterers in the flow that is being monitored. The scatterers could be turbulent eddies or vortex shedding for liquid single-phase flows, and solid particles for solid/fluid mixed-phase flows. The basic geometry of a Doppler

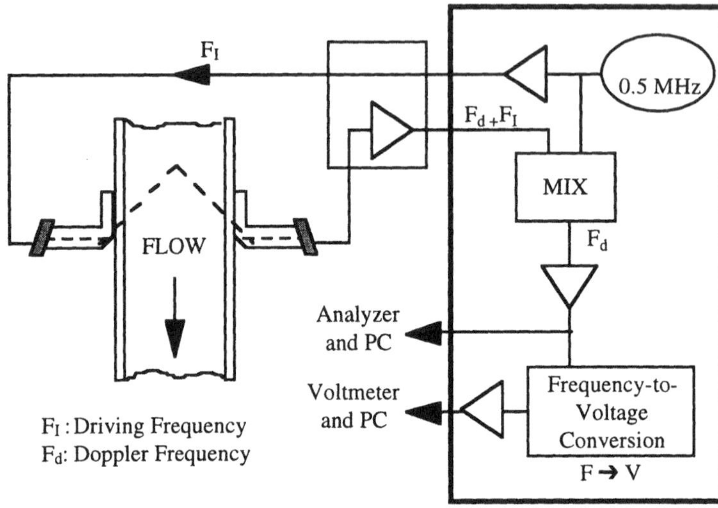

FIGURE 5-7 *Schematic representation of a basic Doppler flowmeter and a block diagram of the signal-processing system.*

flowmeter is shown in Fig. 5-7. In this geometry, we introduce a two-transducer arrangement so the measured Doppler effect represents an averaged value across the pipe flow. Sound waves propagating with sonic velocity C through a fluid are scattered by particles in the fluid moving with a velocity V and a direction θ_1 toward the sound transmitter. The scattered waves propagate to the receiver at an angle of θ_2.

Consider two successive peaks of the sound wave, which are separated by the time interval of one period t (equal to the reciprocal of the frequency f, i.e., $t = 1/f$). During the time interval (t), the distance between the particle and the transmitter has been reduced by $tV\cos\theta_1$, and, similarly, the distance to the receiver is also reduced by $tV\cos\theta_2$, so the distance that the second peak must travel from the transmitter to the receiver is shortened by $tV(\cos\theta_1 + \cos\theta_2)$. Consequently, the arrival time of the successive peaks of the sound wave is separated by $t[1 - V/C\,(\cos\theta_1 + \cos\theta_2)]$. The reciprocal of this time interval is the frequency of the arriving sound wave, and the Doppler frequency (the difference frequency f_d) can be approximated from the equation

$$f_d = f\,(V/C)\,(\cos\theta_1 + \cos\theta_2), \tag{5.13}$$

where f is the carrier frequency of the sound wave. In the special case in which the transmitter and the receiver are positioned symmetrically across the pipe, $\theta_1 = \theta_2$, and Eq. 5.13 reduces to

$$f_d/f = 2V/(C/\cos\theta_1). \tag{5.14}$$

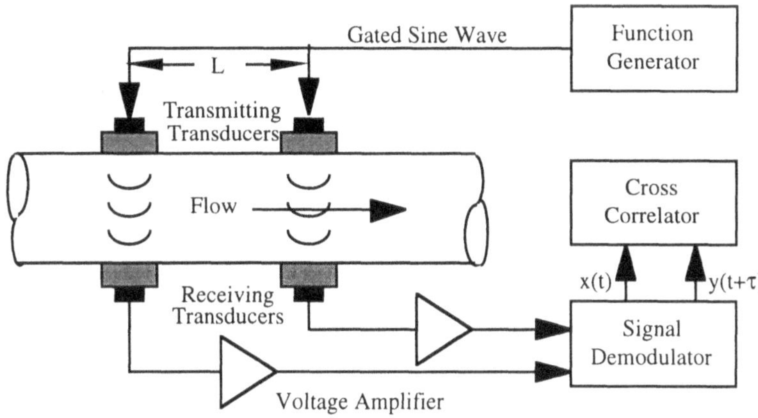

FIGURE 5-8 *Block diagram of basic cross-correlation flowmeter.*

The term $C/cos\theta_l$ in Eq. 5.14 can be determined from the sensing geometry. According to Snell's law, the cosine of the angle θ_l of the sound beam in the fluid is related to the corresponding cosine of the angle in the pipe wall θ_P by the ratio of the relative sound velocities in the fluid and the pipe, i.e.,

$$C/cos\theta_l = C_P/cos\theta_P\ . \tag{5.15}$$

Thus, if one knows the velocity of sound in the pipe wall and the beam direction in the wall, which is generally fixed by the arrangement of the transducer mounting, one does not have to know the velocity of sound in the fluid, which varies with the fluid. Consequently, the calibration of the instrument is independent of the fluid.

5.2.4 Cross-Correlation Technique

The cross-correlation technique measures the time of flight of an inherent flow tag that is passing through two sensors that are separated by a known distance. Figure 5-8 is a block diagram of a basic cross-correlation flowmeter. The sensors are two sets of transducers; one transmits ultrasonic waves to the fluid and the other receives the signals. As the flow tag passes through the interrogating ultrasonic beam, it modulates the beam by attenuation and scattering; thus, the received signals carry the specific tag signature. If the same tag remains unchanged when it reaches the downstream sensor, the signals from the two sensors can be cross-correlated to give the transit time of the tag traveling between the sensors. The flow tag can be turbulent eddies or solid particles. If $x(t)$ and $y(t)$ are the two sensor signals, the cross-correlation function of the signals $R_{xy}(t)$ is given by

$$R_{xy}(\tau) = \frac{1}{T}\int_0^T x(t)y(t+\tau)dt. \tag{5.16}$$

The cross-correlation function represents the transit-time probability distribution, the maximum peak of which corresponds to the most probable transit time, and the width of which is related to the velocity profile and the decay of flow perturbation due to the tag. Thus, the most probable flow velocity can be obtained from $V = L/\tau$, where L is the separation between sensors. In practice, the flow-related signals can be mathematically modeled as

$$x(t) = S(t) + N_x(t) \tag{5.17a}$$

and

$$y(t) = a\, S(t + D) + N_y(t), \tag{5.17b}$$

where $N_x(t)$ and $N_y(t)$ are noise signals that generally are not correlated with the real flow-related signal $S(t)$; and a and D represent decay and delay, respectively, of the signal y(t) with respect to $x(t)$. Ideally, the noise portion of the signal will be averaged out to reveal a cross-correlation peak given as

$$R_{xy}(\tau) = a\, R_{ss}(\tau) \cdot \delta(\tau - D), \tag{5.18}$$

where $R_{ss}(\tau)$ is the autocorrelation function of the signal $S(t)$, and $\delta(\tau - D)$ is a delta function at $\tau = D$. The cross-correlation function is simply the autocorrelation function of the signal convoluted with a delta function, which shifts the autocorrelation function by a delay time D. The RMS error in estimating the time delay by the cross-correlation technique depends on the S/N ratios, averaging time, and signal bandwidths. The relationship can be expressed by

$$\mathrm{E}(\hat{\mathrm{D}} - \mathrm{D})^2 = 2\mathrm{G}_{\mathrm{nn}} / \left[\mathrm{T}\mathrm{B}_{\mathrm{s}}^2 \mathrm{R}_{\mathrm{ss}}(0)\right], \tag{5.19}$$

where G_{nn} is the power density spectrum of the noise, $R_{ss}(0)$ is the autocorrelation function of the signal at $t = 0$, T is the averaging time, and B_s is the signal bandwidth. Clearly, the larger the bandwidth, the smaller the measurement error.

5.3 MEASUREMENT OF SOLID/LIQUID FLOW

Solid/liquid flows are commonly found in industrial processes; to avoid flow obstruction, nonintrusive flowmeters are generally preferred. Flowmeters based on ultrasonic techniques are ideal nonintrusive instruments because, in most applications, the ultrasonic transducers are simply clamped on the outside pipe wall. In this section, we describe two ultrasonic flowmeters based on the Doppler and cross-correlation methods. Both require an inherent flow tag; thus both are directly applicable to solid/liquid flows because of the presence of solid particles. Both flowmeters measure mainly particle velocity; liquid-phase velocity, if different from the particle velocity, is not determined.

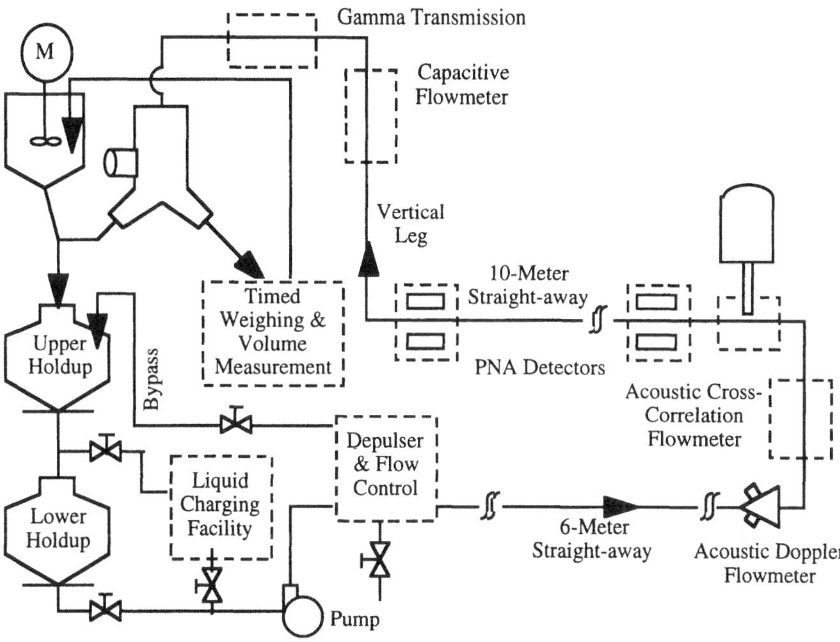

FIGURE 5-9 *ANL solid/liquid flow test facility.*

5.3.1 Volumetric Flow Rate

Determination of the volumetric flow rate of a pipe flow requires only a measurement of average flow velocity. Both Doppler and cross-correlation flowmeters provide such an averaged flow rate measurement. Designs and performance of the two types of flowmeters are described here. Results from calibration tests conducted at an Argonne National Laboratory (ANL) flow facility and prototype instrument demonstration tests at coal-conversion pilot plants are presented.

Figure 5-9 shows a schematic diagram of the ANL solid-liquid test facility (SLTF), which was designed and constructed at ANL to promote development, testing, evaluation, and calibration of flow instruments and to provide an understanding of flow regimes encountered in industrial coal slurry systems. Flowmeters undergoing evaluation were installed on the 6-m straight-a-way and in the vertical leg. The flow rate was adjusted by adjusting the pump speed, bypass valve, and main throttle valve settings. Flow tests were run by gradually increasing coal concentration and establishing stable flows at various flow rates for each concentration.

The facility is equipped with an on-line, timed, weight/volume diversion system that diverts the mainstream into the weighing and volumetric tanks while the content of a reserve tank is dumped into the upper holdup tank. In this way, the sudden loss of flow was compensated for and a constant pump head was maintained. The density of the medium can be monitored to within 1% and was obtained by combining the flow speed and weight readout. A solid weight percent- vs. density curve was produced from carefully obtained measurements of a series of standard samples. The coal was finely ground Ohio #9, a high-ash, highly volatile bituminous coal that is fairly typical of coal used in commercial gasifiers. The major constituents are 59 wt.% carbon, 4 wt.% hydrogen, 4 wt.% sulfur, and 24 wt.% ash. A sieve analysis disclosed that ~86% had a 63-125 m diameter. The oil was representative of organic liquids used in feed lines of some pilot coal conversion plants. Oil density at 20°C was 0.868 g/cm^3.

5.3.1.1 Doppler Flowmeter

The classic Doppler effect is the shift produced by the relative motion between transmitter and receiver. In Doppler flowmeters, the transmitting and receiving transducers are fixed with respect to each other, and the relative motion is produced by the particles or bubbles carried in the stream. The particles act as intermediate receivers and retransmitters. Commercially available acoustic Doppler flowmeters can operate in a temperature environment of up to 300°F (150°C), which is far below the requirements for coal-conversion plants. The temperature restriction on the Doppler flowmeters is mainly due to the temperature capabilities of the commercially available transducers.

Another reason for the erratic behavior of commercially available acoustic Doppler flowmeters is the electronic circuits that are used to process the Doppler-shift signals. In accordance with theoretical predictions, the output Doppler-shift signal exhibits single-tone characteristics and can be counted by the zero-crossing method. This type of method assumes turbulent-flow characteristics. However, in coal-conversion streams at high temperatures, low velocities, and high viscosities, the flow characteristics of the streams are laminar and the spectrum of the Doppler signal is not like a tone spectrum, but is distributed over a wide range of frequencies.

To overcome these two problems, ANL has developed a high-temperature acoustic Doppler flowmeter with the following characteristics: (a) waveguides or standoffs are used to transmit the sound through the temperature gradient, and (b) an electronic circuit determines the corner frequency of the Doppler-signal spectrum. Both of these characteristics are unique to the ANL system and have resulted in a high-temperature acoustic Doppler flowmeter that operated successfully in the recycle lines of the SRC-I, SRC-II and H-coal pilot plants (Karplus et al., 1985].

5.3.1.1.1 High-Temperature Acoustic Doppler Flowmeter

Figure 5-7 is a schematic representation of the installation of the transducers of the acoustic Doppler flowmeter and a block diagram of the signal-processing system. The silver standoffs are 1 ft (305 mm) long and soldered to the pipe. A wedge for mode conversion is attached at the end of each standoff, and the transducers are attached to the wedges. The transmitter driven by an oscillator at a frequency f of 0.5 MHz, transmits through the standoff and the pipe wall, and then into the flow medium. The reflected signal is shifted by a frequency f_d, received by the second transducer, amplified and fed into one of the inputs of a mixer. The other input of the mixer is connected to the original signal. The

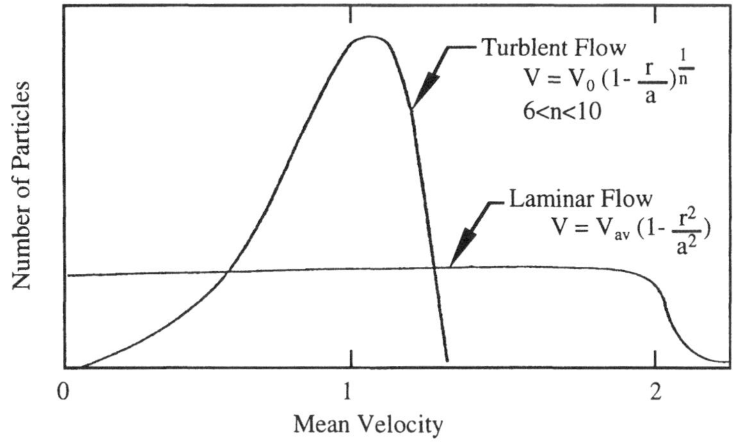

FIGURE 5-10 *Velocity of particles normalized to mean velocity vs. number of particles.*

output of this demodulation process is a Doppler signal that is proportional to flow. The Doppler signal is then coverted to voltage and fed into a spectrum analyzer for further analysis or to a voltmeter for display.

The Doppler frequency can be determined by the zero-crossing method. This method is well suited to turbulent flows that typically show a single-tone frequency spectrum. At the recycle slurry line of the SRC-II pilot plant, the output of the frequency-to-voltage converter gave a flow rate that agreed very closely with the material balance of the plant for most of the time during stable operating conditions. But, drifting of flowmeter readings, which caused measurement uncertainty, was also observed. Upon checking the Doppler- shift signal, we observed that this signal was contaminated by low-frequency signals that caused the Doppler signal to rise above and below the reference line of the zero-crossing counter used to count the frequency content of the Doppler waveform. These low-frequency signals may be attributed to either laminar flow or the presence of gas bubbles in the line. Small quantities of hydrogen may be present during the SRC-II process. Hydrogen in the line greatly attenuated the received signal and placed the Doppler signal well below the background noise.

5.3.1.1.2 Flow-Profile Effects

The low-velocity and high-viscosity conditions in the SRC-II pilot plant clearly indicate that the flow is laminar. The design of the flowmeter initially tested at the SRC-II was based on experiments performed in a laboratory loop, where the flow is turbulent. To explain the effect of the laminar-flow conditions, consider Fig. 5-10, where the velocity of particles normalized to mean velocity, is plotted against the number of particles. The figure shows that, for turbulent flow, the particles are concentrated at the mean of the range, whereas, for laminar flow, they are uniformly distributed over a range that is twice that of the mean. This observation agrees with the results obtained by taking the spectra of Doppler signals obtained from turbulent flow in the laboratory and from the

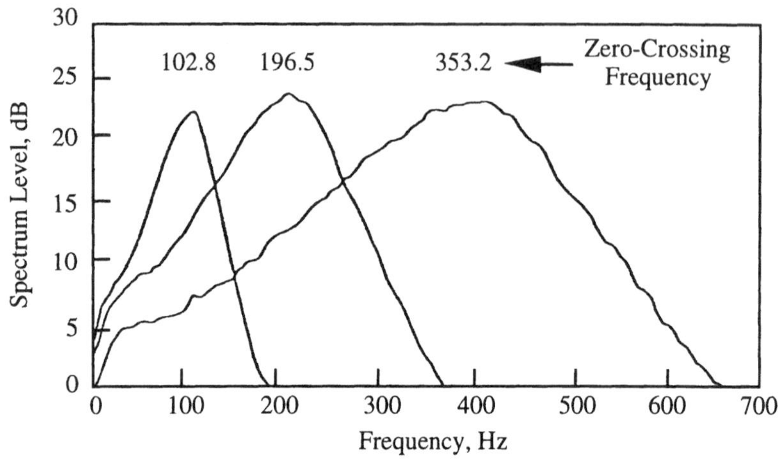

FIGURE 5-11 *Spectrum of Doppler-shift signal for turbulent flow at various velocities in the laboratory loop.*

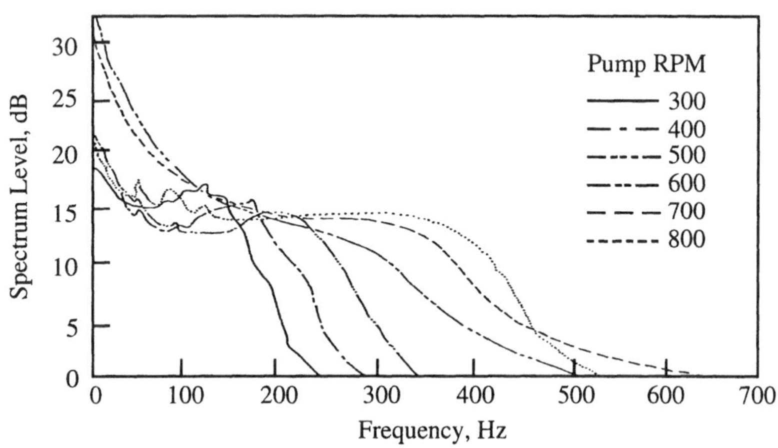

FIGURE 5-12 *Spectrum of Doppler-shift signal for luminar flow at various pump speeds at in the SRC-II pilot plant.*

laminar flow at the recycle slurry line of the SRC-II pilot plant. These spectra are shown in Figs. 5-11 and 5-12 for turbulent and laminar flows, respectively.

Figure 5-11 indicates that, for turbulent flow, the energy of the signal is concentrated about the Doppler-shift frequency and therefore closely resembles a single-tone signal. The zero-crossing counting mechanism will work and yield the Doppler frequency that is proportional to flow velocity.

Figure 5-12 displays the spectra of the Doppler signal for laminar flow at the recycle slurry line at the SRC-II pilot plant at various pump speeds (flow velocities). The characteristics of these spectra are (a) considerable energy concentration at low frequencies (the nature of this high-energy concentration

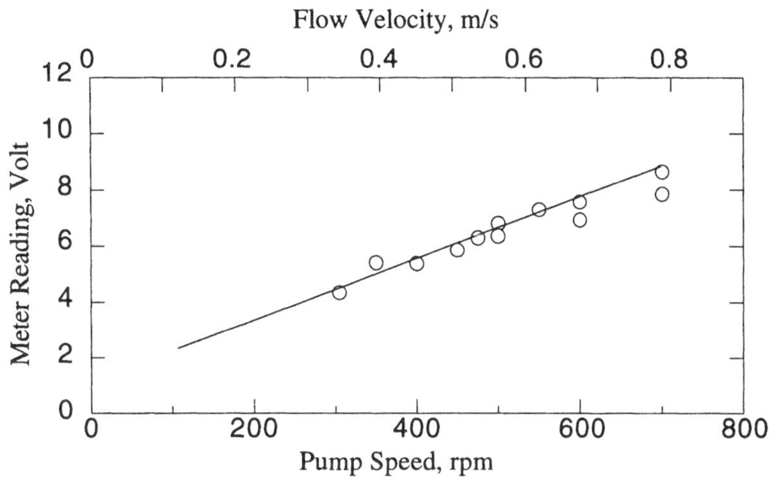

FIGURE 5-13 *Flowmeter reading vs. pump speed, obtained when pump speed was changed from 300 to 800 rpm during SRC-tests. Line represents a linear fit to the data.*

was later found to be due to hydrogen bubbles in the line, (b) a flat region, and (c) a rapid decrease after attaining a frequency that is twice that of the mean Doppler flow frequency. Spectra of this nature were observed repeatedly, except when very intense low-frequency components completely masked the Doppler signal.

If an adjustable-gain high-pass filter has been used to eliminate the low-frequency contamination, the resulting spectrum would look like a band-pass filter spectrum. But, such a filter introduces a limit in the low range of the meter, and a nonlinearity at the lower end of the meter range. The corner frequencies or 3-dB points of such a filter will yield a frequency at the high end that can be related to the lowest possible measurable flow.

To determine the corner frequency of the waveforms shown in Fig. 5-12 and relate it to flow, the property of a uniform spectrum is used. This property states that, for a uniform spectrum, the mean-square amplitude of the total signal energy is the product of the signal energy level and its bandwidth. In other words, the area under the curve is the product of its length and its height above the abscissa. The electronics circuit based on this concept was developed and successfully tested at the SRC-II pilot plant.

5.3.1.1.3 Experimental results

During the SRC-II tests, the flow velocity in a vertical recycle line was varied by changing the pump speed from 300 to 800 rpm. The spectra of the Doppler signals are shown in Fig. 5-12 and the processed data are shown in Fig. 5-13. The voltage output of the flowmeter, which is related to the corner frequency of the spectra in Fig. 5-12 and hence to flow velocity, is plotted against pump rpm.

At another vertical recycle line, the Doppler signals displayed characteristics completely different from those observed at the first line. The observed spectra had a monotonically declining slope, which was very rich in low-frequency content. Obviously, such spectra cannot be processed with the electronic system because they do not have an apparent corner frequency. The reason for these variations in the Doppler spectra is related to the operation of the snubber installed behind the charge pump and in front of the Doppler flowmeter. A small amount of hydrogen was leaking from the snubber into the line; this excessively attenuated the Doppler signal to a point below the background noise. However, when the hydrogen flow was turned on, the signal strength increased by ~35 dB, but it did not change in an obviously repeatable manner with changes in flow. In fact, the spectra characteristics showed a monotonically decreasing slope with no apparent corner frequency. This result may be attributed to back-scattering of signals from hydrogen bubbles that are stagnant near the pipe wall. The back-scattered signals, however, can be transmitted through the pipe wall to the receiver on the opposite side of the pipe and thus increase the signal strength.

Based on the above results, we conclude that an ultrasonic Doppler flowmeter applicable to flows that contain suspended solids can be designed for both turbulent and laminar flows, but requires further development if excess gas bubbles are present.

5.3.1.2 Cross-Correlation Flowmeter

The cross-correlation technique measures the time of flight of an inherent flow tag passing through two sensors separated by a known distance. The technique has been used successfully to monitor single-phase fluid flows in which turbulent eddies modulate the interrogating ultrasonic beams. This type of correlation flowmeter has also been developed for solid/liquid and gas/liquid flows, in which the density fluctuation, caused by clusters of solids and by gas bubbles, is the prime inherent flow tag.

A coal slurry flow is a practical mixed-phase flow that requires close monitoring to ensure safe and efficient transport of the coal slurry. In addition to the flow rate measurement, detection of gas bubbles and settling of solids is equally important in coal slurry lines. In this section, we will describe a cross-correlation flowmeter that can reliably measure coal slurry flow rates over wide ranges of coal concentration and flow velocity. We will also illustrate how the flowmeter can detect settling of solids and recognize the presence of gas bubbles. Both laboratory and pilot plant flow tests are included.

5.3.1.2.1 Basic Configuration

The basic design of an ultrasonic cross-correlation flowmeter (Sheen and Raptis, 1983), as shown by Fig. 5-8, consists of two pairs of transducers positioned in parallel and separated a known distance. The transducers can be mounted by either a clamp-on arrangement or use of special windows that provide a better impedance match to the flow so that more acoustic energy can be transmitted. "Pitch-catch" is the typical mode of operation. The transducers on one side generate ultrasonic waves in the form of sine tone bursts and the

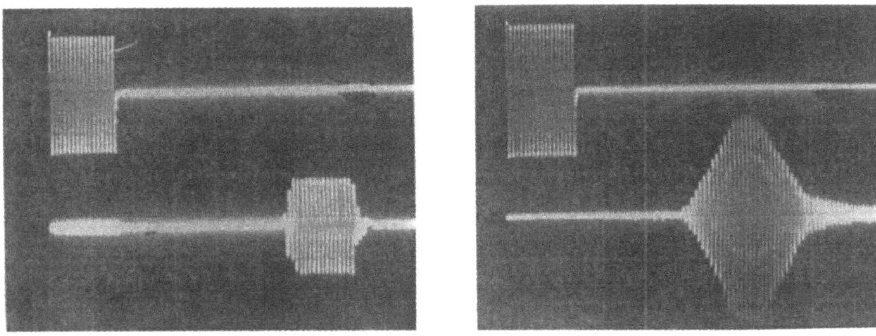

Window Arrangement | Clamp-on Arrangement

FIGURE 5-14 *Typical transmitted and received signals for acoustic cross-correlation flowmeter.*

transducers on the opposite side act as receivers that detect the attenuated signals. The received signals from the two receivers are demodulated and cross-correlated in a time domain to produce a correlation function, from which the transit time, thus the flow velocity, of solid particles traveling between the two sensing volumes can be determined.

Piezoelectric transducers are typically used for ultrasonic flowmeters. Commonly available transducers are resonance types that produce narrow-band high-intensity compression waves. Wide-band transducers can be designed at the expense of signal strength. To select a proper transducer one must consider operating frequency, process temperature, and impedance matching. If the wavelength is compatible with or smaller than the particle size, ultrasonic waves will be mostly scattered by the particles in the path. Consequently, received signals under the pitch-catch arrangement will be strongly attenuated, resulting in a poor signal-to-noise ratio *S/N*. On the other hand, if the wavelength is much larger than the particle size, the absorption cross section becomes small and the received signals lack sensitivity. The typical working frequency of an ultrasonic flowmeter is in the 1-5 MHz range.

The operating temperature of a piezoelectric transducer is dictated by the Curie temperature of the piezoelectric ceramic and its front-end coupling material. Commonly used ceramics are lead zirconate titanates (*PZTs*) whose Curie points are in the range of 300-360°C. In practice, once the process temperature exceeds 100°C, a waveguide is used to thermally isolate the transducer from the process stream. Impedance matching is needed to maximize energy transfer between process stream and transducers. The front-end coupling material often serves the impedance matching purpose.

The matching material is typically a single quarterwave plate of a material with an acoustic impedance of $\left(Z_L^2 Z_c\right)^{1/3}$, where Z_L is the acoustic impedance of the process stream and Z_c is the impedance of the piezoelectric ceramics. If the process stream is a water-based medium, the impedance of the desired material is ~4 MRayls (1 MRayl = 10^6 Kg/m^2/sec). Most plastics and polymers are suitable for this application. However, when the flowmeter is of the direct

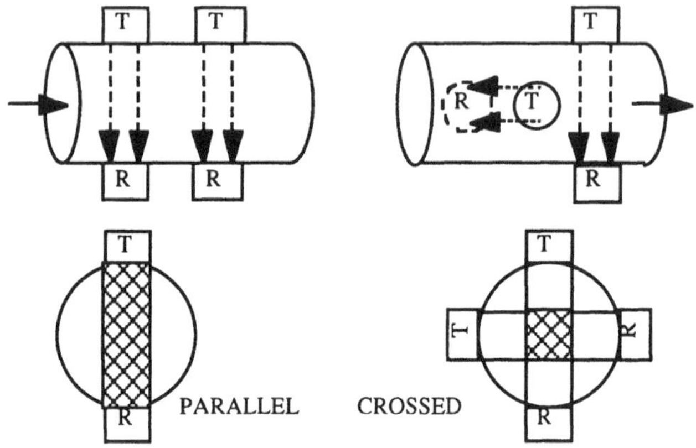

FIGURE 5-15 *Parallel and crossed acoustic-beam orientations of cross-correlation flowmeter (shaded areas are intersecting crosssections)*

clamp-on configuration, a soft metallic film, such as silver, is often used as the coupling between the transducer and the pipe wall. Both, the direct clamp-on and the acoustic windows generally focus the ultrasonic beam in the flow, but some window materials (e.g., Teflon), in which sound propagates at a speed close to that in the fluid, can deliver a parallel beam in the far field. Another advantage of using acoustic windows is noise reduction. In Fig. 5-14 we show the received signals for a two transducer mounting arrangement. The direct clamp-on arrangement shows interference caused by signal paths through the pipe wall.

Other important design parameters of an active acoustic cross-correlation flowmeter are transducer size, spacing, and orientation, because they define the overlap volume between the two sets of transducers, from which the flow correlation is derived. The transducer size determines the beam width and field (near or far field). Commonly used transducers have a diameter of 2.54 cm, which gives a nearfield distance of 11 cm in a water-based medium for a frequency of 1 MHz. Thus, the nearfield beam is the beam profile for pipes with a diameter of <5 in. The profile converges slightly and can be assumed to have the same width as the transducer. To obtain the maximal correlation, the parallel-transducer configuration is chosen. The other extreme orientation is the cross-beam arrangement. Figure 5-15 illustrates the two beam orientations.

In principle, the two orientations give differing velocity measurements; the cross-beam measures the centerline flow velocity, whereas the parallel beam gives an average flow velocity across the pipe. On a vertical pipeline, the transducers can be mounted in any orientation with respect to the pipe axis, but on a horizontal line, the common mounting arrangement is side-to-side. However, if one wishes to detect settling of solid particles in a horizontal line, the transducer mounting arrangement should be top-to-bottom. To choose the proper transducer spacing, one must consider how to minimize signal crosstalk due to beam overlap and short-circuiting through the pipe wall but still preserve the flow correlation. Typically, two to three pipe diameters are selected.

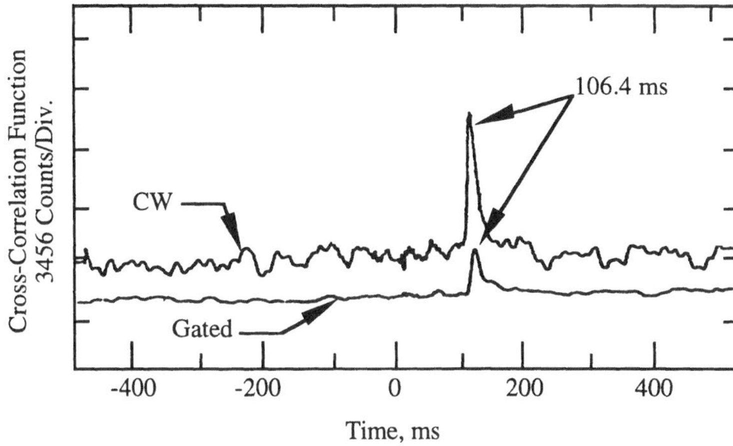

FIGURE 5-16 *Cross-correlation functions in the continuous-wave (cw) and gated-wave excitation modes.*

5.3.1.2.2 Flow Measurements

To illustrate the capabilities of the cross-correlation technique to measure flow velocity, we used two coal-slurry tests. During the tests, various sensing parameters and geometries were evaluated. For example, the examined sensing geometries were direct clamp-on vs. acoustic windows. With the clamp-on geometry, we were able to examine the effects of sensor separation and orientation on velocity. Other instrument variables that were evaluated were operating frequency and mode, demodulation schemes, and signal filtering.

Before we present the test results, we describe some general features of the cross-correlation technique. First, the velocity data obtained by this technique are typically derived from the maximum cross-correlation peak, which evidently differs from the actual average velocity over the pipe cross section because of the finite acoustic beam width. Thus there appears to be a meter factor that is due to the velocity profile. Second, in the cross-correlation system, we use two parallel sets of electronics; any phase difference caused by the electronics will lead to an error in measurement.

To eliminate such an error, the filter bandwidth must be much greater than the content of the signal frequency. The choice of the low-pass filter can be determined by examining the power spectrum of the transmitted signals. Power spectra of a coal/water slurry flowing at differing velocities between 0 and 1.3 m/s. showed that the frequency content of the signals is mainly in the low-frequency range (<l kHz). Hence, a 1-kHz low-pass filter is adequate for most applications. Either continuous waves or gated sine waves can be used to excite the transmitters. The two excitation modes yield the same velocity measurement; the only difference is the signal strength (Fig. 5-16). Nevertheless, the gated mode has two advantages: less power is used so less heat is generated in the transmitters, and, in addition to velocity measurements, information about relative sound attenuation and the time delay between the driving and received signals can be obtained. The latter can then be used to calculate sound speed in the slurry.

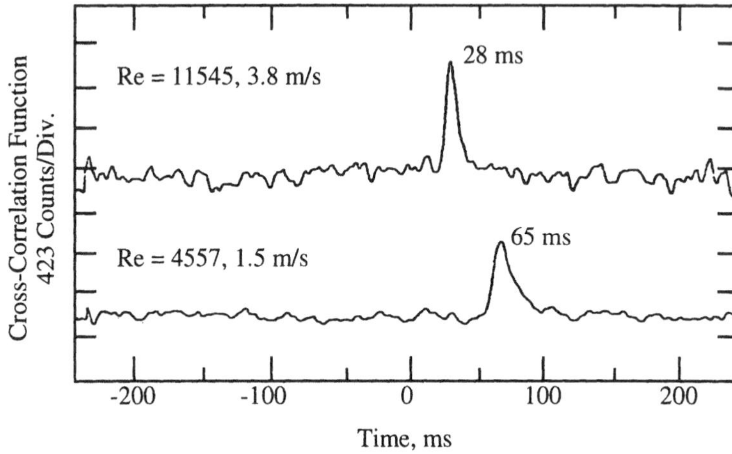

FIGURE 5-17 *Cross-correlation functions at two velocities for the oil flow at two Reynolds numbers(Re).*

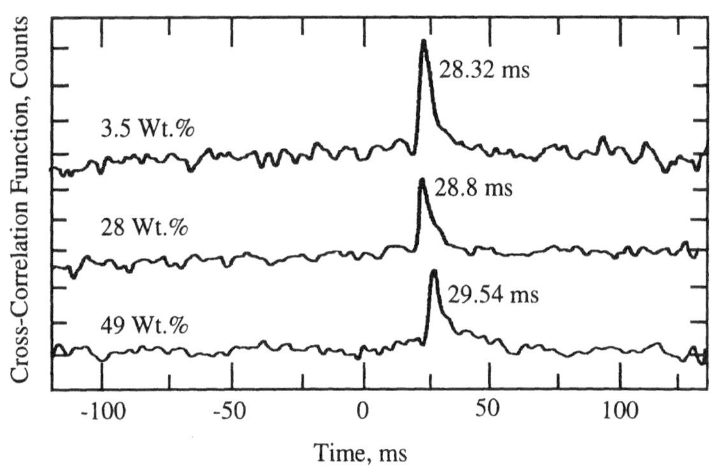

FIGURE 5-18 *Cross-correlation functions for coal/oil slurries with differing solids concentrations but similar velocities.*

Coal/Oil Slurry Tests. Coal/oil flow tests were conducted at ANL's SLTF in 1982. The cross-correlation flow instrument that was used contains Teflon windows and broad-band transducers (0.5-2 MHz). The transducer separation was 12.7 cm, i.e., 2.42 times the pipe diameter (2-in. Schedule 40 pipe). Amplitude demodulation was used to analyze the correlation signals. Ohio #9 coal, which has a mean particle size of 100 μm, was used. The oil was a commercial transformer oil (Shell Diala-Ax) that has a density of 0.868 g/cm^3 and a viscosity of ≈15 cP at 25°C.

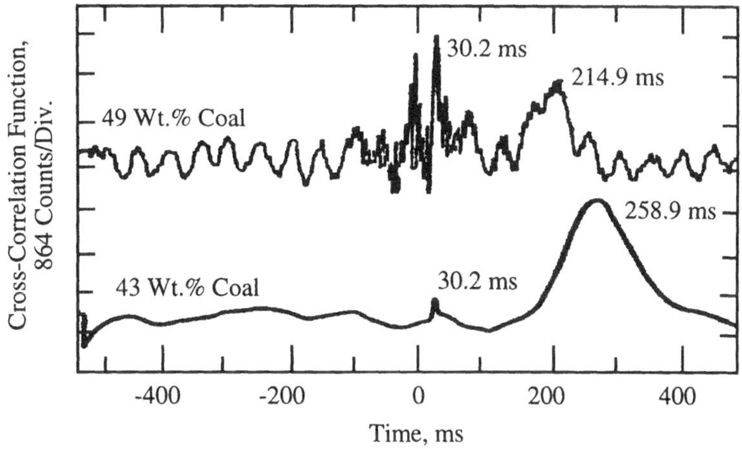

FIGURE 5-19 *Separation detected for two coal/oil slurry concentrations at 2.39 m/s in a 2-in. horizontal pipe.*

Figure 5-17 shows the correlation functions for oil flows (<1 wt.% particle concentration) at flow velocities of 3.8 and 1.5 m/s. The cross-correlation function of the higher velocity displays a sharper peak and higher amplitude; the peak of the lower velocity tends to be broader. The peak broadening is a distinct indication of changing velocity profile toward the laminar regime. In principle, the flow regime is also affected by the viscosity of the slurry or the concentration of coal (at 60 spindle rpm, viscosity varies from 15 to 95 for coal/oil slurries with 0-40 wt.% solids). But this effect is not obviously detectable from the shape of the cross-correlation function. Figure 5-18 shows the cross-correlation functions of three flows with differing concentrations of coal but similar flow rates. Neither the peak position nor the shape of the peak is altered significantly, except at 49 % wt., where the peak broadens slightly.

To examine the effects due to transducer orientation and spacing, we adopted the clamp-on transducer arrangement on a horizontal line. Between 43 and 60 wt.% coal, cross-correlations were computed for three transducer orientations: horizontal, 45°, and vertical with respect to the axis of the horizontal pipe. Measured velocities were the same for the horizontal and 45° orientations, but the vertical orientation produced multiple peaks for certain coal concentrations. Figure 5-19 shows the cross-correlation plots for 43 and 49 wt.% coal. Two well-separated peaks are obtained for the 43 wt.% slurry and the slow-moving part of the slurry dominates; particle settling is clearly observed. For the 49 wt.% slurry, more phase separation but less velocity spread was measured. The average transit time of the two peaks for both cases predicts a velocity close to the average velocity measured by flow diversion. For a 60 wt.% slurry, phase separation disappears.

The persistence of flow disturbance can be measured through space correlation, with increasing spacing between the sensors. For the coal/oil slurry with 60 wt.% coal, cross-correlations were measured with various sensor spacings. Figure 5-20 is a composite of three cross-correlation functions. A

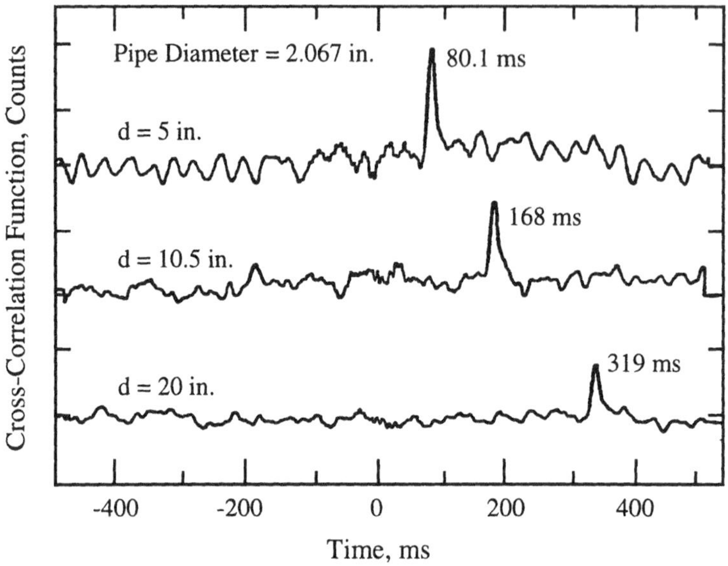

FIGURE 5-20 *Cross-correlation functions for three transducer spacings for a 43-wt.% coal/oil slurry at 1.33 m/s.*

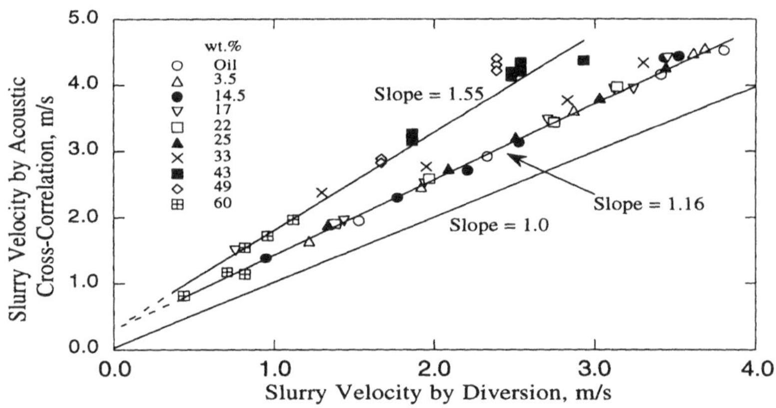

FIGURE 5-21 *Slurry velocities obtained from cross-correlation peaks vs. velocities measured by flow diversion for coal/oil slurries.*

cross-correlation peak still can be resolved, even at a spacing as large as ten times the pipe diameter, and the shape of the cross-correlation peak, as well as the width of the peak, remains unchanged. Flow velocities derived from the three correlation peaks agree within 1%. The only noticeable difference is the change of amplitude. One possible interpretation is that the correlating physical phenomenon results from specific, persistent, flow nonuniformity, such as air bubbles and large coal clusters.

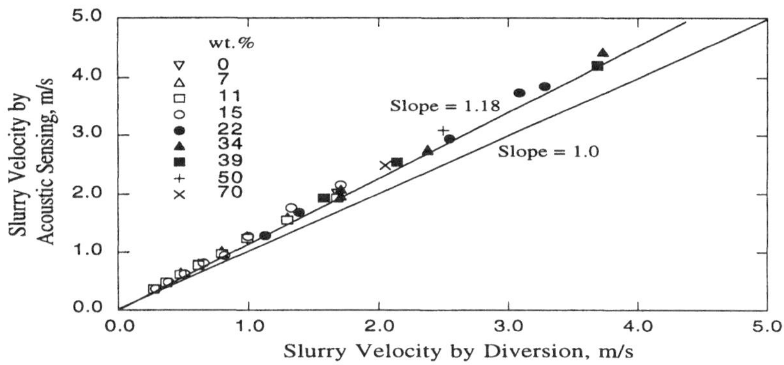

FIGURE 5-22 *Uncorrected slurry velocities by acoustic sensing vs. corresponding velocities obtained by flow diversion.*

Figure 5-21 shows the coal/oil slurry velocity derived from the peak of the correlation function plotted against the velocity measured by flow diversion. The flow diversion gives basically the mass flow rate divided by density to give the volumetric flow rate and then the flow velocity. The data can be fitted by two linear relationships with slopes of 1.16 and 1.55, corresponding to meter factors of 0.86 and 0.64. The meter factor may be directly related to the flow profile effect (Sheen et al., 1985).

Coal/Water Slurry Tests. Coal/water slurry tests, which covered a range of coal concentrations (22-70 wt.%). were conducted at the ANL SLTF. The clamp-on configuration was used throughout the tests and transducers were operated at 0.83 MHz. Cross-correlation functions were resolved over the whole concentration range, (0-70 wt.%). The lowest measurable velocity in a 2-in. pipe was ~0.3 m/s. Because of the low viscosity of water or the low coal-concentration/water mixtures, the flow at 0.3 m/s (reflection coefficient R ~ 15,000) still behaves as a turbulent flow. Thus, turbulent eddies may be the prime source of flow modulation. At high concentrations, phase separation was not detected; this is different from the coal/oil slurry. Figure 5-22 shows uncorrected slurry velocities derived from the maximums of peaks obtained by acoustic sensing plotted against corresponding averaged velocities obtained by flow diversion. Similar to the coal/oil results, velocity data are independent of coal concentration; a linear relation with a slope of 1.18 fits most data points within 5% accuracy. For high concentrations or at low flow rates, the acoustic sensing peaks show some asymmetry, which may represent the velocity profile across the pipe. The centroid of an asymmetric peak yields a slower (about 5% slower in the present tests) flow velocity than that derived from the peak maximums.

During the tests, we also examined how the cross-correlation functions were affected when air is injected into the slurry and when cross-beam geometry is used. Table 5-1 shows the velocity when crossed-beam geometry was used. (Fig. 5-15). When compared with the normal parallel geometry, the crossed-beam geometry measures a velocity that is greater by as much as ~20%. This is expected, because the cross-beam geometry correlates primarily the centerline velocity that readily determines one of the two parameters in a typical velocity profile function. But in practice, to resolve a cross-correlation function by the crossed-beam method requires a long averaging time because the flow information that can be correlated is reduced significantly with this geometry. Figure 5-23 shows the cross-correlation functions obtained with the two geometries. The averaging time for the crossed-beam geometry is 16 times that of the parallel geometry, and the peak magnitude is smaller by a factor of four. Presence of air bubbles in the slurry enhances the density contrast and thus dominates the signal modulation. If large bubbles, whose sizes are compatible with the acoustic beam width, are flowing with the slurry, the bubble-modulated signal exhibits a pattern like a step function that leads to a triangular-type correlation function. Figure 5-24 displays such an example. Due to the broad shape of the peak, the peak maximum contains a large uncertainty; thus, the difference in the position of peaks with and without bubble injections varies without a definite trend. We believe that the measured velocity represents the bubble velocity.

Table 5-1 *Coal/water slurry velocity data for 0-15 wt.% coal.*

Coal Concentration, wt.%	Slurry Velocity by Diversion, m/s	Ultrasonic Parallel, m/s	Cross-correl. Crossed, m/s	Meter Factor Parallel Geometry	Meter Factor Crossed Geometry
~0	0.37	0.47	-	0.79	-
	0.62	0.78	-	0.80	-
	0.99	1.22	1.39	0.81	0.71
	1.68	2.01	-	0.84	-
7	0.27	0.36	0.38	0.75	0.71
	0.37	0.48	0.52	0.77	0.71
	0.48	0.65	0.70	0.74	0.69
	0.62	0.81	0.86	0.77	0.72
	0.80	1.02	1.10	0.78	0.73
	0.99	1.28	1.30	0.77	0.76
	1.30	1.60	1.70	0.81	0.76
11	0.27	0.36	-	0.75	-
	0.37	0.47	-	0.79	-
	0.48	0.61	-	0.79	-
	0.62	0.77	-	0.81	-
	0.80	0.97	-	0.82	-
	0.99	1.24	-	0.80	-
	1.30	1.55	-	0.84	-
	1.68	1.93	-	0.87	-
15	0.363	0.285	-	0.785	-
	0.48	0.385	-	0.80	-
	0.617	0.51	-	0.83	-
	0.805	0.655	0.814	0.81	0.77
	0.946	0.82	-	0.87	-
	1.27	1.00	-	0.79	-
	1.76	1.33	-	0.76	-
	2.15	1.71	-	0.80	-

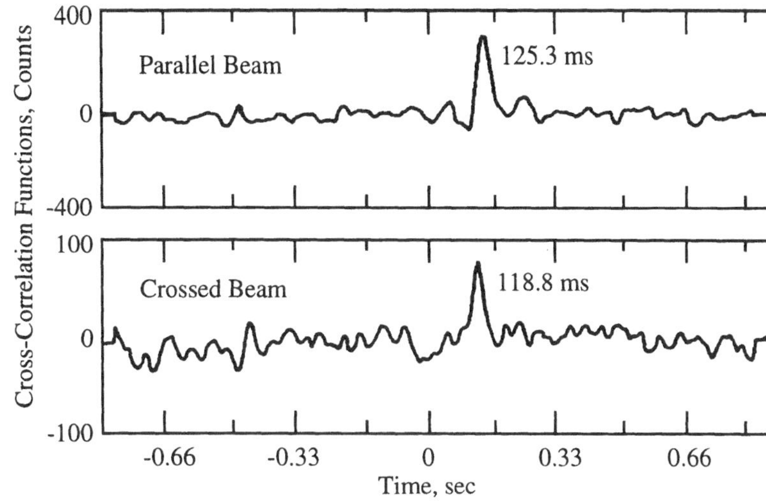

FIGURE 5-23 *Cross-correlation functions obtained under parallel- and cross-beam geometries.*

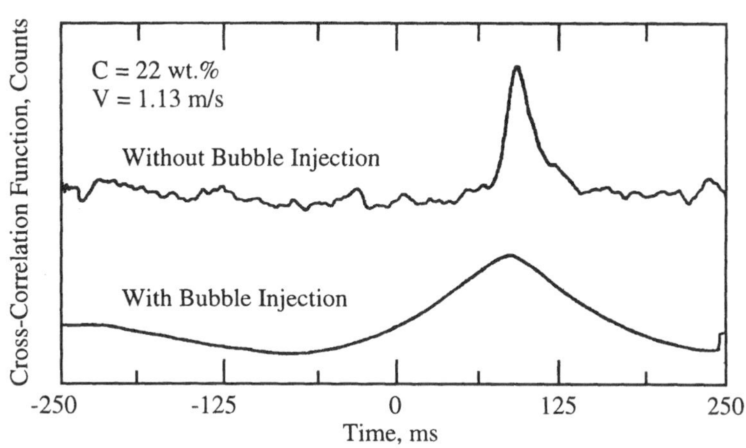

FIGURE 5-24 *Cross-correlation function for flowing coal slurry with many air bubbles at 22 wt.% and 1.13 m/s.*

5.3.2 Mass Flow Rate

To measure mass flow we must determine slurry density and volumetric flow rate. A Coriolis flowmeter may be considered a true mass flowmeter because it directly responds to flow momentum (i.e., ρV). Ultrasonic flowmeters only measure volumetric flow rate; thus, a second density measurement is necessary to obtain mass flow rate. For single-phase fluid flows, ultrasonic methods such as the impedance method have been developed to accurately measure fluid density. But for mixed-phase flows, in principle, one must measure the density, the flow velocity of each phase, and the phase distribution. The only technique that may provide such a total flow characterization is flow imaging, which has not been fully developed and will be discussed latter in this chapter. Currently, particle concentration of a solid/liquid flow is often inferred from sound attenuation or measurements of sound velocity.

5.3.2.1 Measurement of Sound Attenuation

A classical problem in acoustics is the absorption of sound in solid suspensions. Sewell (1910) first conducted a theoretical study of the case of small rigid spherical particles suspended in fluids. The condition of immobility in this case is satisfied by water droplets in air; thus, Sewell's treatment can be applied to sound propagation in fogs and clouds.

Rayleigh (1894) laid out the foundation for the scattering theory of sound wave propagation in fluids that contain suspended solids. He discussed the plane-wave disturbance produced by small obstacles and observed that (a) the zero-order term in the partial wave expansion of the disturbed field is a manifestation of the compressibility difference between the particles and the suspending fluid and (b) the first-order term is determined from the density difference as well as from the relative motion of particles (viscous drag losses).

Urick (1948) measured ultrasonic attenuation in aqueous kaolin as well as sound dispersion. The results were in good agreement with the losses predicted from viscous drag at the particle surface. Sound propagation in a suspension can also produce temperature gradients at the particle/suspending-fluid interface, and thus, results on attenuation via thermal diffusion.

Another process that attenuates sound waves is wave scattering. A theory that describes wave propagation in solid suspension (Ishimaru, 1978) has been well established for a medium in which uniform particles are homogeneously suspended. The attenuation can be determined from the coefficients of the reflected compression wave, viz.,

$$\alpha = -\frac{3\varepsilon}{2k^2 r^3} \sum_{n=0}^{\infty} (2n+1)\,\mathrm{Re}\, A_n\,, \tag{5.20}$$

where α is the attenuation coefficient, ε is the volume fraction of the suspended particles, k is the compression wave number, r is the particle radius, and A_n is the n-th partial wave reflection coefficient. In general, only the first few terms must be considered. Equation 5.20 does not consider multiple scattering; thus, the attenuation is linearly dependent on the concentration of solids. However, the use of attenuation measurements to estimate the concentration of solids is very much process dependent; empirical relationships may be established. In practice, to eliminate system attenuation due to acoustic window, transducer coupling material, and beam dispersion, one would measure relative attenuation with reference to the attenuation in the liquid phase of a medium that contains suspended solids.

The relative attenuation α_r is defined as

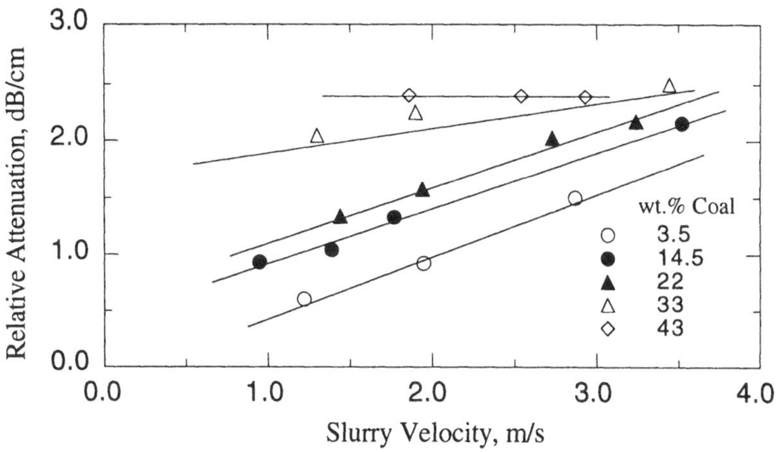

FIGURE 5-25 *Relative attenuation vs. slurry velocity for various coal concentrations.*

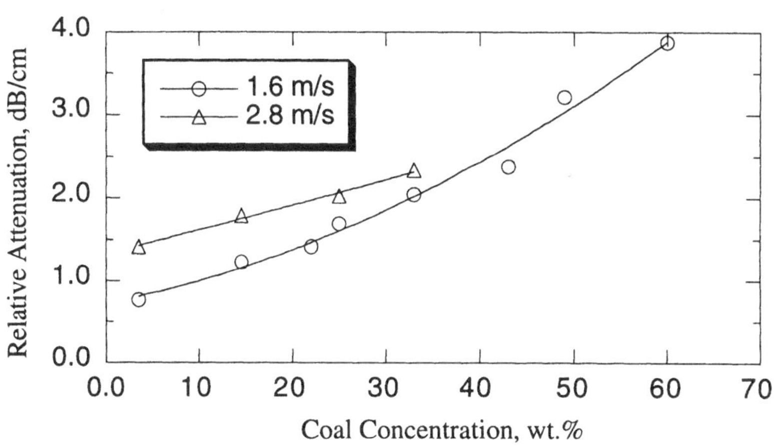

FIGURE 5-26 *Relative attenuation vs. coal concentration for two slurry velocities.*

$$\alpha_r = \frac{20}{d} \log \frac{I_f}{I_s}, \tag{5.21}$$

where d is the pipe diameter and I_f and I_s are the received signals that are being transmitted through the fluid and slurry. Figure 5-25 shows a plot of the relative attenuation vs. slurry velocity for various coal/oil concentrations. In the figure, measured attenuations are slightly dependent on velocity, particularly at low coal concentrations. Figure 5-26 provides the best estimated curves of attenuation vs. concentration for two flow velocities based on the data in Fig. 5-25. The increase in attenuation is more exponential than linear. Important factors that contribute to an exponential increase in attenuation are slurry viscosity and particle scattering, because both change nonlinearly with particle concentration. To date, no quantitative model is available to estimate the extent of these effects. However, attenuation measurements might still be a useful tool for monitoring particle concentration in slurries because the change in attenuation at a frequency near 1 MHz is due primarily to particle scattering. The accuracy of attenuation measurements is ~10%. The flow-dependent attenuation can be attributed to a beam-drifting effect and increasing turbulent eddies at high velocities (the former is a relatively small effect).

5.3.2.2 Measurement of Sound Velocity

Phase velocity and attenuation are the real and imaginary parts, respectively, of the complex wave number. Phase velocity can generally be measured more easily and accurately than attenuation. Several theoretical models have been proposed, however, because measurement data are lacking, none of the models can provide a quantitative measure of phase velocity. We describe three models to illustrate the complexity of the problem.

5.3.2.2.1 Effective-Medium Approach

The effective-medium approach is a phenomenological approach that assumes that, in a suspension, there will be a well-defined phase velocity V, which depends on an effective density ρ_{eff} and an effective compressibility β_{eff}, given as

$$V = (\rho_{eff}\beta_{eff})^{-1/2}. \tag{5.22}$$

The effective compressibility commonly used is a simple averaging,

$$\beta_{eff} = \varphi\beta_2 + (1 - \varphi)\beta_1, \tag{5.23}$$

where *1* and *2* represent fluid phase and solid, respectively. There are, however, different ways of averaging the density, depending on the assumption that is chosen. The simplest expression, used by Urick (1948), is

$$\rho_{eff} = \varphi\,\rho_2 + (1 - \varphi)\,\rho_1. \tag{5.24}$$

Ament (1953) considered the effect of fluid viscosity and particle size a, and obtained

$$\rho_{eff} = \varphi\rho_2 + (1-\varphi)\rho_1 - 2(\rho_2 - \rho_1)^2\varphi(1-\varphi)Q/(Q^2 + U^2), \tag{5.25}$$

where

$$Q = 2(\rho_2 - \rho_1)(1-\varphi) + \frac{9\delta}{2a}\rho_1 + 3\rho_1, \tag{5.26}$$

and

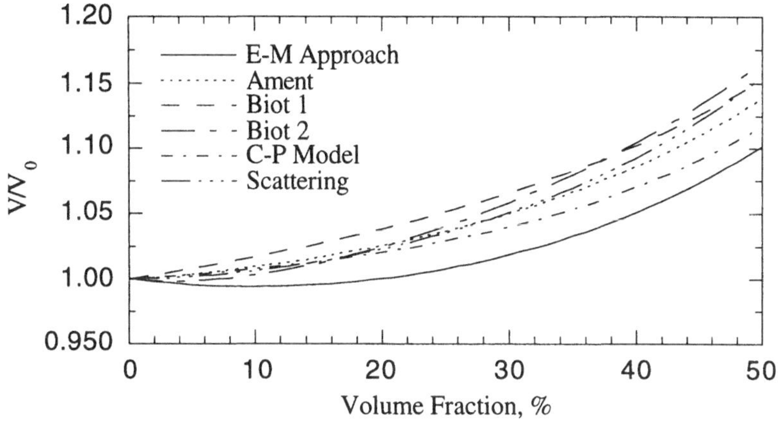

FIGURE 5-27 *Phase velocity vs. volume fraction for glass beads, obtained from various models.*

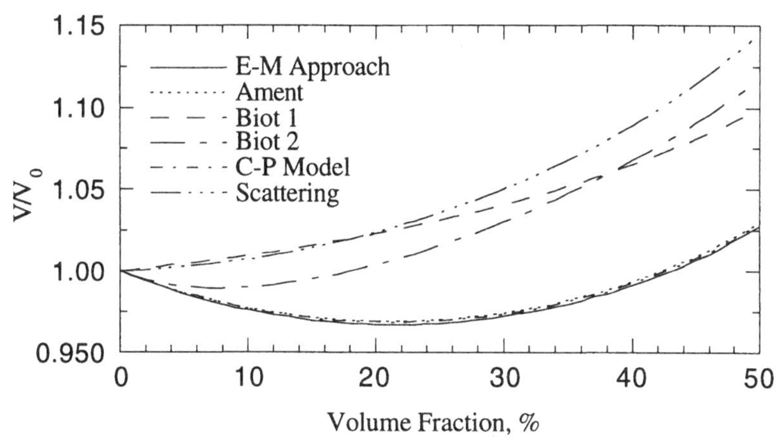

FIGURE 5-28 *Phase velocity vs. volume fraction for spherical kaolins, obtained from various models.*

$$\delta = \sqrt{2\eta / \omega\rho_1},$$
$$U = \frac{9}{2}\rho_1\left[(\delta / a) + (\delta / a)^2\right]. \tag{5.27}$$

For a suspension with a large volume fraction of suspended particles, Biot (1956) developed a theory by treating the medium as a porous solid. The theory gives for the effective density

$$\frac{1}{\rho_{eff}} = \frac{(1-\varphi)\varphi\rho_2 + (\tau - 2\varphi + \varphi^2)\rho_1}{\tau\rho_1\left[\varphi\rho_2 + (1 - 1/\tau)(1-\varphi)\rho_1\right]}, \tag{5.28}$$

where τ is a parameter called "tortuosity". Two expressions are suggested for τ: $(2-\varphi)/2(1-\varphi)$ and $(3-\varphi)/2$.

5.3.2.2.2 Coupled-Phase Model

The coupled-phase c-p model is based on analysis of a two-phase fluid by solving four differential equations that govern the motion of the two-phase mixture. For the effective wave number k, the solution gives

$$k^2 = \omega^2\left[(1-\varphi)\beta_1 + \varphi\beta_2\right] \times \frac{\rho_1\left[\rho_2(1-\varphi+\varphi S) + \rho_1 S(1-\varphi)\right]}{\rho_2(1-\varphi)^2 + \rho_1\left[S + \varphi(1-\varphi)\right]}, \tag{5.29}$$

where S, defined in Eqs. 5.30 and 5.31, is a complex quantity that corresponds to an attenuated wave:

$$S = R + iU/2\rho_1 \tag{5.30}$$

and

$$R = \frac{1+2\varphi}{2(1-\varphi)} + \frac{9\delta}{4a}. \tag{5.31}$$

From the effective wave number, one obtains the effective wave velocity $V = Re(\omega/k)$ and attenuation $\alpha = Im(k)$.

Variation of phase velocity over a range of solid-volume percentages are calculated for the above models for two types of particles, i.e., glass beads and kaolins (with acoustic impedances of 21.12 x 10^5 and 10.66 x 10^5 g/cm^2-s, respectively). Calculated results are shown in Fig. 5-27 for glass beads and in Fig. 5-28 for kaolins. All models, except Biot-2, show decreasing phase velocity at lower volume fractions, then increasing phase velocity at higher volume fractions.

5.3.2.2.3 Multiple-Scattering Treatment

The widely used multiple-scattering treatment was developed by Waterman and Truell (1961) and Twersky (1962). The treatment is based on an approximation in which the exciting field seen by a scatterer may be represented by the total field that would exist at the scatterer if the scatterer were not present. Furthermore, it assumes that scatterers are statistically independent, i.e., the probability of finding a scatterer at one point is independent of other scatterers. The treatment yields the following expression for the effective wavenumber in terms of single-particle scattering amplitudes:

$$\left(\frac{k}{k_0}\right)^2 = \left[1 + \frac{2\pi N}{k_0^2} f(0)\right]^2 - \left[\frac{2\pi N}{k_0^2} f(\pi)\right]^2, \tag{5.32}$$

where k is the effective wave number, k_0 is the propagation wave number in the fluid, and $f(0)$ and $f(\pi)$ are the single-particle scattering amplitudes in forward and backward directions, respectively.

If we consider scattering by a spherical particle, the scattering amplitude can be given as

$$f(\theta) = \frac{i}{2k_o} \sum_{m=0}^{\infty} (2m+1)(1+\Re_m) p_m(\cos\theta), \tag{5.33}$$

where R_m is a reflection coefficient at the particle surface, and is defined as

$$\Re_m = \frac{h_m'^*(k_o a) + i\chi_m h_m^*(k_o a)}{h_m'(k_o a) + i\chi_m h_m(k_o a)}, \tag{5.34}$$

where $h'_m = j'_m + i\,\eta'_m$; $h_m = j_m + i\,\eta_m$; h'^*_m and h_m^* are the complex conjugates of h'_m and h_m , respectively; and j_m and η_m are the m-th order spherical Bessel and Neumann functions, respectively. χ_m, is a constant derived from the boundary conditions at the particle surface, has the form (Morse and Ingard, 1968)

$$\chi_m = i \frac{\rho C}{\rho_e C_e} \frac{j_m'(k_e a)}{j_m(k_e a)}, \tag{5.35}$$

where ρ_e, C_e, and k_e are the density, phase velocity, and wave number of the spheres having a diameter of a; and ρ and C are density and phase velocity of the fluid. The phase velocity of the spheres can be obtained from Eq. 5.22. In effect, Eq. 5.35 considers only the acoustic properties of the spheres; thermal and viscous effects are not included. For rigid spheres, χ_m approaches zero.

In general, k in Eq. 5.32 is a complex value defined as

$$k = k_R + i\alpha, \tag{5.36}$$

where $k_R = \omega/V$ (the real part of the modified wave number), and α is attenuation. Substituting Eqs. 33 and 35 into Eq. 32, we obtain two coupled equations, Eqs. 36 and 37, which can be solved for sound velocity V and attenuation α as follows:

$$\frac{k_R^2 - \alpha^2}{k_o^2} = 1 + \frac{3\varphi}{k_o^2 a^3} f_R(0) + \frac{9\varphi^2}{4k_o^4 a^6}[f_R^2(0) - f_I^2(0) - f_R^2(\pi) + f_I^2(\pi)] \tag{5.37}$$

$$\frac{k_R \alpha}{k_o^2} = \frac{3\varphi}{2k_o^2 a^3} f_I(0) + \frac{9\varphi^2}{4k_o^4 a^6}[f_R(0) f_I(0) - f_R(\pi) f_I(\pi)], \tag{5.38}$$

where f_R and f_I represent the real and imaginary part of the scattering amplitude, respectively.

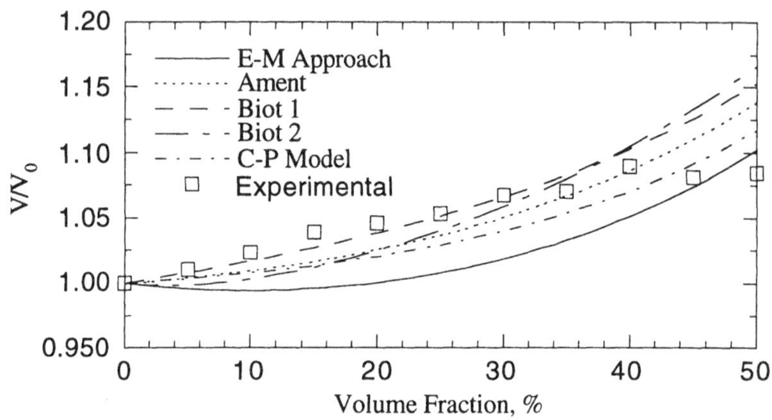

FIGURE 5-29 *Experimentally obtained and theoretically predicted sound velocity vs. solids concentration for 8 μm hollow glass beads suspended in Echogel.*

The model predicts both attenuation and phase velocity in solid suspensions of uniform spherical particles. Nonuniform particles of show very little effect on wave propagation. However, the size distribution can be treated with statistical averaging.

5.3.2.2.4 Experimental Results

To verify model predictions of phase-velocity variation over a range of solids concentrations, we conducted laboratory measurements with hollow glass beads suspended in Echogel, a water-based gel that is commonly used as a transducer coupling material. The measured speed of sound in Echogel is 0.153 cm/μs, which gives a wavelength of 0.153 cm for 1-MHz longitudinal waves. The nominal diameter of the glass beads is 8 μm, i.e. they are much smaller than the wavelength; thus, we are primarily measuring Rayleigh scattering, the principle of which is that the absorption cross section is inversely proportional to the wavelength and directly proportional to the volume of the particle. Measurement of the absorption cross section is still in progress. The data we present here correspond to the phase velocity measurement. Phase velocities of 1-MHz longitudinal waves were measured for solids concentrations up to 50% by volume. Figure 5-29 shows the data and model predictions. The Biot-1 model (Eq. 5.28) provides the best fit to the data up to ≈30 vol.% solids. An increase of ~5% in phase velocity is observed as the particle concentration is increased to 30 vol.%. However, for concentrations >30%, the measured phase velocities show very little change.

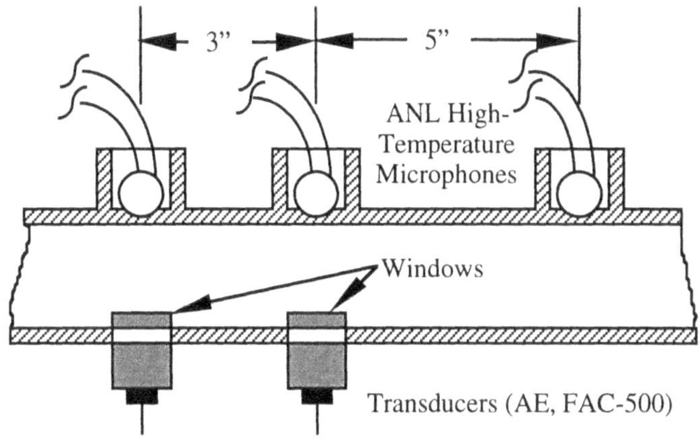

FIGURE 5-30 *Acoustic test section used to measure solid/gas flow noise.*

5.4 MEASUREMENT of SOLID/GAS FLOW

Monitoring of solid/gas flow is important to safe and efficient operation of pneumatic transport that is used in many industrial processes such as coal mining and powder transport. Commonly employed techniques to measure particle velocity are a radioactive tracer method (Somerscales, 1981), optical techniques (Lee & Srinivasan, 1978), electromagnetic methods (Bobis et al., 1986) and conventional mechanical approaches (Soo, 1990).

Radioactive tracers measure particle velocity directly, but require a nuclear radiation source, which limits the technique to research applications. Laser Doppler anemometry has made the optical technique a promising diagnostic method for two-phase flows. However, the requirement of an optical window has hampered its industrial use. Conventional mechanical methods are typically intrusive. For example, an isokinetic sampling technique for measuring particle velocity applies only to a very dilute solid suspension with small particles because of intrusion and plugging problems. Electromagnetic methods are based on either electrostatic induction of charges on particles or changes of electrical capacitance as particles pass through an electrical field. The electrostatic-induction method often includes the use of a probe, which, again, limits its application to dilute solid suspensions. The capacitance technique measures the change in the dielectric constant of the medium. One difficulty of the capacitance technique is that a uniform electrical field must be maintained across the sensor spool, which must be nonmetallic.

In this section, we introduce acoustic techniques that may have practical applications to real-time on-line monitoring of solid/gas flows.

5.4.1 Flow Noise and Flow Rate

Use of acoustic flow noise to monitor solid/gas flow rate is probably the simplest method, but it lacks accuracy. The technique has been used to measure mass flow rate of a powder flow line. A linear dependence was observed between the mass flow rate and the total mean square voltage of the flow noise. To measure the acoustic flow noise, only a wide-band transducer (or

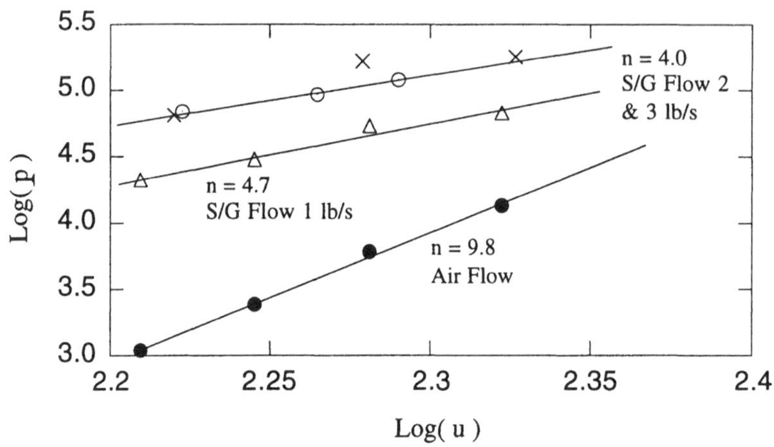

FIGURE 5-31 *Logarithmic plot of autocorrelations of flow noise vs. logarithmics of solid/gas feed rates of 0, 1, 2, and 3 pounds per second (lb/s).*

microphone), which can be mounted on the outside of the pipe wall or in direct contact with the flow, is required. Figure 5-30 shows a test section used to acquire solid/gas flow noise. The test section consists of three ANL high-temperature microphones (Gavin et al), which have a flat frequency response up to 100 kHz when exposed to the 7/8-in. flowthrough openings and to two clamp-on transducers that are isolated from the flow by acoustic windows (Teflon or stainless steel).

Signals from the transducers and microphones are amplified, filtered, and analyzed. Typically, charge amplifiers are used for the microphones, and voltage amplifiers, for the transducers. High-pass filters are applied to the signals to eliminate low-frequency mechanical noise, such as pipe vibration. The signals are analyzed by either a true RMS voltmeter or a wide-band spectrum analyzer.

A series of limestone/air flow tests was conducted with the test section at an ANL solid/gas test facility. The average particle size of limestone is <1 mm. The tests covered three mass loadings (1, 2, and 3 lb/s) at various air flow rates. In general, the microphones detected several tones that might have been the result of direct impacts of particles on the microphone. The tones were not observed with the clamp-on configuration. Instead, a wide-band noise spectrum was typically detected. The clamp-on configuration, was therefore, used for flow analysis. The flow noise levels were measured in the 10-50 kHz range; they were given in RMS voltages or autocorrelation values. Data, as shown in Fig. 5-31, can be fitted into an empirical relationship, given as

$$\bar{p} = cu^n, \tag{5.39}$$

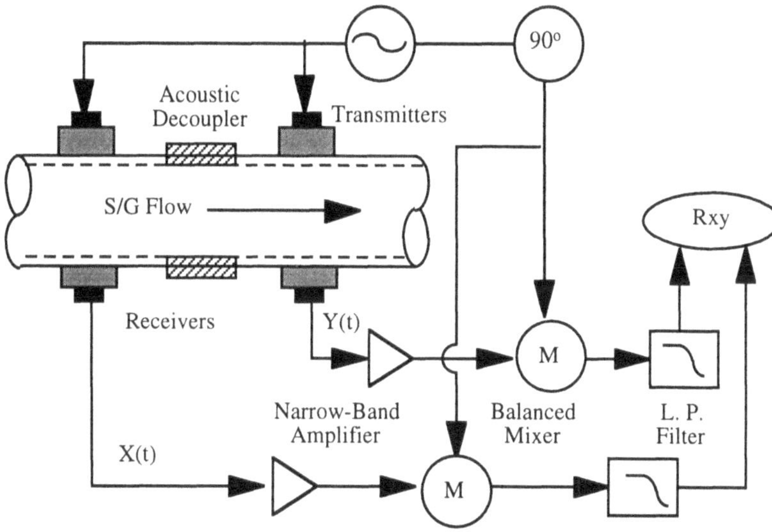

FIGURE 5-32 *Block diagram of the active cross-correlation system for solid/gas flow.*

where $\bar{p}$ is the averaged acoustic power, c a constant, u the air flow rate, and n is a power factor. The power factor, as estimated from the plots, ranged from 4.0 to 5.0 for solid/gas flows and was close to 10 for air flows. The acoustic power increases with solid loading up to 2 lb/s. The results indicate that flow noise level may be used to monitor solid loading in a solid/gas flow. Although the technique of measuring flow noise level is not accurate, it is simple and inexpensive.

5.4.2 Cross-Correlation Method

In principle, the ultrasonic techniques described for solid-liquid flow measurement can be applied to measure air flow rate and particle velocity. Direct measurement of air flow rate by measuring upstream and downstream transit times has been demonstrated. But, the Doppler and cross-correlation techniques have never been applied to solid/gas flow because the attenuation of ultrasound in the air is high. Recent developments have shown that high-frequency (0.5-MHz) air-coupled transducers can be built and 0.5-MHz ultrasound can be transmitted through air for a distance of at least 1 in. Thus, the cross-correlation technique should be applicable to monitoring of solid/gas flow. Here, we present a new cross-correlation technique that does not require transmission of ultrasonic waves through the solid/gas flow. The new technique detects chiefly the noise that interacts with the acoustic field established within the pipe wall. Because noise may be related to particle concentration, as we discussed earlier, the noise-modulated sound field in the pipe wall may contain flow information that is related to the variation in particle concentration. Therefore, cross-correlation of the noise modulation may yield a velocity-dependent correlation function.

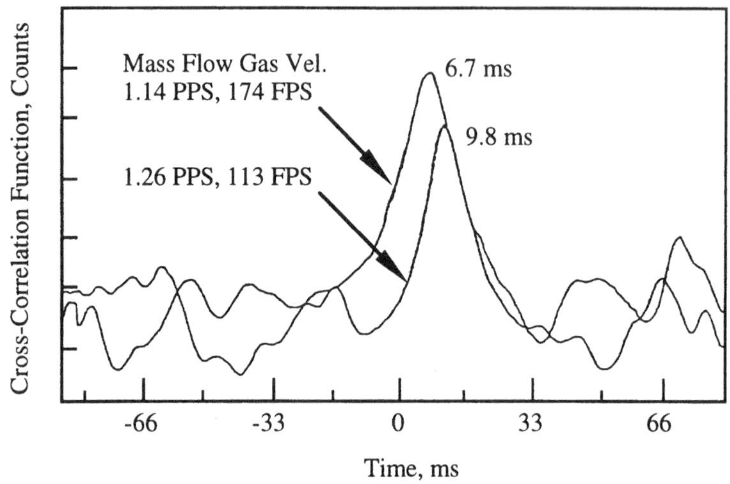

FIGURE 5-33 *Cross-correlation functions for two solid/air flows obtained from demodulated signals in 30-500-Hz bandwidth.*

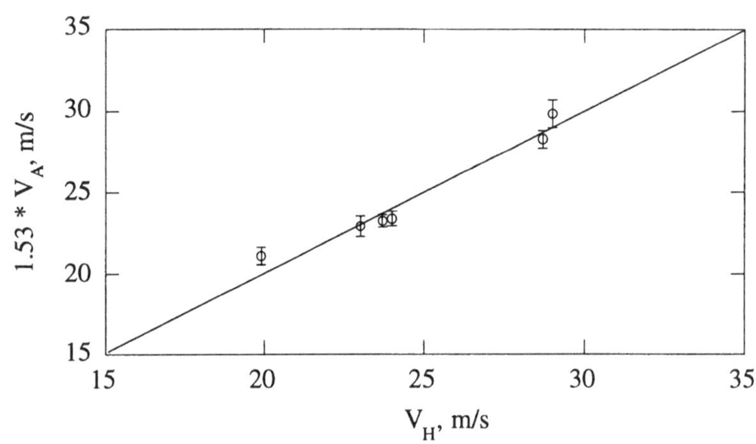

FIGURE 5-34 *Particle velocities (V_A) sensed by the acoustic method vs. partial velocity determined with the radioactive-particle injection method.*

Figure 5-32 is a block diagram of the ANL acoustic cross-correlation test section (Sheen and Raptis, 1986) for monitoring solid/gas flows. Three pairs of wide-band transducers are clamped directly on the pipe. In each pair, one transducer acts as a transmitter and the other as a receiver. To avoid acoustic crosstalk, each pair of transducers is acoustically isolated from the others. The isolation is achieved by using viton gaskets between flanges. During operation, the transmitters delivered a continuous wave with a frequency of ~1 MHz. Receivers were conditioned with amplifiers and band-passed filters set at 100-1

MHz. Received signals were demodulated, and cross-correlation functions were calculated between two pair of receivers. The instrument was tested with a series of glass bead/air flow tests. During the tests, a radioactive tagging technique was used to obtain the true particle velocity for direct comparison. Figure 5-33 shows two cross-correlation functions obtained from the demodulated signals in a narrow bandwidth of 30-500 Hz. The cross-correlation functions displayed clear peaks whose maximums yielded the particle velocities. The particle-velocity data derived from the correlation peaks differed significantly from the velocities measured by the radioactive tagging method. The difference, however, remained the constant over the measurement range. This is illustrated in Fig. 5-34, in which a multiplier of 1.53 has been applied to the cross-correlation data. The source of the multiplier may be the narrow filter bandwidth and the particle velocity profile effect. Tagging methods sense particles mainly in the turbulent region, whereas the acoustic technique detects particles near the pipe wall.

5.5 MEASUREMENT of LIQUID VISCOSITY/DENSITY

To measure liquid density and viscosity, we must characterize a solid/liquid flow. In this section, we describe an in-line nonintrusive ultrasonic technique for determining liquid density and viscosity.

5.5.1 The ANL Ultrasonic Viscometer

ANL's ultrasonic viscometer is a nonintrusive in-line device that measures both fluid density and viscosity. The design of the viscometer is based on a technique that measures acoustic and shear impedance. The technique was first applied by Moore and McSkimin (1970) to measure dynamic shear properties of solvents and polystyrene solutions. The reflections of incident ultrasonic shear (1-10 MHz) and longitudinal waves (1 MHz), launched toward the surfaces of two transducer wedges that are in contact with the fluid, are measured. The reflection coefficients, along with the speed of sound in the fluid, are used to calculate fluid density and viscosity. Oblique incidence was commonly used because of better sensitivity, but mode-converted waves often occur in wedges that do not exhibit perfect crystal structure and lack well-polished surfaces. For practical applications, we use the normal-incidence arrangement.

5.5.1.1 Longitudinal Waves and Acoustic Impedance of Fluid

Acoustic impedance of a fluid Z_l is the product of fluid density ρ and phase velocity V of sound in the fluid; it can be determined by measuring the reflection coefficient R at the boundary of the fluid and transducer wedge. If we select the normal-incidence configuration, R is given by

$$R = \frac{Z_l - Z_w}{Z_l + Z_w} , \tag{5.40}$$

where Z_W is the acoustic impedance of the wedge in which longitudinal waves propagate from transducer to fluid. If the phase velocity in the fluid can be determined accurately by other measurements (e.g., time-of-flight of longitudinal waves traveling in the fluid), the fluid density can be derived from

$$\rho = \frac{Z_w(1-|R|)}{V(1+|R|)}, \tag{5.41}$$

where the absolute value of the reflection coefficient is used because, in principle, R is a complex number. However, in practice, if we assume that wave attenuation in the wedge and fluid can be neglected, and only the real parts of R and Z_W are used in the density calculation.

5.5.1.2 Shear Waves and Shear Impedance of Fluid

Use of the ultrasonic shear reflectance method to obtain the shear mechanical properties of fluids has been the subject of many studies of Newtonian (Cohen-Tenoudji et al., 1987) and non-Newtonian (Harrison and Barlow, 1981) fluids. Consider that gated shear-horizontal (*SH*)-plane waves propagate in a wedge at an incidence angle that is normal to the polished surface in contact with the fluid, and that these waves are reflected back. The shear reflection coefficient can be expressed as given in Eq. 5.40, with shear impedances replacing acoustic impedances. The shear impedances of the wedge Z_{WS} and fluid Z_{ls} are given as

$$Z_{ws} = \sqrt{\rho_w C_{44}} \tag{5.42}$$

and

$$Z_{ls} = \sqrt{j\omega\rho\eta}\ , \tag{5.43}$$

where ρ_W is the density of the wedge material, C_{44} is the stiffness constant of the wedge, ω is the radial frequency of the shear wave, and η is the fluid viscosity. By using Eq. 5.43, we have assumed that the fluid behaves as a Newtonian fluid; more complex expressions are expected for non-Newtonian fluids (Sheen et al., 1997). The shear impedance of a fluid is a complex value that consists of amplitude and phase. The phase change is very small for a single reflection, so we consider only the variation in amplitude. The shear reflection coefficient R_s, which is a measurable quantity, can be used to calculate the product of density x viscosity as follows:

$$\sqrt{\rho\eta} = \sqrt{\frac{\rho_w C_{44}}{2\omega}}\,\frac{1-\sqrt{1-2K^2}}{K}, \tag{5.44}$$

where $K = \dfrac{1-|R_s|^2}{1+|R_s|^2}$.

Equation 5.44 predicts the sensitivity of the measurement and the range of the shear reflectance method. Figures 5-35 and 5-36, show the dependence of the reflection coefficient on the product of density x viscosity for various

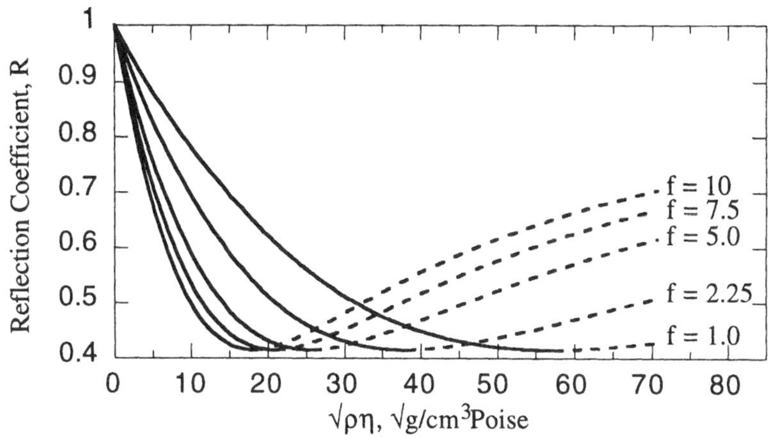

FIGURE 5-35 *Reflection coefficient as a function of fluid density viscosity for various operating shear wave frequencies.*

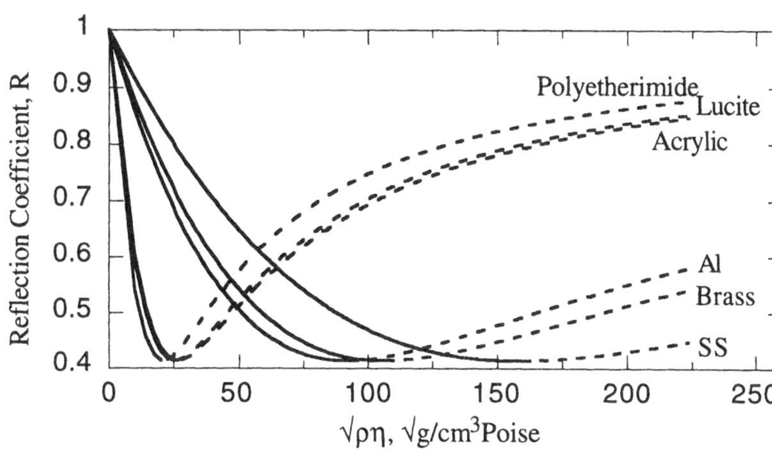

FIGURE 5-36 *Reflection coefficient vs. square root of fluid density viscosity product for various wedge materials (S.S. = stainless steel).*

operating shear frequencies and wedge materials, respectively. In principle, lower shear-impedance materials and higher operating shear frequencies provide better sensitivity but a smaller measurement range. However, for tank-waste applications, the choice of Lucite and 10 MHz is not sufficient to achieve the desired sensitivity.

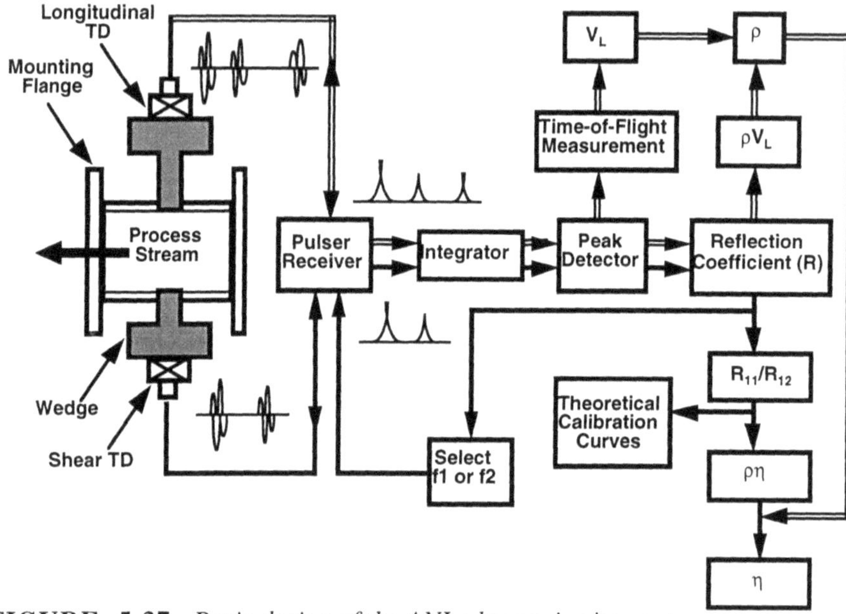

FIGURE 5-37 *Basic design of the ANL ultrasonic viscometer*

5.5.1.3 Viscometer Design

Figure 5-37 shows the basic design of the ultrasonic viscometer and its signal-processing scheme. The basic design consists of two transducer wedges mounted on a pipe, opposite one another and flush with the inner surface of the pipe. The wedges have an offset surface to provide the reference reflection, which is compared with the reflection from the sensing surface to give the reflection coefficient. In effect, the offset surface provides a continuous reference signal for self-calibration. Two types of transducers, shear horizontal (*SH*) and longitudinal, are used; both operate in the pulse-echo mode. Three major reflections are detected for longitudinal-wave operation, corresponding to reflections from the offset surface, the sensing surface that is in contact with the fluid, and the pipe wall on the opposite side. The amplitude ratio of the first two reflections is a measure of the reflection coefficient, whereas the time-of-flight between the second and third reflections allows us to deduce the phase velocity of the longitudinal wave in the fluid. Thus, longitudinal-wave operation gives a direct measure of fluid density. Shear-wave operation detects only two reflections because most fluids do not support shear waves. The amplitude ratio of the two reflections allows us to calculate the reflection coefficient, from which we can deduce the product of the density x viscosity.

5.5.2 Laboratory Tests and Results

The wedge material determines the sensitivity and accuracy of the density and viscosity measurements. Table 5-2 lists the tested wedge materials and their acoustic properties. For the tests, transducers (longitudinal and shear) were attached to wedges with epoxy glue. They are excited by a wideband pulse and generate pulses with a center frequency of 1 MHz for longitudinal waves and 5 MHz for shear waves. Reflections from the two surfaces of each wedge are rectified and integrated. The integrated reflection amplitudes are used to calculate the reflection coefficients. Typically, 500 averages are applied to the signals to reduce the signal-to-noise ratio.

Table 5-2 *Characteristics of various wedge materials.*

Material	Density g/cm^3 ρ	Longitudinal Velocity V_L, cm/μs	Longitudinal Impedance ρV_L	Shear Velocity V_{SH}, cm/μs	Shear Impedance ρV_{SH}	Working Temperature T_W, °F
ABS[a]	1.5279	0.2330	0.3560	——	——	185
Acrylic (Cast)	1.1800	0.2731	0.3222	0.1369	0.1615	200
Acrylic (Extruded)	1.1800	0.2525	0.2979	0.1369	0.1615	200
Delrin	1.0341	0.2137	0.2210	0.0931	0.0963	180
Lucite	1.2800	0.2335	0.2989	0.1119	0.1432	200
Plexiglass	1.1897	0.2701	0.3214	0.1621	0.1928	200
Polyetherimide	1.2700	0.2403	0.3052	0.1041	0.1323	338
Polystyrene	1.0279	0.2042	0.2099	——	——	170
WTD[b]	1.2624	0.2352	0.2969	0.1016	0.1283	350
HTD[b]	1.4038	0.2591	0.3637	0.1127	0.1582	500
VHTD[b]	1.4315	0.2309	0.3305	0.0985	0.1409	900

[a]ABS = Acrylonitrile-butadiene-styrene.

[b]Delay lines were supplied by Panametric, Inc., for high-temperature applications. WTD = moderate-temperature delay line; HTD = high-temperature delay line; VHTD = very-high-temperature delay line.

5.5.2.1 Measurement of Density

The longitudinal-wave reflectance method is used to measure fluid density. Table 5-3 lists the density of standard liquids that were used in the test to calibrate density. The longitudinal-wave phase velocity in each liquid, deduced from the time-of-flight measurement, is also given. Note that variation in phase velocity of the standard liquids does not correlate with their density change; thus, phase velocity alone cannot be used to predict liquid density. However, by combining phase velocity and acoustic impedance measurements, we can obtain an accurate measure of liquid density. Figure 5-38 shows the density calibration results for two wedge materials, polyetherimide and aluminum. The polyetherimide wedge gives an accuracy better than 0.5% for the test liquids, but results from the aluminum wedge are significantly lower than the actual values. The discrepancy of the aluminum wedge may be due to wetting

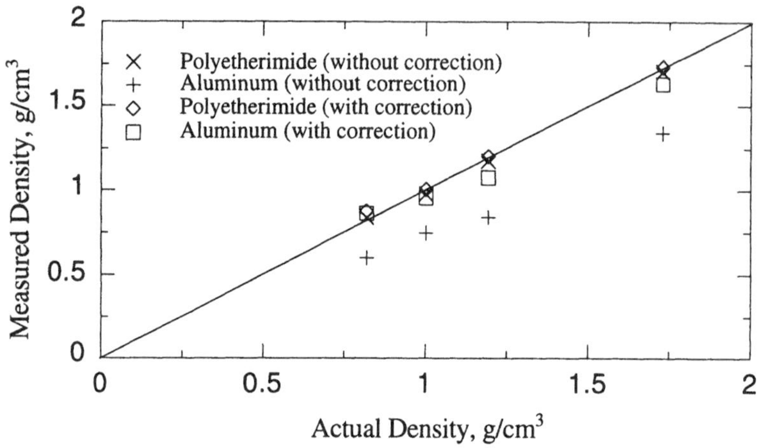

FIGURE 5-38 *Density calibration results for two wedge materials, polyetherimide and aluminum. A wedge correction function of 4% is determined.*

Table 5-3 *Liquids used for density calibration tests.*

Liquid*	Chemical Constituents	Density, g/cm³	Longitudinal-Wave Phase Velocity, cm/μsec
R-827	Kerosene Chloronaphthalene Naphthol	0.818	0.12766
G-1000	2-Butoxy Ethanol 51.9% Ethylene Glycol 47.2% BASACID Green <1%	1.002	0.15906
Y-120	Chloronaphthalene 99% Kerosene <1% Mono Azo Dye <1%	1.194	0.14272
B-175	Diazene-42 99% *Diazene*-200 <1% Solvent Blue 36 <1%	1.730	0.11452

*Supplied by ALTA Robbins, Anaheim, CA.

problems and wedge geometry, which consistently give a 4% higher reflection coefficient. If we apply the 4% correction to both wedges, which have the same design, the discrepancy is significantly reduced (as shown in Fig. 5-38).

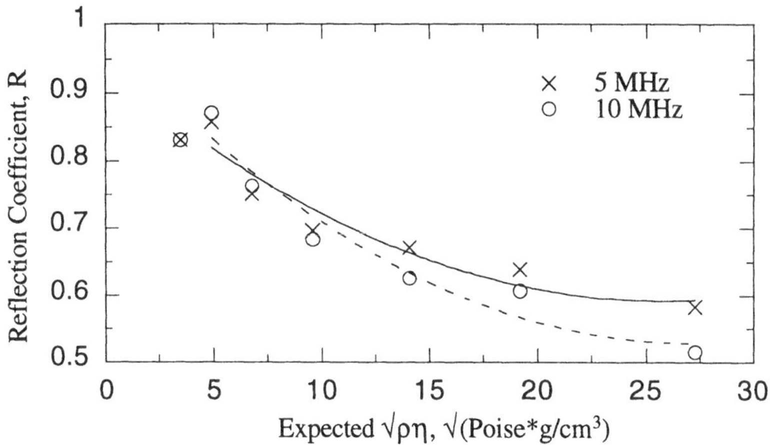

FIGURE 5-39 *Measured reflection coefficients as a function of viscosity with a polyetherimide wedge at two operating frequencies.*

5.5.2.2 Measurement of Viscosity

When compared with the density wedges, the design of the wedge for measuring viscosity is scaled down in a ratio of longitudinal to shear velocities. An aluminum wedge gives an ~1 % change in reflection amplitude for a 250-cP viscosity change. This sensitivity is very poor, especially for low-viscosity measurements; however, if a low-impedance wedge and high operating frequency are used, it can be improved. Figure 5-39 shows the reflection coefficients as a function of viscosity for the polyetherimide wedge at two frequencies. In the high-viscosity range, better sensitivity is obtained at the higher frequency (10 MHz).

We performed calibration tests for three low-impedance wedge materials, polyetherimide, acrylic (Lucite), and polystyrene. Table 5-4 lists the liquids used for the calibration tests; note that all of the liquids have similar densities but vary in viscosity. The calibration results are shown in Fig. 5-40, which reveals that Lucite is the best of the three as a wedge material for measuring viscosity. However, all of the measured viscosities are lower than their expected values. The discrepancies may be attributed to non-Newtonian fluid behavior, surface wetting, and poor sensitivity.

For low-viscosity liquids, the detection sensitivity of the technique based on impedance measurements (or reflection coefficient) must be improved. One improvement would be to monitor multiple reflections because each echo represents one interaction at the wedge/liquid boundary. To obtain multiple echoes, the wedge design must be modified. Two design factors should be considered, echo interference and signal attenuation. The simplest design is to use a thin-plate configuration. Thus, a thin polyetherimide plate was fabricated and tested in glycerol/water solutions. In Fig. 5-41, we show the results that were derived from the second and third echoes over a viscosity range of 1 to

Table 5-4 *Liquids used for viscosity calibration tests*

Liquid*	Chemical Constituents	Density, g/cm^3	Viscosity, cP
N600	Mineral oil	0.8876	1381
N1000	PAO oil	0.8479	2823
N2000	Poly(1-butene)	0.8753	5248
N4000	Poly(1-butene)	0.8812	10450
N8000	Poly(1-butene)	0.8873	22390
N15000	Poly(1-butene)	0.8919	41360
N30000	Poly(1-butene)	0.8954	83040

*Supplied by Cannon Instrument Company, State College, PA.

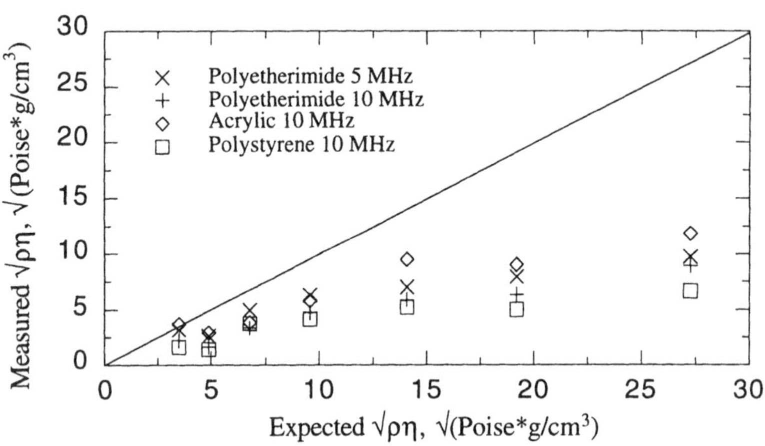

FIGURE 5-40 *Viscosity calibration data for various wedge materials.*

≈600 cP. The derived viscosities in Fig. 5-41 were calculated from the measured reflection coefficients by using Eq. 5.44, in which R_s is replaced by $(R_s)^{1/n}$, where n is the echo number. It is evident that multiple reflections improve measurement sensitivity; thus, low-viscosity liquids can be monitored with this technique.

5.6 SUMMARY AND FUTURE DEVELOPMENT

This chapter gives an in-depth overview of ultrasonic techniques for monitoring fluid/particle flows. Ultrasonic instruments are generally favored by industrial

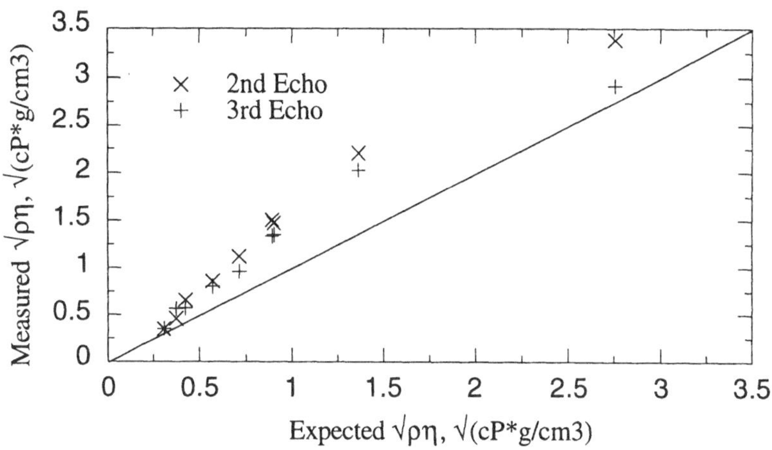

FIGURE 5-41 *Viscosity calibration data in low-viscosity range-obtained with polyetherimide wedge and multiple reflection technique.*

process engineers because of their low cost and nonintrusiveness. For both solid/liquid and solid/gas flows, ultrasonic techniques or instruments have been developed to measure average velocities of particles and fluids. The measured velocities are averaged over the width of the ultrasonic beam; therefore, it is possible to obtain a velocity profile in a pipe by properly arranging ultrasonic beams. For example, the cross-correlation technique can measure both the centerline and average velocities, from which a velocity profile can be derived. It is also possible to extract velocity profile information from the shape of the cross-correlation function or from the frequency spectrum of a Doppler flowmeter.

We have presented an innovative technique to measure liquid density and viscosity of a solid/liquid flow. The technique, which gives an accurate and quantitative measurement of liquid density, is the only technique that provides liquid density and viscosity separately. The drawback of this technique is that it does not measure bulk phenomena, instead, it senses the liquid properties near the pipe wall. The presence of solid particles has little effect on the measurements.

To characterize a solid/fluid flow, particle-size distribution, solid concentration, and effects of particles on fluid viscosity must be determined. At present, no technology is available to accomplish these tasks. Optical methods can conceivably be used to determine particle-size distribution in dilute flows. Ultrasonic techniques suffer from high attenuation in solid suspensions, particularly in the frequency range of interest (>10 MHz for typical particle sizes <100 μm). Thus, how to measure these parameters is a challenge for future research in this field.

5.7 NOTATION

A_n	*n*-th partial wave reflection coefficient
a	particle size
B_s	signal bandwidth
C	sound speed or phase velocity
C_{44}	stiffness constant
c	constant
D	separation between two Teflon windows
DG	receiving transducer directional gain
d	pipe diameter or transducer diameter
f	carrier frequency of sound wave or scattering amplitude
f_d	Doppler frequency, i.e., difference frequency
G	power density spectrum of noise
I	received transmitting signal
j_m	*m*-th order spherical Bessel function
k	wave number
L	sensor separation
L_0	separation between intersections of pipe wall and acoustic path
N(t)	noise signal
NL	background noise level
PG	processing gain
PSL	processed signal level
$\bar{p}$	averaged acoustic power
R	reflection coefficient
R(t)	cross-correlation function of signals
r	particle radius
S(t)	decay
SL	source level at inner wall
SLD	square law director
T	averaging time
TL	transmission loss through fluid
t	travel time
t_m	travel time in dead space $(L- L_0)$
u	air flow rate
V	sound or phase velocity
V_0	ratio of received voltage and driving voltage
x(t)	sensor signal
y(t)	sensor signal
Z	acoustic impedance
ΔS	averaged measurement parameter (Δf, $\Delta\phi$, Δt, etc.)

Subscripts

d	downstream
eff	effective
f	fluid
I	imaginary part

l	liquid or fluid
m	at particle surface
p	pipe
r	relative parameter
R	real part
S	slurry
u	upstream
w	wedge
0	toluene/benzene slurry
1	fluid phase or coal/toluene/benzene slurry
2	solid phase

Superscripts

n	power factor
'	complex conjugate

Greek Symbols

α	relative attenuation in dB/cm
β	compressibility
ε	volume fraction of suspended particles
χ_m	constant
η	fluid viscosity
η_m	*m*-th order spherical Neumann function
λ	wavelength
ρ	density
θ	angle between pipe axis and acoustic path
φ	volume concentration
τ	leap time or tortuosity
ω	radial frequency of shear wave

5.8 REFERENCES

Ament, W. S., "Sound Propagation in Gross Mixtures," J. Acoust. Soc. Am., Vol. 25, 1953, 638.

Biot, M. A., "Theory of Propagation of Elastic Waves in a Fluid-Saturated Porous Solid -- II. Higher Frequency Ranges," J. Acoust. Soc. Am., Vol. 28, 1956, 179.

Cohen-Tenoudji, F., Pardee, W. J., Tittmann, B. R., Ahlberg, L. A., and Elsley, R. K., "A Shear Wave Rheology Sensor," IEEE Trans. Ultrason. Ferroelect. Freq. Conf., Vol. UFFC-34(2), 1987, 263.

Foster, G. A., Karplus, H. B., and Mulcahey, T. P.,"Multipath Ultrasonic Flow Measurements in Water, A Final Report," Argonne National Laboratory Report, 1985; ANL-85-14.

Gavin, A. P., Anderson, T. T., and Janicek, J. J., "Sodium-Immersible High-Temperature Microphone Design Description," Argonne National Laboratory Report, 1975; ANL-CT-75-30.

Harrison, G. and Barlow, A. J., "Dynamic Viscosity Measurement," Meth. Exp. Phys., Vol. 19, 1981, 138.

Ishimaru,A., Wave Propagation and Scattering in Random Media, Vols. 1 & 2, Academic Press, New York, 1978.

Karplus, H. B., Raptis, A. C., Lee, S., and Simpson, T., "The ANL Doppler Flowmeter," Argonne National Laboratory Report, 1985; ANL/FE-85-14.

Moore, R. S., and McSkimin, H. J., "Dynamic Shear Properties of Solvents and Polystyrene Solutions from 20 to 300 MHz," Phys. Acoust., Vol. 6, 1970, 167.

Morse, P. M. and Ingard, K. U., Theoretical Acoustics, Chap. 8, McGraw-Hill, New York, 1968.

Raptis, A.C., Sheen, S. H., Roach, P. D., and Mech, J. F., "Acoustic Noise Background and Sound Transmission Tests in a Slurry Line at the HYGAS Pilot Plant," Argonne National Laboratory Report, 1979; ANL-FE-49622-TM04.

Sewell, C. J. T.,"The Extinction of Sound in a Viscous Atmosphere by Small Obstacles of Cylindrical and Spherical Form," Phil. Trans. Roy. Soc. London Vol. A210, 1910, 239.

Sheen, S. H. and Raptis, A. C. "Active Acoustic Cross-Correlation Technique Applied to Flow Velocity Measurement in a Coal/Liquid Slurry," Proc. of Sonics/Ultrasonic Symp., 1983, 591.

Sheen, S. H., Raptis, A. C., Bobis, J. P., Lee, S., and Simpson, T., "Evaluation of Active Ultrasonic Cross-Correlation Technique in Coal/Liquid Pipe Flow Measurements," Argonne National Laboratory Report, 1985; ANL/FE-85-12.

Sheen, S. H. and Raptis, A. C., "Development of Acoustic Flow Instruments for Solid/Gas Pipe Flows," Argonne National Laboratory Report, 1986; ANL/FE-85-07.

Sheen, S. H., Chien, H. T., and Raptis, A. C., "Measurement of Shear Impedances of Viscoelastic Fluids," Proc. 1996 IEEE Int. Ultrasonic Symp., Vol. 1, 1997, 453.

Somerscales, E. F. C., "Measurement of Velocity -- Tracer Methods," Chap. 1, Methods of Experimental Physics, Vol.18 Part A, Academic Press, New York, 1981.

Soo, S. L. Multiphase Fluid Dynamics, Science Press-Gower Technical, Aldershot, England, 1990.,

Twersky, V., " On Scattering of Waves by Random Distributions -- I. Free-space Scattering Formalism," J. Math. Phys., Vol. 3 number 4, 1962, 700.

Urick, R. J., "Absorption of Sound in Suspensions of Irregular Particles," J. Acoust. Soc. Am.., Vol. 20 number 3, 1948, 283.

Waterman, P. C. and Truell, R.," Multiple Scattering of Waves," J. Math. Phys., Vol. 2 number 4, 1961, 512.

6

Instrumentation for Fluid-Particle Flow: Electromagnetics

Shu-Haw Sheen, Hual-Te Chien, and Apostolos C. Paul Raptis

6.1 INTRODUCTION

Electromagnetically based flow metering encompasses electrical, magnetic, and optical techniques, where optical techniques are not limited to the use of visible light, but cover the entire electromagnetic (EM) spectrum, from microwaves to γ rays. Electrical techniques involve measurement of the electrical impedance, capacitance, or conductivity of the flow medium. When applied to the flow of suspended solids, each measured electrical property generally is not related directly to flow rate but provides an estimate of solids concentration. Similarly, most optical techniques that measure attenuation of light or EM waves in a medium (Van De Hulst, 1957) sense percent solids in the region where the light is interrogating. To measure flow rate, more than one sensor or sensing station is required; for example, cross-correlating the outputs of two-sensors to obtain the flow rate is a common practice.

Instruments that measure primarily flow rate have been developed. Examples of these instruments are the magnetic flowmeter for fluid flow and the hot-wire anemometer for gas flow. Although magnetic flowmeter that measures the induced electrical field strength when a conducting fluid flows through a magnetic field has been developed and applied to single-phase conducting flows, its application to the monitoring of solid/liquid flows is still infrequent. Application of a hot-wire anemometer is limited to gas flows. In fact, an anemometer must be protected from solid particles in the gas stream because of potential erosion problems. A discussion of the hot-wire anemometer can be found in Blackwelder (1981), thus it will not be discussed here.

In today's industrial applications, Coriolis mass flowmeters are widely used by process engineers to monitor mass flow rate. This meter measures the Coriolis force that depends on the mass momentum of the flow, and, in principle, it can be applied to both single- and mixed-phase flows. Magnetic or optical detectors are generally used to detect mass-flow-related Coriolis acceleration. A brief description of the Coriolis flowmeter will be presented because it is widely used in industrial processes.

A laboratory technique that is commonly used to calibrate particle velocity is the tracer technique. The tracers can be optically illuminating particles or short-lived radioactive tracers. We will briefly discuss two techniques that can be used to produce radioactive tracers: on-line pulsed neutron activation (PNA) and off-line irradiation.

To conclude this chapter, we will present a brief introduction to some emerging technologies for monitoring mixed-phase flow.

6.2 MEASUREMENT PRINCIPLES

Electrical techniques primarily measure the electrical impedance of a mixed-phase medium. Because the dielectric constant or electrical conductivity of a solid phase differs from that of the fluid, one can measure electrical conductance or capacitance to determine phase distribution. To attain better sensitivity, conductance flowmeters are usually applied to conducting media, such as aqueous solutions or solids/water slurries, whereas capacitive flowmeters are applied to solid/gas flows and solid/nonconducting-liquid flows. Capacitance measurements are generally more reproducible because they are not affected by the ion concentration of the solution, which is difficult to control during processing.

Optical or radiometric techniques for measuring solids concentration are based on the dependence of attenuation and scattering of an optical beam or radiation on the number of particles in the optical path. The theory that guides these techniques assumes a single-scattering process; thus, it is only valid for low concentrations of solids. These techniques require special optical windows, which sometimes are impractical and limiting.

In industry, it is often more important to measure mass flow rate than to monitor volumetric flow rate. To determine the mass flow rate of a mixed-phase flow, one must measure the velocity and concentration of each phase. This becomes a very difficult task and none of the current techniques can accomplish it. In practice, most mass flowmeters measure the relative change of a physical parameter like capacitance. To relate this measurement to solids concentration, the flowmeters must be calibrated against other, direct, methods

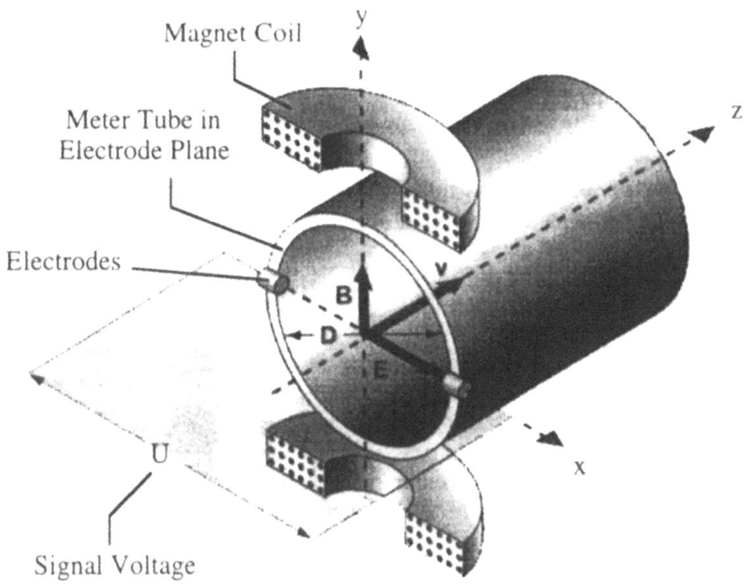

FIGURE 6-1 *Schematic representation of conventional electromagnetic flowmeter. (Source: Hoske, 1998)*

such as timed flow diversion. In this section, we will outline the fundamental principle of each electromagnetic technique, with a focus on what each technique can measure. In most cases, the theories assume homogeneous pipe flows; effects due to velocity and concentration profiles are not addressed.

6.2.1 Electromagnetic Methods

Electromagnetic flowmeters are mainly applied to single-phase conducting fluids, for example liquid metals, water-based industrial liquids, and blood. However, they are also being used to measure solid/liquid flows such as water-based slurries and sludge as long as conductivity of the liquid medium exceeds 50 microsiemens per centimeter (μS/cm). Output signals from an EM flowmeter are basically proportional to the volumetric flow rate, which is applicable to solid/liquid flows only if the solids particles are uniformly distributed in the liquid. Moreover, to obtain the mass flow rate of a solid/liquid flow, the EM flowmeter requires other means to measure solids concentration.

The measurement principle of the EM flowmeter is well established (Shercliff, 1962; Bevir, 1970). Figure 6.1 shows the basic configuration of a conventional cylindrical flowmeter (Haske, 1998). The electrical conducting fluid passes through a magnetic field, producing an electrical field, the strength of which is measured by a pair of electrodes that are in contact with the fluid and insulated from the conductive pipe. If we let V be the vector velocity of the conducting fluid and B, the flux density of the magnetic field, we will obtain the generated electrical potential across the sensing electrodes U by solving the equation

$$U = \int \vec{V} \cdot \vec{W} d\xi, \tag{6.1}$$

where dξ integrates over the volume within the flowmeter, and the relationship between W, the weight vector and B is $W = B \times J$. Here J is called the virtual current, a hypothetical current density that would be set up if unit current were passed between the electrodes while the fluid is stationary. The flowmeter sensitivity signal S per unit flow rate, can be determined from

$$S = \frac{U}{\int V d\tau}. \tag{6.2}$$

The sensitivity depends on the spatial distribution of the magnetic field, the electrode arrangement and the flow velocity distribution. Ideally, one would like to design a flowmeter that is independent to flow velocity distribution. This may be achieved by arranging the magnetic field and the electrode configuration so that the output signal is proportional only to flow rate. Theoretically, Bevir (1970) showed that the necessary and sufficient condition for output to be independent of the flow pattern is that curl W = 0, with W$\rightarrow$ 0 at ∞. If we assume that the induced magnetic field is small and the conductivity uniform, curl B and curl J = 0. We may then set $B = \nabla$F and $J = \nabla$G, to obtain $W = \nabla$F x ∇G, where both F and G are solutions of Laplace's equation. The condition of curl W = 0 implies either ∇F or ∇G must be constant. Thus, a simple class of ideal flowmeter is obtained by making ∇F constant; this means that magnetic field B is constant.

In the conventional EM flowmeter, point electrodes are used in a circular pipe. This flowmeter is normally overly sensitive to flow near the two electrodes; it may accurately measure flow rate if the magnetic field is uniform and flow is rectilinear (i.e., parallel to the flowmeter axis) and axisymmetric. For a rectilinear, axisymmetric flow, Eq. 6.2 can be written as

$$S = \frac{\int_0^a W'(r) V(r) r dr}{\int_0^a V(r) r dr}, \tag{6.3}$$

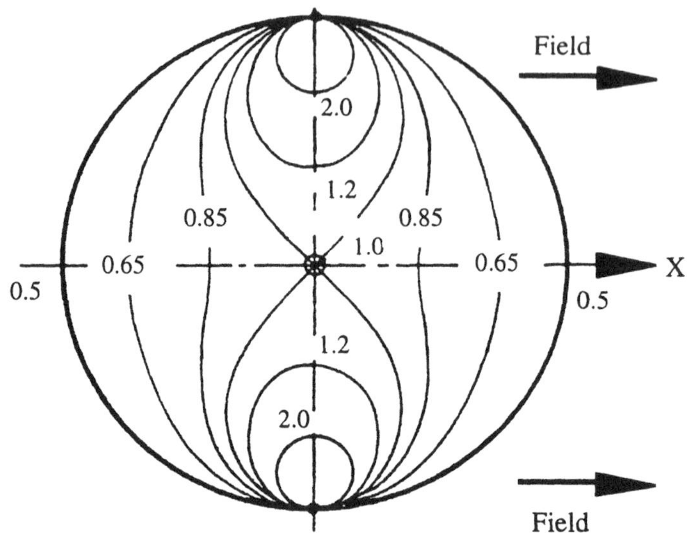

FIGURE 6-2 *Weight function contours for the uniform-field point electrode flowmeter. (Source: Cox and Wyatt, 1984)*

where a is the pipe radius and

$$W'(r) = \frac{1}{2\pi} \int_0^{2\pi} W(r,\theta) d\theta . \tag{6.4}$$

Equations 6.3 and 6.4 show that the sensitivity to rectilinear flow will be independent of flow profile only if $W(r,\theta)$ is constant over the cross section of the flowmeter. Unfortunately, for the point electrode configuration, $W(r,\theta)$ tends to infinity at the electrodes because the virtual current J increases without limit as the electrode size decreases. Therefore, to obtain a uniform $W(r,\theta)$, one needs to modify the magnetic field and the electrode configuration. The method to compute the weight functions was described by O'Sullivan and Wyatt (1983). Figure 6.2 shows the contours of equal-weight functions for the conventional uniform-field, point electrode flowmeter (Shercliff, 1962). A practical EM flowmeter generally does not use point electrodes; multiple electrodes (O'Sullivan, 1983) and electrodes of large area (Cox and Wyatt, 1984) have been suggested, and the magnetic field is generated by using magnet coils operated at a frequency of < 1 kHz and of either continuous waves or pulses. Figure 6.3 shows the weight function for the configuration of six-point electrodes. Performance tests of such a multiple-point-electrode flowmeter showed that it improved the signal-to-noise ratio and thus improved the

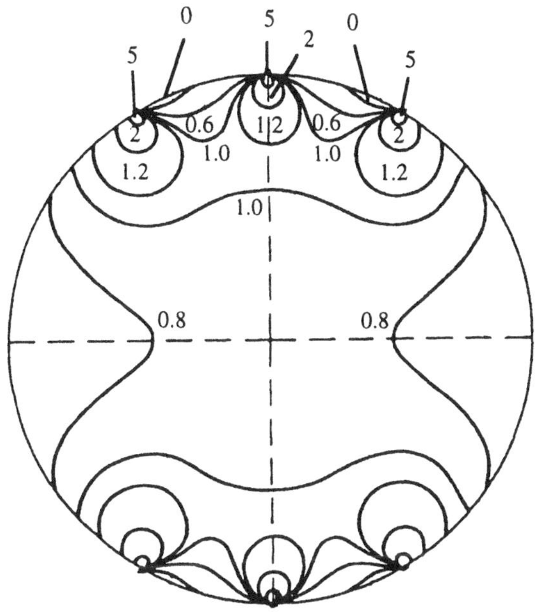

FIGURE 6-3 *Rectilinear weight function with six electrodes at $\theta = \phi$, $180^o + \phi$, where $\phi = 60$, 90, and 120°, and $\phi = 0.243$. (Source: O'Sullivan 1983)*

consistency of the flowmeter response; however, it did not correct the nonuniformity in magnetic field.

Large-area electrodes, in principle, are less sensitive to flow velocity distribution. Cox and Wyatt (1984) constructed and tested such a flowmeter using saddle-shaped electrodes that subtend 120° at the pipe axis. The flowmeter configuration and its weight function are shown in Fig. 6.4. The tests of Cox and Wyatt (1984) showed that the flowmeter was 14 times less sensitive to velocity distribution than the conventional point-electrode flowmeter. However, large-area electrodes suffer from many engineering problems, e.g., defects in the insulating film and seals which cause electrical leakage; variations in the electrochemical state of the surface that lead to nonuniformity of the electrode/fluid interface; and deposition of solids on the electrodes that cause additional nonuniformity. As a result, the large-area-electrode flowmeter may suffer from calibration uncertainty and severe baseline instability.

6.2.2 Capacitive Methods

A capacitive sensor measures the dielectric properties of a medium; it is applied mainly to a medium of low electrical conductivity. The basic design of

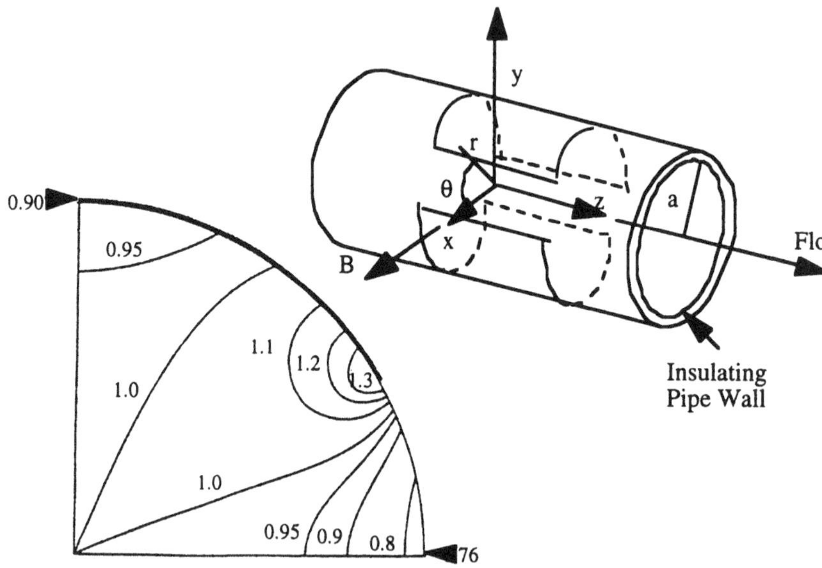

FIGURE 6-4 *Configuration and weight function contours of a flowmeter with saddle-shaped electrodes. (Source: Cox and Wyatt, 1984)*

a capacitive flow sensor is similar to that of a capacitor; it consists of two parallel metal plates separated by the dielectric flow medium. The capacitance of this capacitor is given by

$$C = \varepsilon_0 \varepsilon A / d, \tag{6.5}$$

where ε_o and ε are the dielectric constants in vacuum (8.85 pF/m) and in the dielectric medium, respectively; A is the overlap area of the plates; and d is their separation.

The capacitive method for measuring solids concentration is based on the change in capacitance in the presence of solids within the sensing volume. For example, in a solid/gas flow, because solids normally have a higher dielectric constant than the gas, the measured capacitance increases with solids concentration. As long as the solids concentration is low, a linear expression for the capacitance of a mixed-phase flow is generally assumed. For mixed-phase flows, Eq. 6.5 can be rewritten as

$$C_m = \varepsilon_0 \left\{ \varepsilon_s \phi + \varepsilon_f (1 - \phi) \right\} A / d, \tag{6.6}$$

where C_m is the capacitance of the mixed-phase medium, ϕ is the solid volume fraction, and ε_s and ε_f are dielectric constants of solids and fluid, respectively. In practice, for better sensitivity and larger dynamic range, the instrument

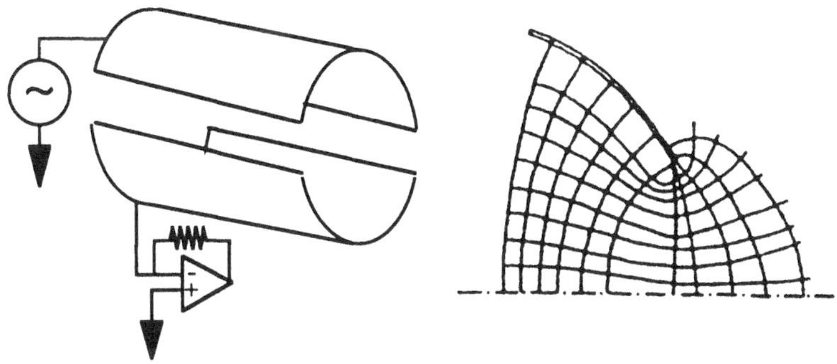

FIGURE 6-5 *Wrapped-electrode capacitor.*

measures the relative change in capacitance with respect to the pure fluid, from which it deduces the volume fraction ϕ according to

$$\phi = \frac{(C_m - C_f)/C_f}{(\varepsilon_s - \varepsilon_f)/\varepsilon_f}, \tag{6.7}$$

where the subscripts m, f, and s represent mixed-phase, fluid, and solid medium, respectively.

Again, uniformity of the electrical field determines the accuracy of the capacitive technique; therefore, electrode geometry is a major design parameter. The most common geometry is that of two parallel plates wrapped around a pipe, as shown in Fig. 6.5. The electrodes, however, must be insulated from the conductive pipe wall. Typically, the parallel plates cover a large pipe area, and thus produce a relatively homogeneous field within the pipe, except at the electrode edges. Fig. 6.6 shows the equipotential field lines near the edges of the parallel-plate electrodes. Overall, this electrode geometry provides good accuracy.

Another common electrode geometry is annular, and consists of at least three axial-ring electrodes, as shown in Fig. 6.7. The central ring is the live electrode that is electrically insulated from the other two ground electrodes. Fig. 6.8 shows a plot of the equipotentials of the annular geometry. The field of this geometry is axial and obviously nonuniform in sensitivity to solids distribution, and more sensitive to the regions near the wall. However, this geometry is relatively easy to manufacture; hence it is practical for industrial applications.

Thus far, we have discussed capacitive techniques for sensing solids concentration in a mixed-phase flow. We must now discuss how to measure flow velocity by the capacitive method, because the ultimate goal is that of measuring mass-flow-rate. Two methods are commonly suggested for

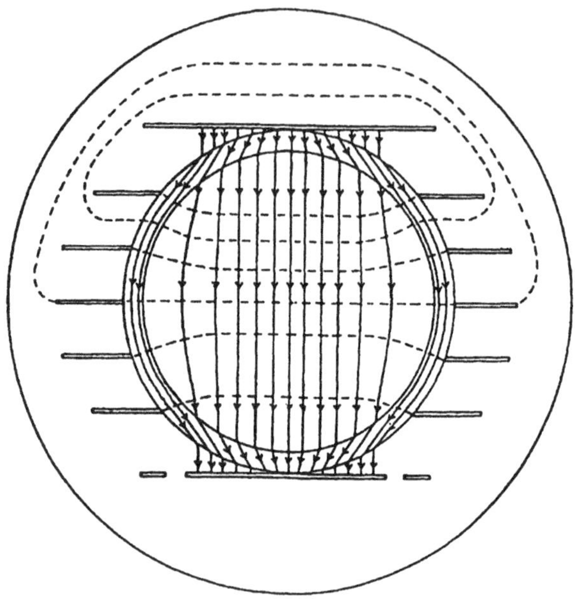

FIGURE 6-6 *Narrow-electrode capacitor, ANL-I version.*

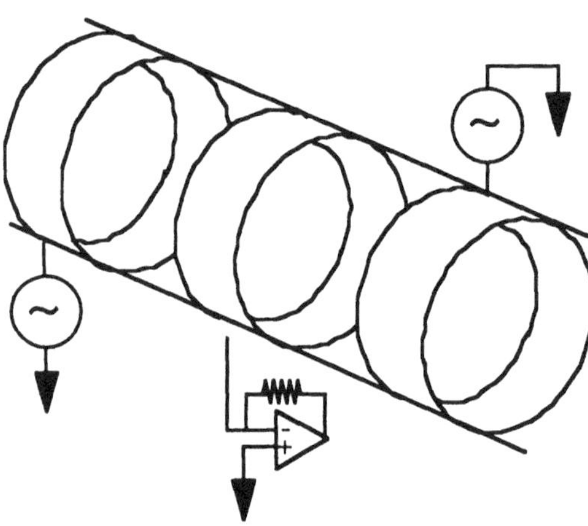

FIGURE 6-7 *Annular-electrode capacitor.*

measuring flow rate. One is based on a cross-correlation technique (Beck et al., 1990), which requires two sets of capacitive electrodes, separated by a known distance D, and measures the transit time τ necessary for a given capacitance fluctuation to traverse from one electrode to the other. The flow velocity V_S (more appropriately called the particle velocity) within the sensing volume is

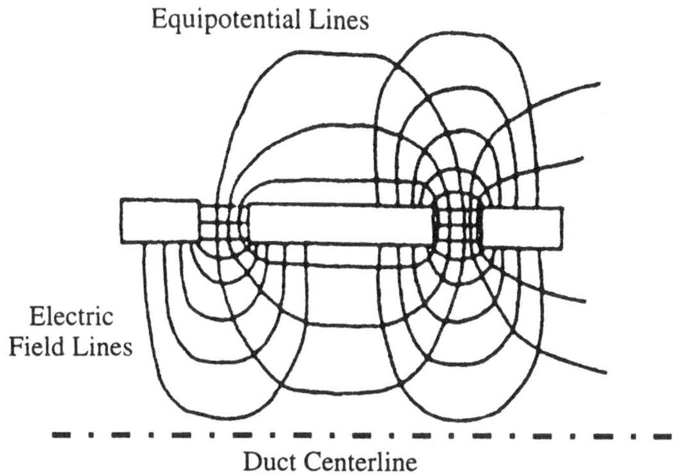

FIGURE 6-8 *Equipotential and electric field lines of annular-electrode capacitor.*

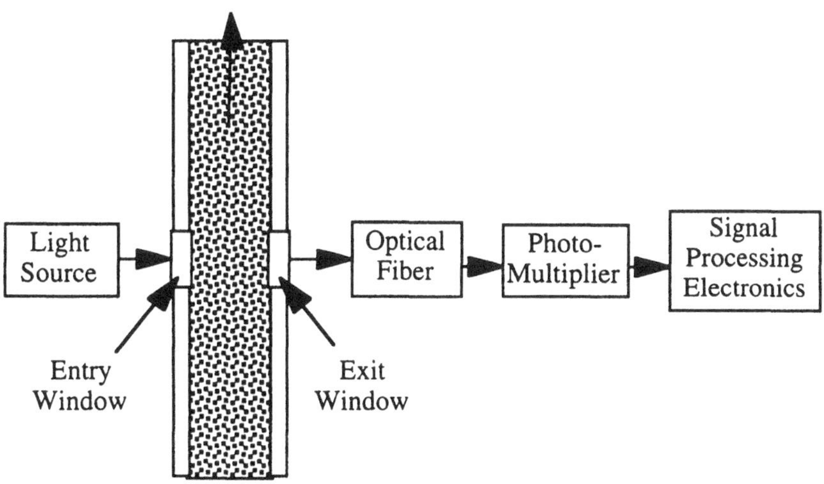

FIGURE 6-9 *Typical cross-correlation flowmeter. (Source: Yan, 1996)*

then calculated from D/τ. Fig. 6.9 shows a typical design of a cross-correlation flowmeter. If one knows the solids density ρ_S, the mass flow rate of the solids F_M is

$$F_M = \rho_S \phi A V_S = \rho_S \phi A \, D/\tau, \tag{6.8}$$

where τ is derived from the peak of the cross-correlation function. In principle, a cross-correlation function represents the particle-velocity distribution convoluted with the sensor response function. Hence, the shape of the cross-correlation function may contain velocity-profile information and its peak is related to the most probable velocity.

The second method of measuring flow rate, the spatial filtering technique (Hammer and Green, 1983), analyzes the frequency content of the capacitance signals and relates the frequency bandwidth to particle velocity. The capacitance signals generally contain rapid fluctuations that are the result of the inhomogeneity of the mixed-phase flow and possess a finite bandwidth that is determined by the time required for a particle to traverse an active range of the electrodes. To illustrate the concept, let us assume that the signal variation is a simple rectangular pulse, the amplitude of which is proportional to the solids concentration and the width of which depends on particle velocity. In this case, the form of the frequency spectrum of the capacitance signals will be expressed as

$$C(j\omega) = K\phi \frac{\sin(\omega a/2V_S)}{\omega a/2V_S}, \tag{6.9}$$

where K is a constant and a is the active length of the capacitance electrode. Equation 6.9 is a sinc function whose maximum at the origin is $K\phi$, and the first zero of which occurs at V_S/a.. This technique seems to be simple and practical because only one set of electrodes is used to measure both the solids concentration and particle velocity; however, it requires calibration to determine the active length of the electrodes, which may vary with electrode geometry and performance. The accuracy of this technique may be a problem.

6.2.3 Optical and Tracer Techniques

The entire EM spectrum range, from microwaves to γ-rays can be used as the optical probe to measure solids concentration and particle velocity. Light attenuation/scattering methods have long been used to determine solids concentration. The theoretical basis of the methods is described by the Lambert-Beer law

$$I = I_0 e^{-\sigma x}, \tag{6.10}$$

where I_0 and I are the intensities of the incident and transmitted light beam, respectively; x is the distance traversed by the beam in the flow medium; and σ is the linear attenuation coefficient. Attenuation of light in a solid/fluid medium is due to absorption and scattering processes. Absorption depends on

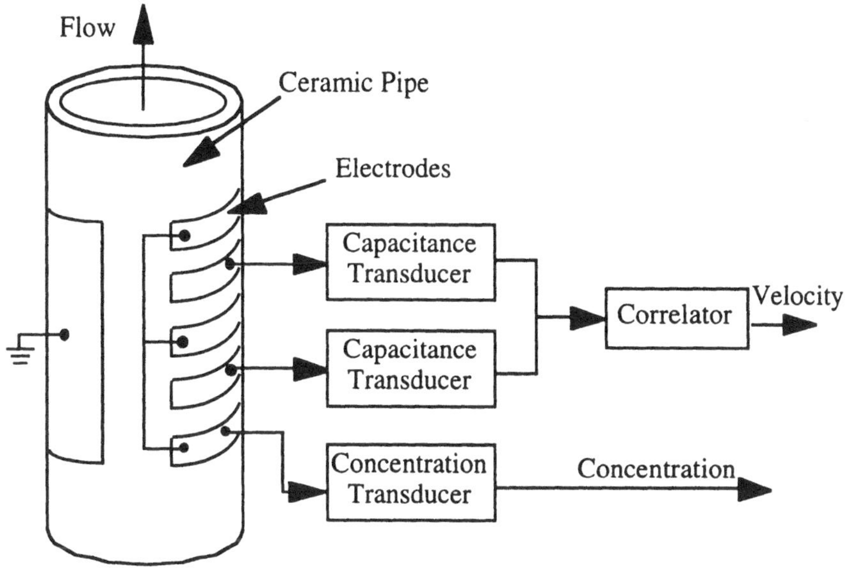

FIGURE 6-10 *Optical solids-concentration monitor. (Source: Yan, 1996)*

optical frequency, whereas scattering is sensitive to particle size. The absorption process involves interactions of the EM waves with the medium at molecular or subatomic levels; it, therefore, depends on the optical frequency and on the types of particles and fluids. For example, coal particles will absorb more light than opaque particles. Hence, an optical instrument for measuring solids concentration requires specific calibration for the solids being measured. Scattering is a process that redistributes the incident beam energy that is due to a change in the refraction index of the beam path. The Mie theory (Van De Hulst, 1957) is the basis for light scattering by spherical particles; it establishes scattering characteristics in terms of the size and refraction index of a particle. The theory predicts that for large particles ($d > 3\lambda$, where d is the particle diameter and λ, the wavelength), scattering in the forward direction is significantly greater than scattering in any other direction. Because in the typical sensor arrangement of an optical solids-concentration monitor, as shown in Fig. 6.10, the detector is directly opposite to the light source, the measured attenuation is, as the theory predicts, higher for smaller particles than for larger particles of the same solids concentration. Furthermore, the derivation of the theory is based on light scattering of a single spherical particle and neglects multiple scattering. Therefore, linear dependence of attenuation on solids concentration only appears in dilute cases; attenuation is

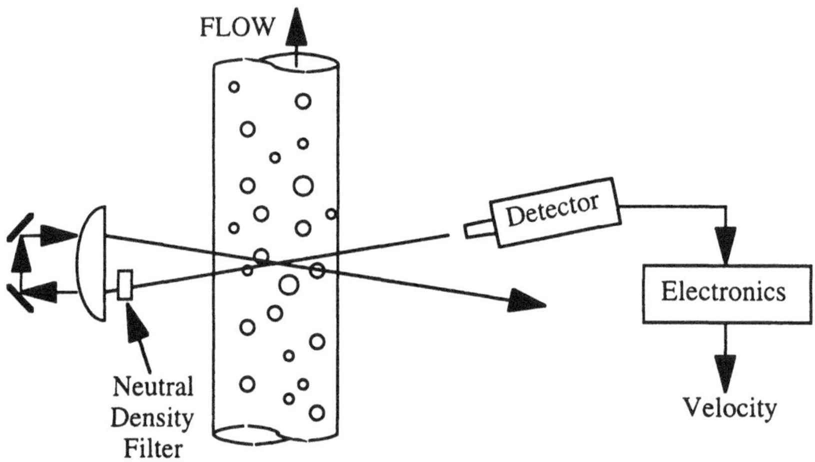

FIGURE 6-11 *Laser Doppler velocimetry in reference beam mode. (Source: Yan, 1996)*

expected to increase exponentially for dense-phase flows. Most commercial optical instruments are therefore suitable for determining solids concentrations below 15% by volume (Hylton et al., 1998). Overall, the optical techniques for measuring solids concentration suffer from the following problems and disadvantages: the measurement depends on particle type and size; the technique is not applicable to dense-phase flows; and contamination and misalignment of sensor windows cause false signals.

The application of optical sensors to measure the velocity of solids can be realized by incorporating Doppler, cross-correlation, or spatial-filtering methods (Yan, 1996). The Doppler technique measures the frequency shift of the signals reflected by the solids particles. Based on the Doppler-shift principle, the difference frequency Δf gives a direct measure of solids velocity V_S via the relationship

$$V_s = \frac{\Delta f}{f} \frac{c}{2\cos\theta}, \tag{6.11}$$

where c is the speed of light; f, its frequency; and θ, the viewing angle of the optical beam transmitted in the flow direction. Laser Doppler velocimetry (LDV) is a well-developed technique that has been applied to mixed-phase flows. In typical operation, application is either in the reference-beam or differential-Doppler mode. The reference-beam mode measures the frequency

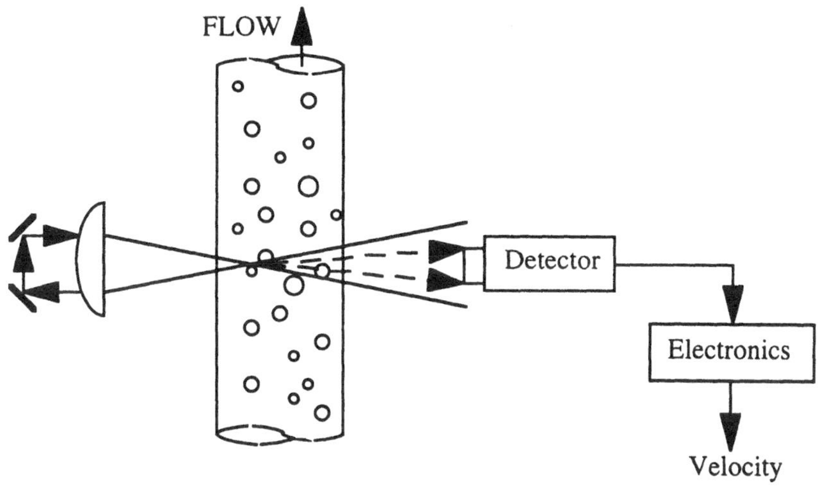

FIGURE 6-12 *Laser Doppler velocimetry in the differential-Doppler mode. (Source: Yan, 1996)*

shift of scattered laser light with respect to a reference laser beam. Figure 6.11 schematically illustrates the setup of the measurement system in the reference beam mode. The differential-Doppler mode, on the other hand, utilizes two coherent, focused laser beams that converge on moving solids from various directions. The photodetector detects the scattered light from both beams and electronically measure the difference in frequency between the two Doppler-shifted signals. The difference provides a measure of the solids velocity in the beam-illuminated region. Figure 6.12 shows this dual- beam setup. Potentially, the LDV technique can be used continuously on line to measure a wide range of solids velocities, ranging from 0.1 mm/s to 100 m/s. However, once again, the technique is only applicable to dilute-phase flow conditions, measurable up to 0.4% alumina powder concentration (Birchenough and Mason, 1976).

The tracer method is the most accurate technique to measure fluid velocity. The method is well developed for fluid flows (Somerscales, 1981). It involves use of light-sensitive particles mixed with the fluid, an optical source of illumination, and a detection system. But for solid/liquid and solid/gas flows, a better approach is to use radioactive tracers. Typically, the tracers are low-level, short-lived radioactive particles that can be either introduced manually or generated on line by a neutron source. Both methods will be described latter in this chapter.

6.3 MEASUREMENT OF SOLID/LIQUID FLOW

Complete characterization of a solid/liquid flow requires that we know phase velocity, solids concentration, particle-size distribution, and liquid density and viscosity. Individual phase velocity and particle-size distribution are difficult to measure, even though, in principle, one may select differing optical frequency bands to track the motion of various phases, and use tomographic techniques to map out particle-size distribution. Instruments to monitor these two parameters are still in the developmental stage. At present, on-line instruments for density and viscosity measurement are available mainly for single-phase fluids; the technologies never address the effects of suspended particles on changes in bulk density and viscosity. For practical purposes, a process engineer must measure mass flow rate accurately, preferably with on-line instruments. In this section, we will limit our discussion to measuring mass flow. A brief description of the Coriolis mass flowmeter will lead the discussion, followed by capacitive techniques and the Pulsed Neutron Activation (PNA) technique for measuring solids velocity. The latter two techniques cover mainly developments at Argonne National Laboratory (ANL).

6.3.1 Coriolis Mass Flowmeter

The Coriolis mass flowmeter measures the mass flow rate of a fluid or slurry that is flowing through the flow tube by detecting the Coriolis force associated with the moving fluid/slurry. The flow tube geometry varies; common tube geometries are U, straight, and Z. The U-tube design has been studied theoretically (Sultan and Hemp, 1989) and widely adopted by industry. The typical arrangement of a U-tube Coriolis mass flowmeter is shown in Fig. 6.13. The arrangement consists of a U-tube lying in one plane and clamped at its ends, an electromagnetic drive that vibrates the tube, and two electromagnetic detectors that sense the relative phase of the limb vibration. In operation, the electromagnetic drive causes the tube to perform an oscillatory rotation about the y-axis. This rotation, in turn, induces a Coriolis force in the straight limbs of the U-tube when fluid or slurry flows in the tube. The Coriolis force is extremely small and can be calculated by solving

$$\vec{F}_c = 2M\vec{\omega} \times \vec{V}, \tag{6.12}$$

where M is the mass of the fluid per unit length of the tube, ω is the angular velocity of the driving rotation, V is the fluid velocity, and x represents the vector cross product operator. The Coriolis force, as defined in Eq. 6.12,

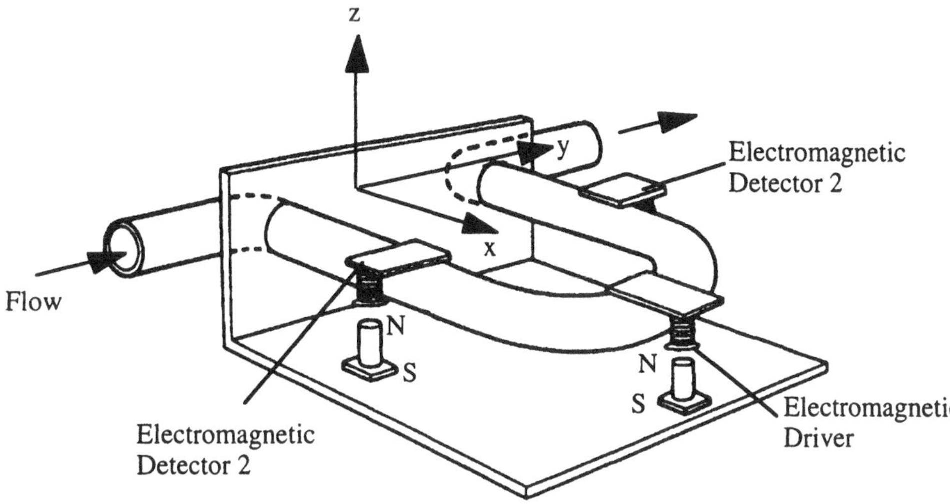

FIGURE 6-13 *Coriolis mass flowmeter in U-tube configuration. (Source: Sultan and Hemp, 1989)*

provides a direct measure of mass flow if the frequency of the oscillatory rotation is constant. Because the fluid/slurry flows in opposite directions in the straight limbs, the Coriolis force causes an oscillatory twisting of the tube about the x-axis. The oscillating moment Γ is given by

$$\Gamma = 4\omega \dot{m} r L = K_s \alpha, \tag{6.13}$$

where K_s is the spring stiffness constant of the tube, L is the tube length, r is the distance from the center of U-tube to the two limbs, $\dot{m}$ is the mass flow rate per unit tube length, and α is the tube deflection detected by sensors that are generally positioned at the midpoint of each limb. The deflection can be measured in terms of the time difference (Δt) between the midpoint crossing time in the two limbs, which is

$$\alpha = \frac{L\omega\Delta t}{2r}. \tag{6.14}$$

Combining Eqs 6.12, 6.13, and 6.14, we obtain

$$\dot{m} = \frac{K_s}{8r^2}\Delta t, \tag{6.15}$$

FIGURE 6-14 *ANL capacitive flowmeter.*

which indicates that a Coriolis flowmeter depends on time difference and pipe geometry constants. Therefore, in principle, the accuracy of Coriolis mass flowmeters is unaffected by changes in slurry temperature, pressure, density, and flow profile.

The measurement range of a Coriolis mass flowmeter varies with tube size, larger tubes being associated with higher flow rates. Also varied in commercial flowmeters are driver arrangements and sensing techniques. For example, the design of the Promass 63 and m-Point Coriolis meters manufactured by Endress and Hauser (Greenwood, Indiana) contains some key differences even though the basic principle involved in the two meters is the same. The Promass 63 has a driver on each tube, whereas the m-Point has a driver on one tube. According to the manufacturer, the additional driver allows the Promass to work with a higher volume of air in the flow stream. In the Promass, a magnetic pickup sensor is used to detect the vibrations in the tube, whereas, in the m-Point, an optical sensor is used. Both meters measure the resonant frequency of the oscillating tubes, which then gives a measure of fluid density in the tube. Performance of the meters, recently evaluated with kaolin/sugar-water slurries (Hylton, 1998), showed less than 1% standard deviation in density measurement.

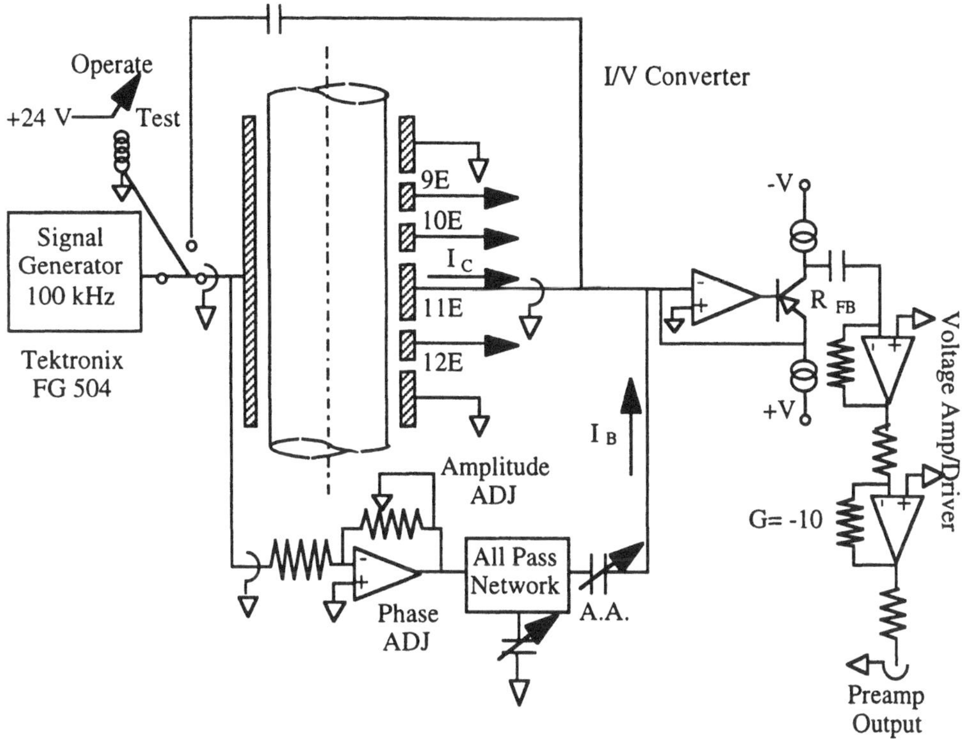

FIGURE 6-15 *Current-sensing preamplifier for capacitive flowmeter electrodes.*

6.3.2 Capacitive Flow Instrument

A capacitive flowmeter, applicable only to slurries with nonconducting fluids, exhibits many attractive features that are often required by industrial processes. For example, it is nonintrusive, has no moving parts, does not disturb the flow, can tolerate hostile temperature and pressure environments, and is resistant to shock and vibration. Figure 6.14 shows the capacitive flowmeter developed at ANL (Bobis et al., 1986). Its electrode arrangement, shown in Fig. 6.15, adopts a parallel-plate configuration. The measuring electrodes are of two types: three short, 0.5 x 0.84-in., electrodes (9E, 10E, 12E) placed in the direction of flow and used to measure flow velocity, and a long, 4.66 x 0.84 in.(11E) electrode that is used to measure density. Both the drive electrode that serves as the other half of the parallel-plate configuration and the measuring electrodes are supported by Teflon insulators and are separated from the flow stream by a 1/8-in. thick ceramic tube with a 2-in. I.D.

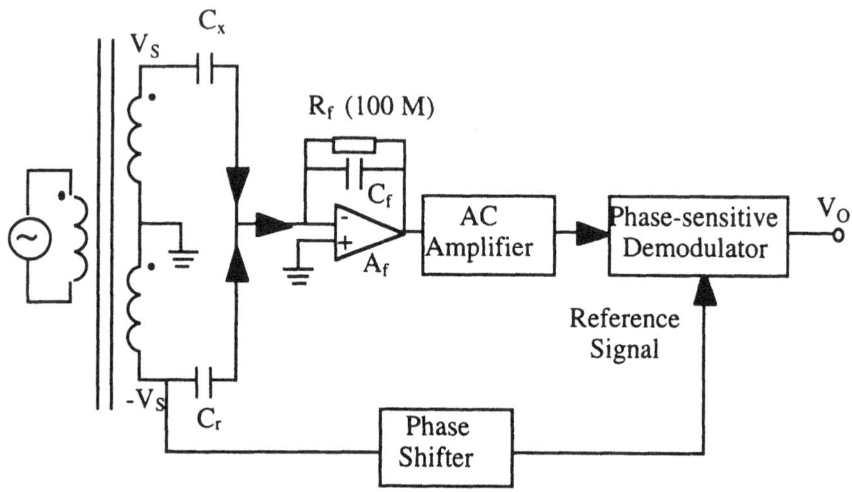

FIGURE 6-16 *Typical transformer-ratio-arm bridge transducer. (Source: Huang, 1988)*

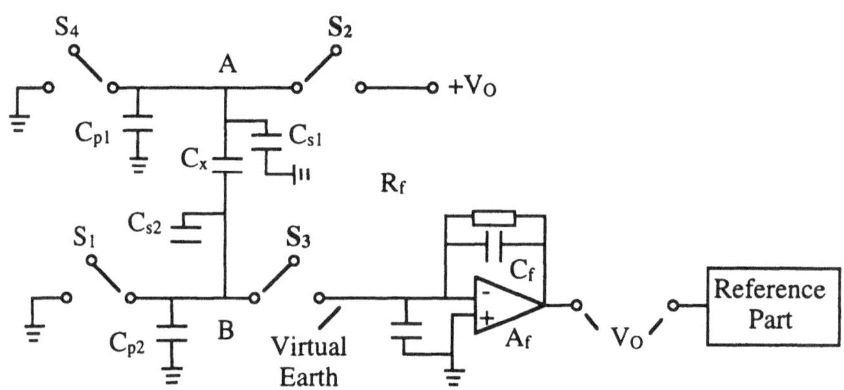

FIGURE 6-17 *Circuit diagram of stray-immune charge/discharge transducer. (Source: Huang, 1988)*

Because the typical capacitance to be measured is in the range of 0.1-10 pF and the required resolution is less than 0.001 pF, the capacitance-measuring circuits become the most critical part of the instrument, which should have high sensitivity and low baseline drift. The circuits must be able to reduce or eliminate the effect due to stray capacitance. Popular capacitance-measuring circuits (Huang et al., 1988) can be categorized into four groups: resonance, oscillation, charge/discharge, and AC bridge methods. Evaluation by Huang et al. (1988) led to the following conclusions. For measurements at frequencies

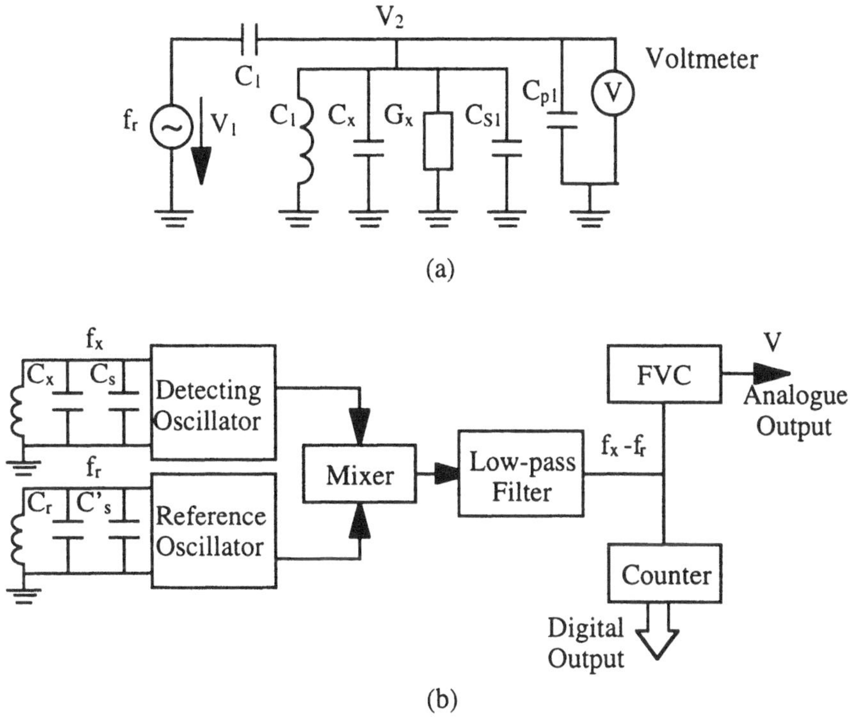

FIGURE 6-18 *Circuit diagrams of (a) resonance and (b) balanced LC oscillator transducers. (Source: Huang, 1988)*

below 100 kHz, the transformer-ratio-arm bridge transducer shown in Fig. 6.16 is recommended; from 100 kHz to 5 MHz, the stray-immune charge/discharge transducer, shown in Fig. 6.17 is suggested; and beyond 5 MHz, LC oscillators or resonance methods, shown in Fig. 6.18 should be used.

6.3.2.1 Density Measurement

To illustrate density measurement by capacitive methods, the performance of the ANL capacitive mass flowmeter is described. Instrument evaluation tests were conducted at the ANL Solid/Liquid Test Facility (SLTF), shown schematically in Fig. 6.19. The SLTF was designed as a specialized instrument-testing and calibration-loop facility for various liquid and liquid/solid media. The facility can provide volumetric flow rates that range from 0 to 10 L/s and flow speeds up to 6 m/s in 2-in. Schedule 40 pipe. It is equipped with an acoustic cross-correlation flowmeter, a PNA system to

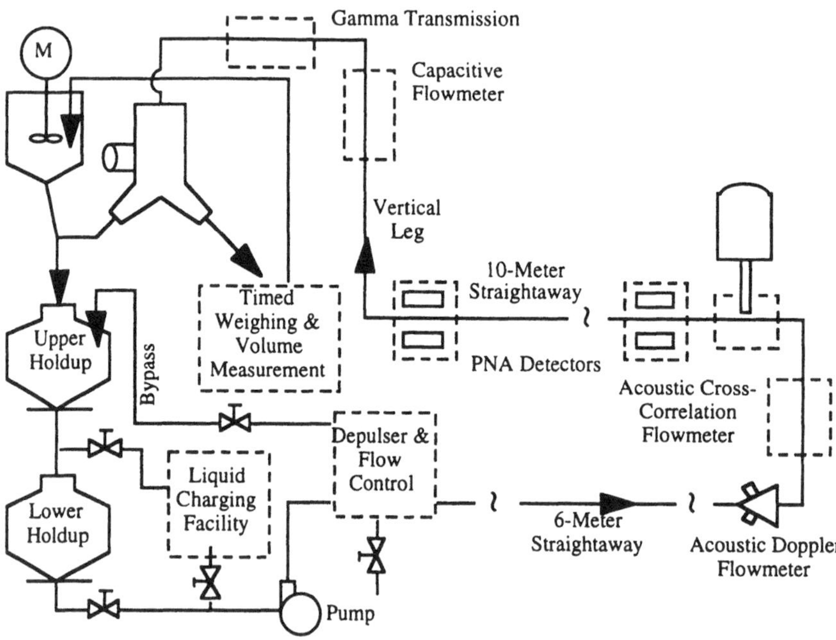

FIGURE 6-19 *Schematic representation of ANL solid/liquid loop test facility (SLTF).*

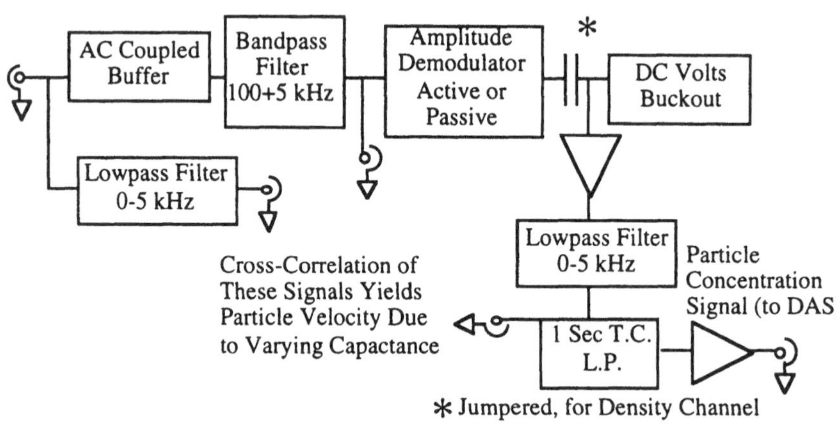

FIGURE 6-20 *Instrumentation of ANL capacitive flowmeter.*

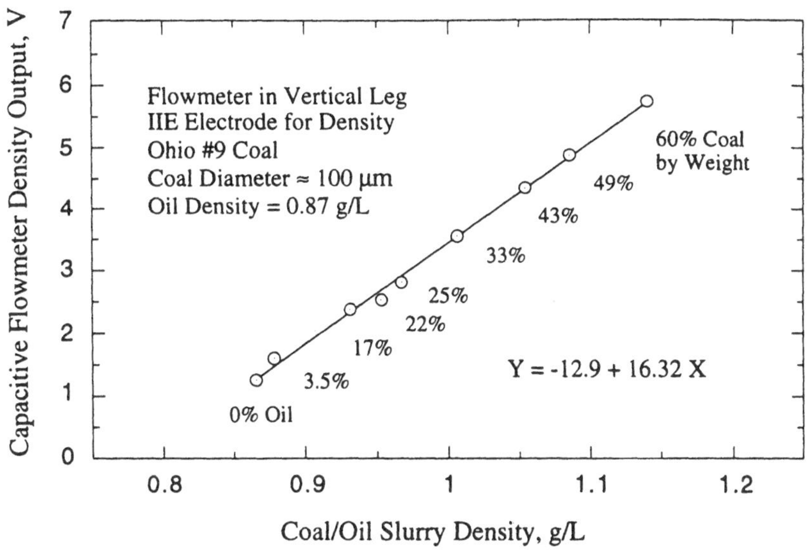

FIGURE 6-21 *Density data of coal/oil slurries obtained by ANL capacitive flowmeter.*

measure particle velocity and an on-line, timed, weight/volume diversion system.

Coal/oil slurries were used in the instrument performance tests. The coal was finely ground Ohio #9, a high-ash highly volatile bituminous coal, the major constituents of which are 59 wt.% carbon, 4 wt.% hydrogen, 4 wt.% sulfur, and 24 wt.% ash. A sieve analysis disclosed that ≈ 86% of the coal particles were of 63-125 μm in diameter. The oil was a representative organic liquid used in the feedlines of some pilot coal conversion plants. Oil density at 20°C was 0.868 g/cm^3.

Schematic diagrams of the instrumentation for the ANL capacitive flowmeter are given in Fig. 6.20. A 100-kHz sine-wave oscillator, with stable frequency and amplitude controls, was used to pulse the drive electrode. Each sensing electrode was connected to a current-to-voltage converter preamplifier. The preamplifier outputs were bandpass filtered at 100 kHz ± 5 Hz and amplitude-demodulated. The demodulated signals were amplified and DC-coupled to a first-order low-pass filter to give density signals.

The density measurements for coal concentrations that ranged from 0 to 60 wt.% are presented in Fig. 6.21. A linear relationship was measured with an accuracy ≤ 5%. The relationship was independent of slurry velocity and particle size. Because of the large contrast between the coal and oil dielectric

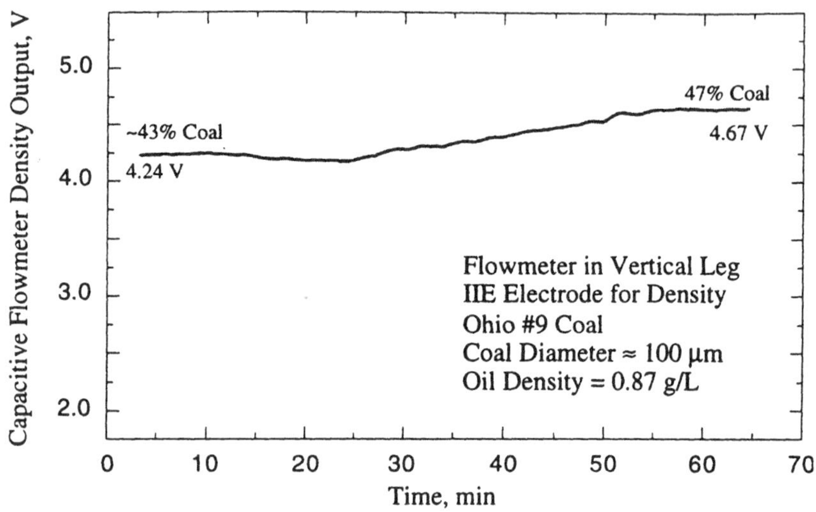

FIGURE 6-22 *Response of density signal to change in coal content of slurry.*

constants, the instrument was highly sensitive when measuring density. Figure 6.22 shows the signal level increase when coal concentration was increased from 43 to 47 wt.%, yielding a change of ≈90 mV in the signal for a 1 wt.% change in coal concentration.

6.3.2.2 Particle Velocity Measurement

The cross-correlation technique is used in the ANL capacitive instrument to measure particle velocity. The outputs from the velocity-sensing electrodes were amplitude-modulated capacitor currents. If one assumes that the output capacitance signals vary sinusoidally ($A_a \sin\omega_a t$), and the applied voltage to the drive electrode is $A_p \sin\omega t$ (where $\omega = 2\pi f$ and f = 100 kHz), the detected signals with amplitude modulation can be given as

$$I(t) = C_0 A_p \omega Cos\omega t + \frac{A_a A_p}{2}\left[\omega(\sin\omega_1 t - \sin\omega_2 t) + \omega_a(\sin\omega_1 t + \sin\omega_2 t)\right], \quad (6.16)$$

where C_0 is the capacitance of the slurry without coal particles, $\omega_1 = \omega + \omega_a$, and $\omega_2 = \omega - \omega_a$. To demodulate the signals, an AC-coupled, low-pass filter (0-5 kHz) was used. Thus, only the terms with ω_2 in Eq. 6.16 remain. The particle velocity or the slurry flow rate is then determined from the cross-correlation function derived from

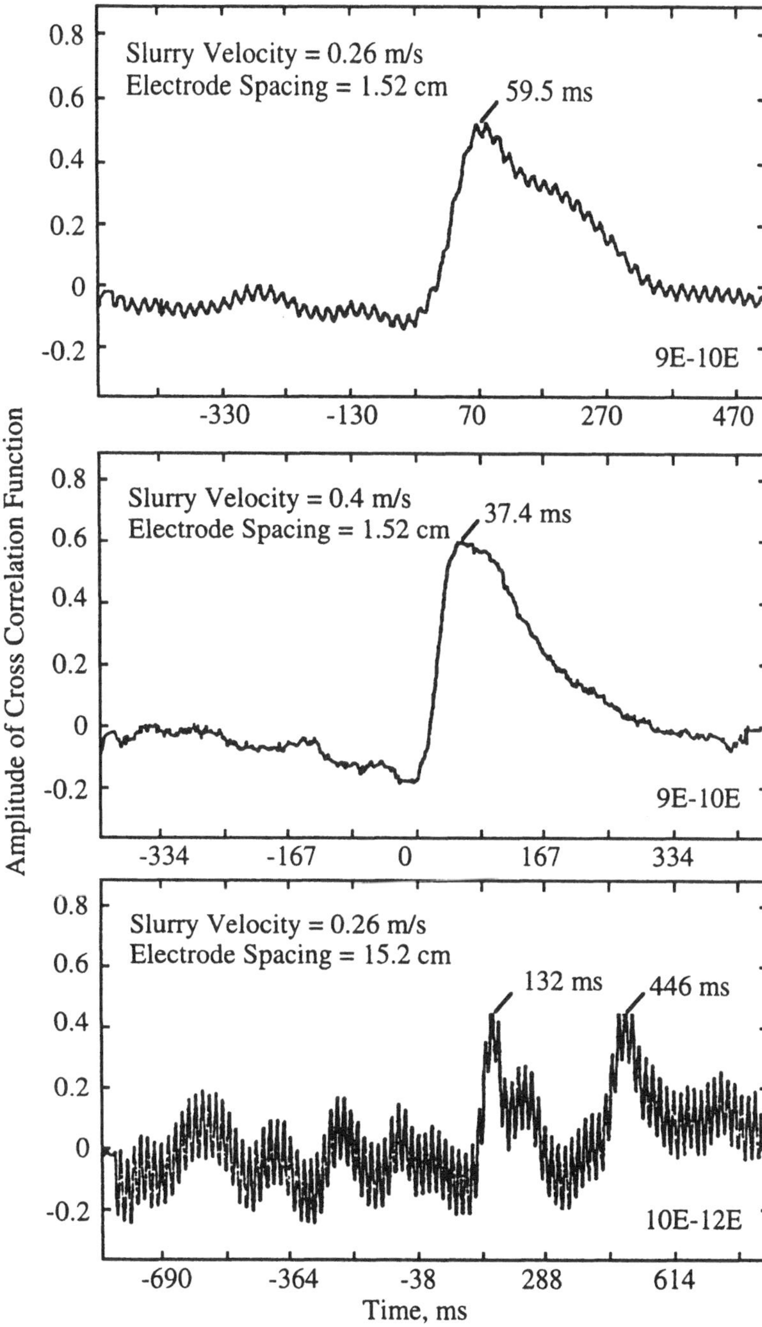

FIGURE 6-23 *Typical cross-correlation function obtained by ANL capacitive flowmeter for two slurry velocities (0.26 and 0.4 m/s) and two electrode spacings (1.52 and 15.2 cm). Coal concentration was 60 wt.%.*

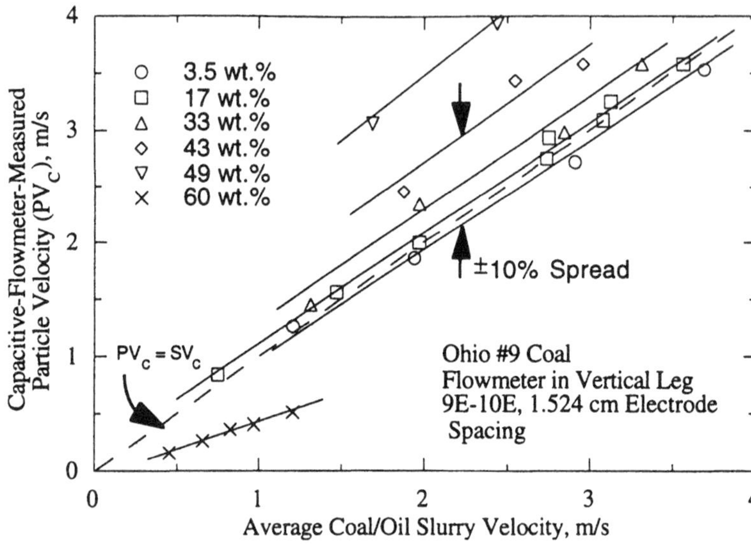

FIGURE 6-24 *Particle-velocity data obtained by flow diversion (S_V) in the ANL capacitive flowmeter over a range of coal concentrations.*

$$F(\tau) = \lim_{T\to\infty} \frac{1}{T} \int_0^T I_1(t) I_2(t+\tau)\, dt \tag{6.17}$$

and

$$U_p = \frac{D}{\tau_m}, \tag{6.18}$$

where D is the separation between sensing electrodes (0.6 in. for the ANL capacitive instrument) and τ_m is obtained from the peak of the cross-correlation function. Figure 6.23 shows typical cross-correlation functions obtained by the ANL capacitive instrument.

The ANL capacitive flowmeter was installed at the vertical line of the SLTF (see Fig. 6.19), thus, particle settling problems would not occur. The only parameter that affected the velocity measurement was the velocity profile. Figure 6.24 shows that particle velocities were measured over a range of solids concentrations. For coal concentrations of 3.5-33 wt.%, the measured velocities were independent of coal concentration and 7% accurate when compared with timed-diversion measurements. For higher coal concentrations, the measured velocities deviate from the diversion measurements; velocity measurements were ≈ 30-80% higher for 43 and 49 wt.% slurries, but 40%

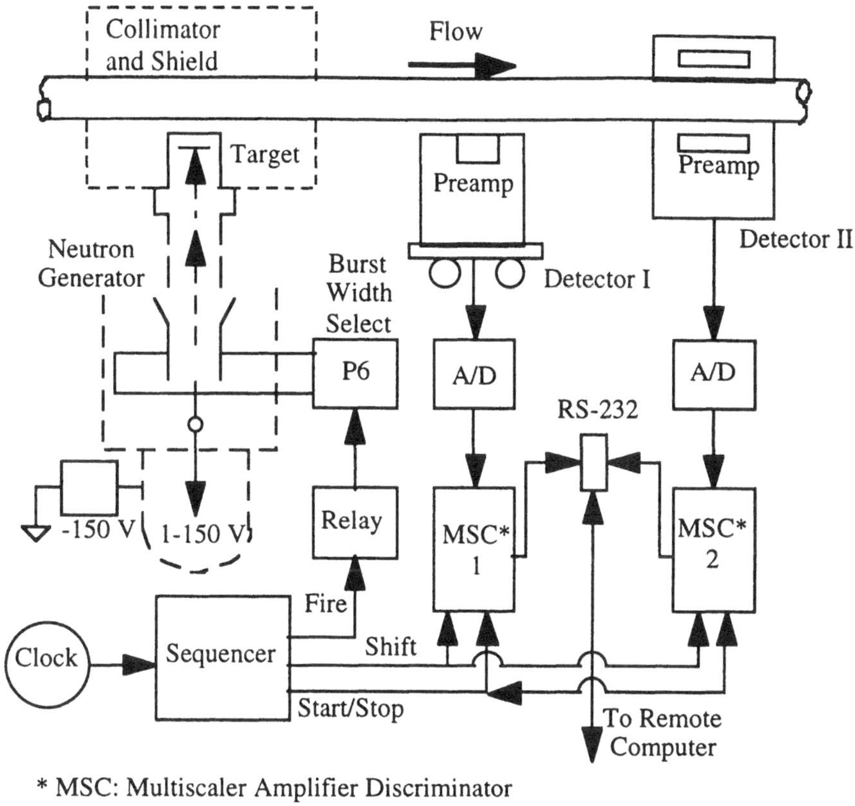

FIGURE 6-25 *Schematic representation of ANL PNA system.*

lower for 60 wt.% slurries. It was suggested that the deviations were caused partly by the velocity profile effect. As flows became more laminar than turbulent, the instrument would give a higher velocity measurement because the electrical field established by the parallel-plate configuration was more intense at the center line than near the pipe wall. But when coal concentration in the slurry is 60 wt.%, the deviation might be due to poor correlation because the slurry showed less fluctuation in density. As shown in Fig. 6.23, the cross-correlation functions for 60-wt.% slurries have a rather broad, skewed shape and also show multiple peaks for electrodes that are further apart.

In conclusion, the capacitive mass flowmeter can be a reliable instrument for measuring the flow of dilute suspended solids. For the flow of high concentrations of suspended solids, the velocity measurement becomes inaccurate. The inaccuracy may be caused by the velocity profile effect, but further study is required to confirm the observation.

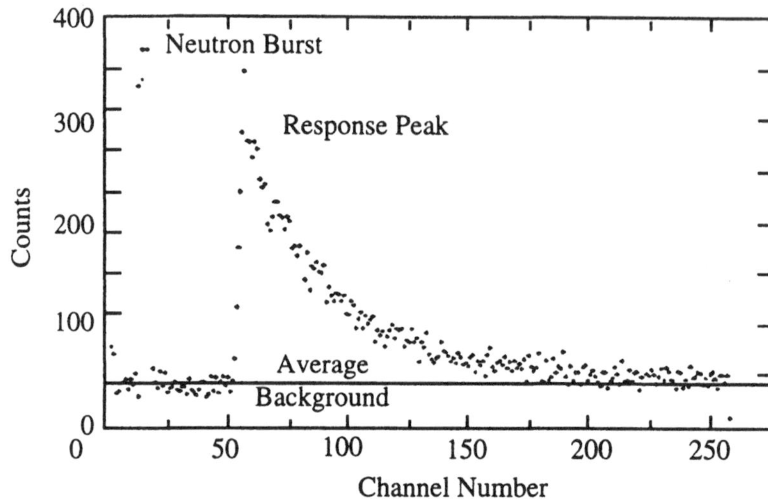

FIGURE 6-26 *PNA profile of 56% solid coal/oil slurry flow at 0.95m/s in a 2-inch pipe, at source-to-detector distance of 3.40 m.*

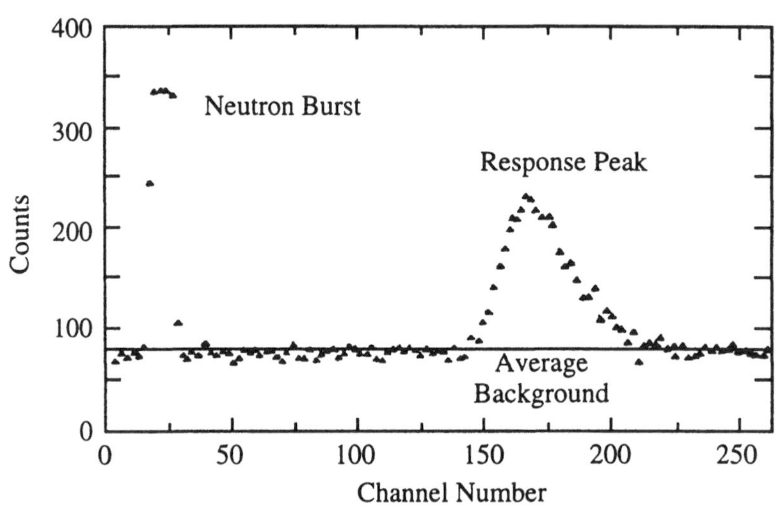

FIGURE 6-27 *PNA profile of 25% solid coal/oil slurry flow at 3.35m/s in a 2-inch pipe, source-to-detector distance of 6.61m.*

6.3.3 Pulsed Neutron Activation Technique

The PNA technique is not commonly used, because it involves radioactivity. In essence, the PNA technique is an on-line tagging method. Short bursts of neutrons are used to irradiate the slurry to produce short-lived γ-emitting elements, the passage of which is registered by downstream γ detectors. Because it is a tagging technique it provides a direct measure of particle velocity.

The overall scheme of the ANL PNA system (Porges et al., 1984) is shown in Fig. 6.25. An RF ion source is used to generate a 150-keV deuteron beam; as a burst of deuterons strikes a tritiated foil, 14-MeV neutrons are produced and fan out into a 4π solid angle. A copper/lead collimator concentrates forward neutrons onto the pipe that passes directly in front of the target snout. Two downstream γ detectors (NaI/Tl crystal scintillators) are used to register the counts.

Typical records obtained from slurries are shown in Figs. 6.26 and 6.27. The variation of the signal profile can be explained by different flow models (Porges, 1984). The averaged flow velocity can be derived from the PNA signal recorded by the detector at a known distance downstream from the PNA source. The PNA velocity results agreed with timed-diversion measurements to within 0.5% and no systematic deviation was found. We believe that the accuracy achieved by the PNA technique can be better than that obtained by other techniques because the PNA system irradiates the entire duct, and the shape of the readout is directly related to the motion of the tags.

6.4 MEASUREMENT OF SOLID/GAS FLOW

Pneumatic transportation of solids is important to many industrial processes, for example transporting coal and powder particles. To an operator of such a pneumatic conveyor, the mass flow rate of the solids is the primary process parameter to be measured accurately. A solid/gas flow is very difficult to control because it behaves quite differently from solid/liquid flows. A recent review (Yan, 1996) discussed several variables that may affect the performance of a flow instrument. The distribution of solids in a pneumatic pipeline can be highly inhomogeneous; consequently, the particle velocity distribution over the pipe cross section can be widespread. Figure 6.28 shows examples in which the "roping" type flow is particularly difficult to understand and monitor.

A solid/gas flow instrument seldom measures mass flow rate directly; instead, it generally measures the volumetric flow rate of solids, which includes measurements of solids velocity and concentration. Every sensing technique basically responds to variations in solids concentration; measuring solids velocity again requires two sensors separated by a known distance so

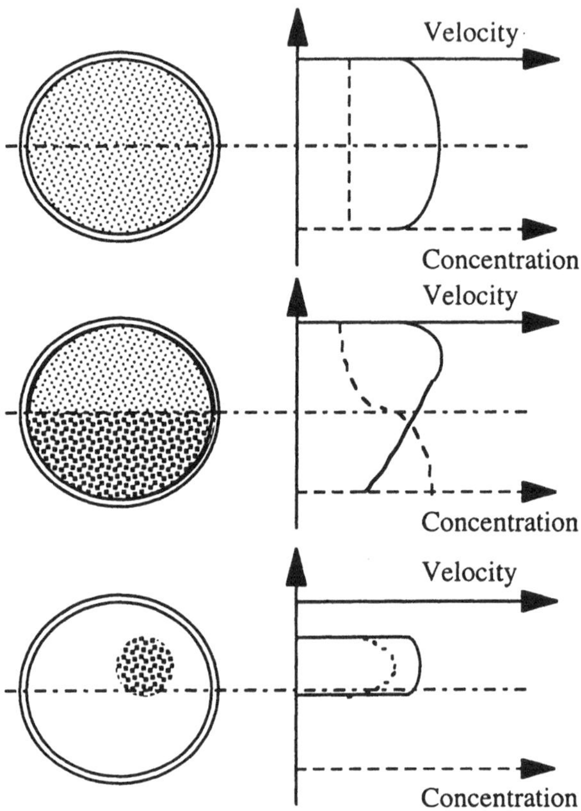

FIGURE 6-28 *Typical solids distribution and velocity profiles of 'roping' type flow over the pipe cross section. (Source: Yan, 1996)*

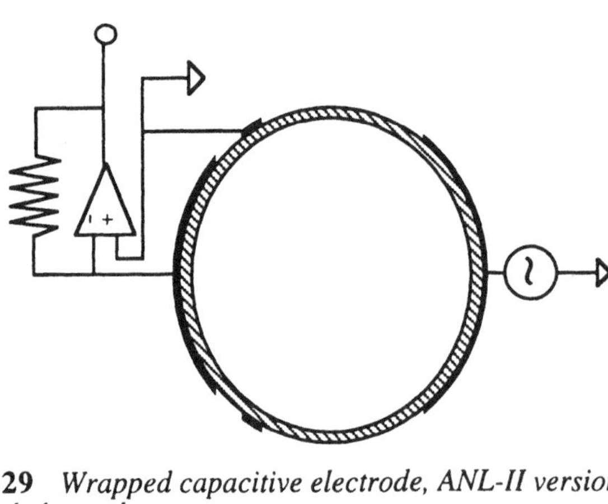

FIGURE 6-29 *Wrapped capacitive electrode, ANL-II version, with guar-equipped signal electrode.*

the solids transit time between the sensors, and thus the solids velocity, can be determined.

Sensing techniques that are applicable to the measurement of solids concentration can be classified into four groups: electrical, attenuation, resonance, and tomographic. The electrical methods utilize the dielectric and electrostatic properties of solids. Typical electrical sensors are capacitive and electrodynamic sensors; the capacitive sensors measure the dielectric property of the solids, whereas the electrodynamic sensors detect the static charges that develop because of collisions between particles, impacts between particles and pipe wall, and friction between particles and gas stream. Attenuation methods are used with optical, acoustic, and radiometric sensors. Both optical and acoustic sensors are applicable to relatively low concentrations of solids. Radiometric sensors, in which γ-rays or X-rays are used, are expensive and may raise safety concerns. They can, however, offer accurate and absolute measurement of particle velocity and thus can be used as calibration tools for other low-cost sensors such as the capacitive sensor. Resonance and tomographic methods, which are still in developmental stages, will be briefly introduced in Section 6.5.

In this section, we will describe only the capacitive technique as a practical instrument, together with a radioactive tracer technique as a calibration method for the capacitive sensor.

6.4.1 Capacitive Instrument

An ANL capacitive solid/gas flowmeter is described to illustrate the design and performance of the capacitive technique. The basic design of the ANL flowmeter for solid/gas flows is similar to the one developed for solid/liquid flows described in Section 6.3.2. Because of the inhomogeneous characteristics of solid/gas flow, the capacitive electrode adopts a wrapped-electrode configuration, as illustrated in Fig. 6.29, to increase coverage of the duct cross section. The ANL capacitive flowmeter, used in the solid/gas flow tests, consists of two measuring units: one to measure solids volume fraction (SVF), the other, to measure particle velocity. The electrodes are copper sheet cutouts with a pressure-sensitive adhesive back. They are bonded onto the external wall of a mullite ceramic tube. The mullite liner, which has a negligible effect on measurement, provides protection against particle erosion. Figure 6.30, the layout of the electrodes, shows two SVF electrodes (D1, D5) and four velocity electrodes (A, B, C, D). The SVF electrodes were placed on both sides of the velocity electrodes to average the SVF over the measuring region.

A schematic diagram of the instrumentation for the capacitive flowmeter is presented in Fig. 6.31. The drive electrode was driven by 100-kHz sine waves

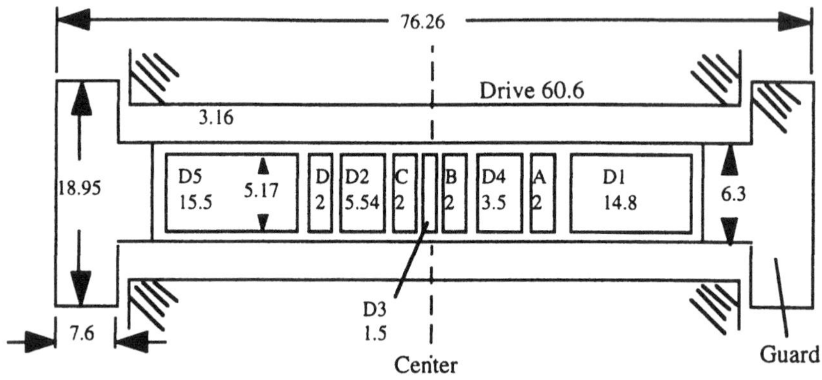

FIGURE 6-30 *Unrolled view of electrode configuration of ANL solid/gas capacitive flowmeter; all dimensions in cm; electrodes are on outside surface of a ceramic tube; electrode gap is 0.23 cm; guard strip width, 0.34 cm; and all electrodes are 0.036-mm copper foil.*

and the sensing electrodes were connected to current-to-voltage converters. The outputs from the converters were amplitude demodulated and cross correlated. The SVF was determined from the amplitude measurement, and the particle velocity was calculated from the peak of the cross-correlation function. To enhance the sensitivity of SVF measurements, an out-of-phase buckout current scheme, with amplitude and phase adjustments, was introduced between the oscillator and converters.

The ANL capacitive flowmeter was tested at the ANL Solids/Gas Flow Test Facility (S/GFTF), which is schematically shown in Fig. 6.32. This facility is a closed system that operates at near-atmospheric pressures and consists of two storage hoppers with electronic scales for direct readout of solids weight; a flow diversion system; sensors for measuring gas flow rate, temperature, and pressure drops; and a radioactive tracer injection system. Typically, solids such as 500 μm polystyrene or 50-, 300-, or 1000-μm glass beads are circulated during the tests. The SVF signals are the voltage outputs from a low-pass (1.6-Hz) filter that follows an amplitude demodulator. The SVF measurements for solids loadings that range from 0 to 1.6 vol.% are shown in Fig. 6.33. The results can be fitted into a linear relationship between the output voltage and the solids volume fraction. The lowest measurable SVF was 0.65% and the measurement accuracy was ≈5%.

Particle velocity was determined by the cross-correlation technique. Figure 6.34 shows typical cross-correlation functions that were obtained from the velocity electrodes. For a solids feed rate of 3.2 lb/s, the cross-correlation functions show well-defined peaks, from which particle velocities can be

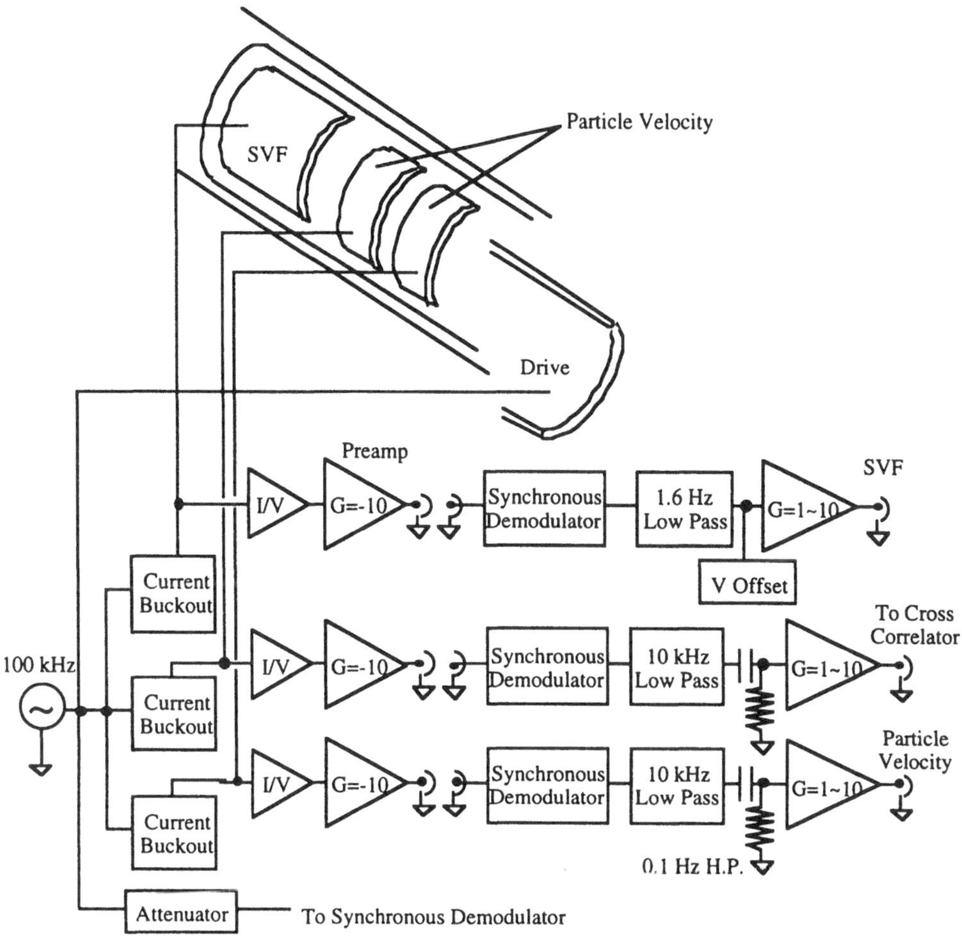

FIGURE 6-31 *Schematic diagram of current-sensing instrumentation of the ANL solid/gas capacitive flowmeter.*

accurately determined. But, for the 4.4 lb/s test, a sharp peak was superimposed on a broad base. The broad base was believed due to an oscillatory delivery phenomenon that developed at the mixing tee for high solids loadings. The particle velocity measured by the cross-correlation technique was compared with the velocity measured by the tracer technique. The results, shown in Fig. 6.35, agreed to within 5% in the velocity range of 21-31 m/s and solids concentration range of 0-1.6 vol.%.

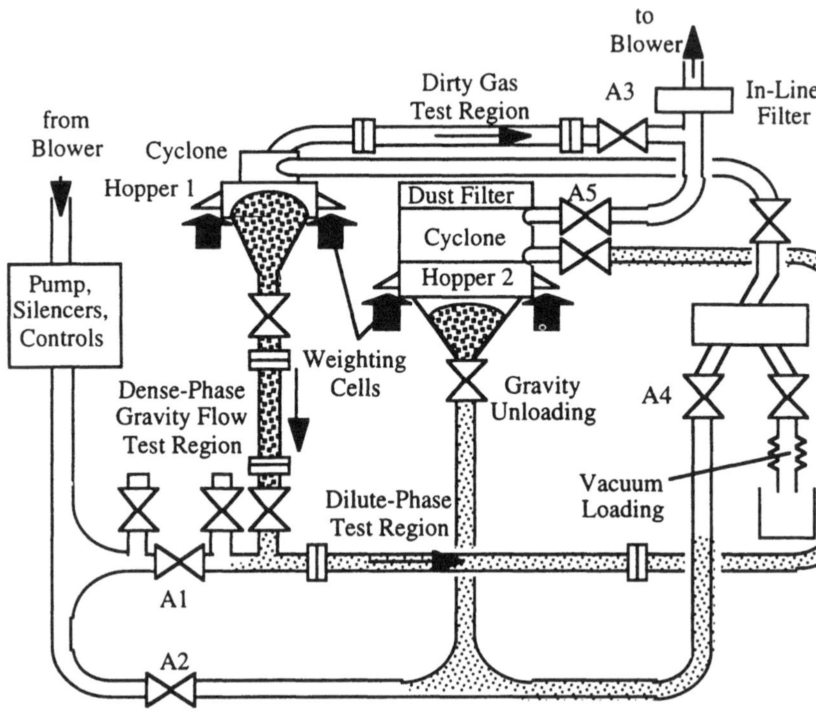

FIGURE 6-32 *ANL Solid/Gas Flow Test Facility.*

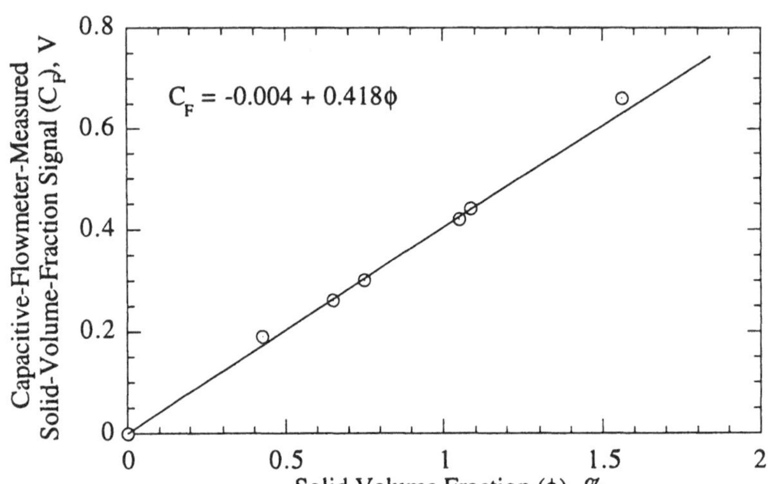

FIGURE 6-33 *Solid-volume-fraction signals obtained by ANL solid/gas capacitive flowmeter.*

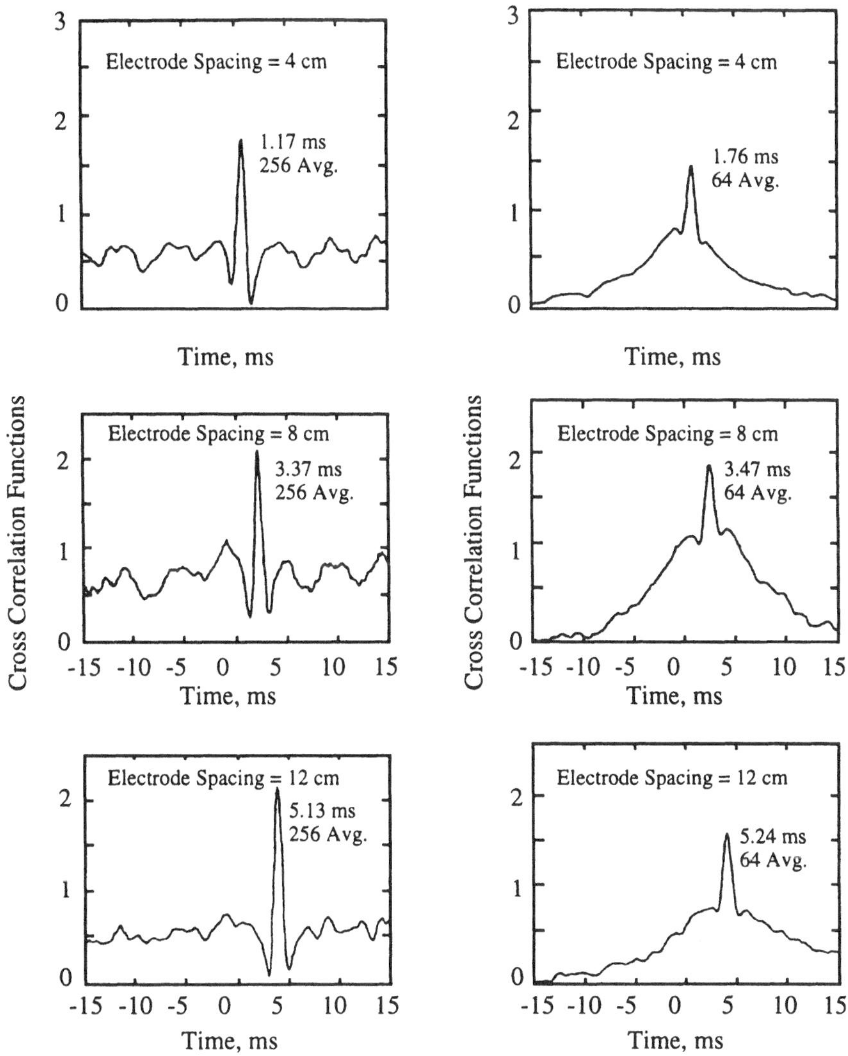

FIGURE 6-34 *Typical cross-correlation functions measured by ANL solid/gas capacitive flowmeter at (left column) $M_S = 3.2$ lb/s and $V_P = 25.1$ m/s and (right column) $M_S = 4.4$ lb/s and $V_P = 24.0$ m/s for three electrode spacings.*

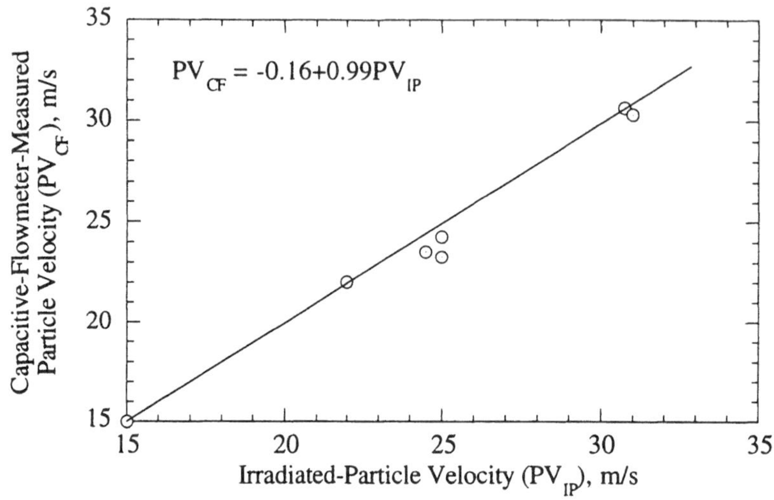

FIGURE 6-35 *Capacitive-flowmeter-measured particle velocity vs. irradiated-particle velocity.*

6.4.2 Radioactive Tracer Technique

The use of irradiated particles as tracers provides an accurate way to measure particle velocity; it allows one to calibrate low-cost flowmeters such as the capacitive flowmeter. Unfortunately, in most industrial environments, it is not possible to produce short-lived radioactive tracers. At ANL, a nuclear research reactor was available for the production of radioactive particles. Ideally, the density and size of the tracer particles and the solids in the flow should be the same so the tracers can be uniformly distributed in the flow and represent the solids velocity distribution. For that purpose, the sample particles to be activated were fabricated from resin, hardener, and indium oxide powder to closely duplicate the size and density of the glass beads used in the flow tests. The particles, after irradiated in the reactor, had a 54-min half-life of γ activity. Twenty particles were injected into the flow stream during the reloading of the feed hopper, becoming randomly mixed with the glass beads.

The passage of each radioactive particle was registered by nine γ detectors (NaI/Tl scintillation detectors), encased in lead shields and placed along the horizontal test section of the S/GFTF. A typical count rate, shown in Fig. 6.36, was in the range of 2×10^5 to 10^4 counts per second.

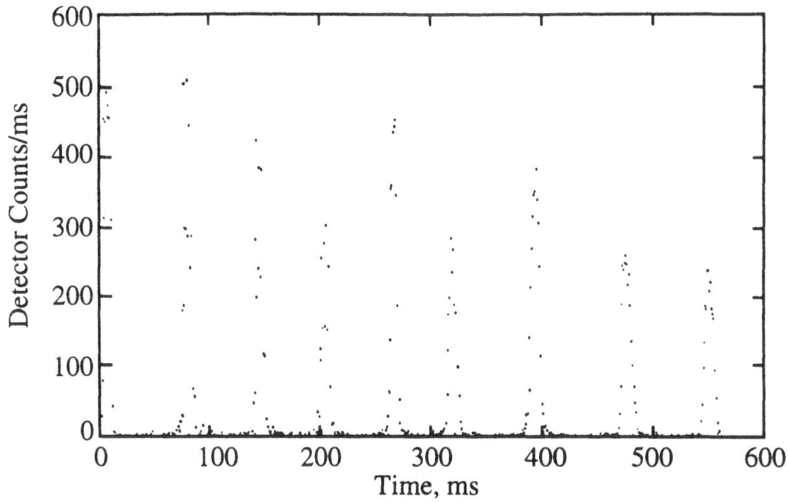

FIGURE 6-36 *Typical irradiated-particle count rate distribution.*

The accuracy of the radioactive tracer velocity measurement was within 1.5%. Detector separation was measured to 0.02%, and the count rate peak locations were within 0.5% of the reading. The averaging of 15 particle velocity measurements reduced the typical velocity dispersion of 5.8% to 1.5%.

6.5 Future Flow Instruments

In our discussion of electromagnetic techniques, we omitted a few available technologies that provide some unique capabilities and, with further development, can attain practical application. One such technique involves the use of a microwave resonance sensor (Kobyashi and Miyahara, 1984) that uses a microwave cavity to measure solids concentration and velocity by monitoring the resonance frequency shift. However, this technique suffers from some shortcomings: the frequency shift may be positive or negative, depending on the dielectric properties of the solids, and the cavity is extremely sensitive to changes in moisture content and temperature.

Another technology involves the use of a flow instrument that is based on magnetic resonance sensors (King et al., 1982). In the presence of an electromagnetic field of suitable frequency, the flow medium that possesses a net magnetic moment either due to atomic nuclei (i.e., nuclear magnetic resonance, NMR) or electrons (i.e., electron magnetic resonance, EMR) will absorb energy from the field and produce a response that is proportional to the number of appropriate nuclei or unpaired electrons per unit volume. The

NMR and EMR techniques provide measurements that are used to determine the pertinent flow parameters, including solids concentration, velocity, and moisture content.

The trend of future flow instruments for mixed-phase flows is toward tomographic flow imaging. In a recent issue of Measurement Science and Technology (Vol. 7, 1996), which was dedicated to process tomography, state-of-the-art developments of various technologies were presented. Tomographic imaging was first developed for X-ray sensing methods, primarily for medical applications. Aside from high cost and safety constraints, radiation-based methods often require long exposure times, which make real-time dynamic measurements of flow difficult. Because of advances in computer technology over the past decade, various imaging techniques have emerged. Tomographic techniques that involve ultrasound, capacitance, electrical impedance, and microwaves, are available for process monitoring. Among the technologies, capacitance tomography has attracted more interest because it is relatively inexpensive and more practical even though it is poor in spatial resolution (Beck and Williams, 1996). A capacitance tomography system for flow imaging consists of several electrodes mounted around the flow pipe, a data acquisition unit that measures all of the capacitances between electrodes, and an image reconstruction unit that basically requires a suitable reconstruction algorithm. The spatial resolution of a capacitance imaging system is determined by the number of electrodes used. For n electrodes, there are $\frac{1}{2}n(n-1)$ independent measurements, and each measurement gives the integrated capacitance within the volume intersected by the electrical field between the two electrodes. In general, the image reconstruction algorithm for capacitance tomography is more complicated than the well-established medical imaging techniques because of the nonlinear nature of the capacitance system. Further development is needed.

6.6 NOTATION

Electromagnetic Methods:

a	=	Pipe diameter
B	=	Magnetic flux density
F	=	Magnetic scalar potential, $B = \nabla F$
G	=	Current scalar potential, $J = \nabla G$
J	=	Virtual current
S	=	Flowmeter sensitivity
V	=	Vector flow velocity

U	=	Electrical potential
W	=	Weight vector, W = B x J
ξ	=	Sensing volume

Capacitive Methods:

A	=	Overlap area of capacitive electrode plates
a	=	Active length of the capacitive electrode
C, C_m, C_f	=	Capacitances, m: mixed phase, f: fluid phase
D, d	=	Separation of capacitive electrodes
F_M	=	Mass flow rate of solids
K	=	A constant proportional to the frequency spectrum
V_S	=	Solid particle velocity
ε, ε_0, ε_f, ε_s	=	dielectric constants, 0: vacuum, f: fluid, s: solid
ρ_S	=	Solid density
ϕ	=	Solid volume fraction
τ	=	Transit time
ω	=	Radial frequency

Optical/Tracer Methods:

c	=	Speed of light
d	=	Particle size
f	=	Frequency
Δf	=	Difference frequency
I, I_0	=	Intensities of incident and transmitted light beam
V_S	=	Particle velocity
x	=	Optical path length
σ	=	Attenuation coefficient
λ	=	Wavelength

Coriolis Method:

$\vec{F}_c$	=	Coriolis force
L	=	Tube length
$\dot{m}$	=	Mass flow rate per unit tube length
M	=	Mass of fluid per unit length
K_S	=	Stiffness constant of the tube
Δt	=	Time difference between the midpoint crossing times in two limbs
α	=	Tube deflection

Γ = Oscillating moment
$\vec{\omega}$ = Angular velocity

6.7 REFERENCES

Beck, M. S., Green, R. G., Plaskowski, A. B., and Stott, A. L., "Capacitance measurement applied to a pneumatic conveyor with very low solids loading," Meas. Sci. Technol. 1, pp. 561-564, 1990.

Beck, M. S. and Williams, R. A., "Process tomography: a European innovation and its applications," Meas. Sci. Technol. 7, pp. 215-224, 1996.

Bevir, M. K. "The theory of induced voltage electromagnetic flowmeters," J. Fluid Mech. 43 (3), pp. 577-590, 1970.

Birchenough, A. and Mason, J. S., "Local particle velocity measurement with a laser anemometer in an upward flowing gas-solids suspension," Powder Technol. 14, pp. 139-152, 1976.

Blackwelder, R. F., "Hot-wire and hot-film anemometers," Methods of Experimental Physics 18, Part A, pp. 259-314, 1981.

Bobis, J. P., Porges, K. G. A., Raptis, A. C., Brewer, W. E., and Bernovich, L. T., "Particle velocity and solid volume fraction measurements with a new capacitive flowmeter at the solid/gas flow test facility," Argonne National Laboratory report, ANL/FE-86-4, 1986.

Cox, T. J. and Wyatt, D. G. "An electromagnetic flowmeter with insulated electrodes of large surface area," J. Phys. E: Sci. Instrum. 17, pp. 488-503, 1984.

Hammer, E. A. and Green, R. G., "The spatial filtering effect of capacitance transducer electrodes," J. Phys. E: Sci. Instrum. 16, pp. 438-443, 1983.

Hoske, M. T., "2-wire magmeter creates freedom from 120V," Control Engineering April, 1998.

Huang, S. M., Stott, A. L., Green, R. G., and Beck, M. S., "Electronic transducers for industrial measurement of low value capacitances," J. Phys. E: Sci. Instrum. 21, pp. 242-50, 1988.

Hylton, T. D., Anderson, M. S., Van Essen, D. C., and Bayne, C. K., "Cmparative Testing of Slurry Monitors," Oak Ridge National Laboratory report, ORNL/TM-13587, 1998.

King, J. D., Rollwitz, W. L., and Santos, A. D. L., "Magnetic resonance coal flowmeter and analyzer," Proc. Symp. Instrumentation and Control of Fossil Energy Processes, Argonne National Laboratory report ANL-82-62, pp. 30-40, 1984.

Kobyashi, S. and Miyahara, S., "Development of microwave powder flowmeter," Instrumentation (Japan) 27, pp.68-73, 1984.

O'Sullivan, V. T. "Performance of an electromagnetic flowmeter with six point electrodes," J. Phys. E: Sci. Instrum. 16, pp.1183-1188, 1983.

O'Sullivan, V. T. and Wyatt, D. G., "Computation of electromagnetic flowmeter characteristics from magnetic field data: Part III. rectilinear weight function," J. Phys. D: Appl. Phys. 16, pp. 1461-1476, 1983.

Porges, K. G. A., "Flow characterization and calibration of slurries by pulsed neutron activation," Argonne National Laboratory report ANL-84-16, 1984.

Porges, K. G., Cox, S. A., Brewer, W. E., and Hacker, D. S., "System description of the ANL slurry loop testing facility (SLTF)," Argonne National Laboratory report ANL-84-20, 1984.

Shercliff, J. A. "The Theory of Electromagnetic Flow Measurement," University Press, Cambridge, U.K., 1962.

Somerscales, E. F. C., "Measurement of velocity: tracer methods," Methods of Experimental Physics, editor, R. J. Emrich, 18, Chapter 1.1, 1981.

Sultan, G. and Hemp, J., "Modeling of the coriolis flowmeter," J. Sound Vib. 132(3), pp. 473-489, 1989.

Van De Hulst, H. C., "Light scattering by small particles," John Wiley & Sons, Inc. New York, 1957.

Yan, Y., "Mass flow measurement of bulk solids in pneumatic pipelines," Meas. Sci. Technol. 7, pp.1687-1706, 1996.

7

Single-Point Laser Measurement

Martin Sommerfeld and Cameron Tropea

7.1 INTRODUCTION

The present review is devoted to recent developments of laser-Doppler and phase-Doppler anemometry (i.e. LDA and PDA) and their application to measurements in particulate two-phase flows. Both measurement techniques are optical non-intrusive single point techniques permitting local instantaneous and ensemble-averaged measurements of particle velocities with high spatial resolution. The velocity is inferred from the Doppler shift of the scattered light caused by a particle moving through the measurement volume. This implies, that for accurate measurements, the probability that two or more particles are in the measurement volume should be small. This can be achieved by adjusting the dimensions of the measurement volume according to the maximum particle concentration expected in the considered flow. Therefore, both methods belong to the class of single particle counting instruments.

To allow measurement of particle diameter or equivalent size on an optical basis, a number of methods may be applied which are summarized in Figure 7-1. Most of these methods (upper row) have been applied in combination with a dual beam LDA-system by extending the optical system or using special signal processing methods. The time of flight method for particle sizing is based on the laser two-focus method and has been recently analyzed by Albrecht et al. (1993). The frequency method is based on a special laser interferometer using two cylindrical waves of incident light (Naqwi et al. 1991). For such an optical configuration, the frequency of the scattered light is composed of two components, the conventional Doppler frequency and an "anisotropic frequency" which is directly dependent on particle size and refractive index. The problem with this approach is however, that the dimensions of the measurement volume are difficult to define.

A method which has received considerable attention over the past 15 years is phase-Doppler anemometry (PDA) which allows determination of the size of spherical particles, droplets or bubbles. Hence, local particle size distributions and size-velocity correlations can be obtained. Moreover, recent developments provide the basis for accurate particle concentration or mass flux measurements and for the estimation of the refractive index, in order to distinguish particles with different optical properties.

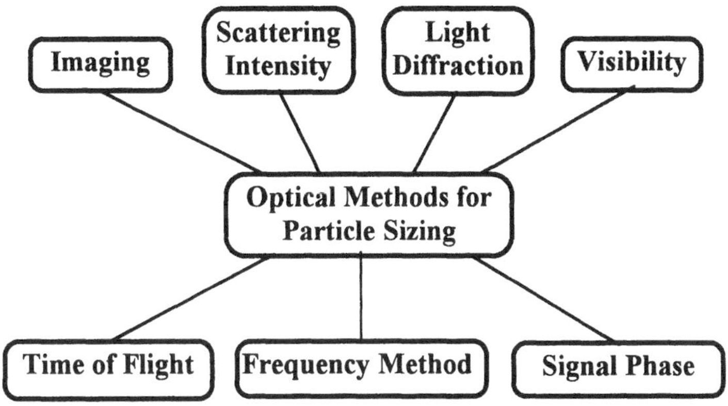

Figure 7-1 Summary of optical methods for particle sizing based on elastic light scattering

Increasing interest in detailed experimental analysis of two-phase flows has led to a number of review papers on single point laser measurements. A more general overview on two-phase flow measurements was given for example by Taylor (1994) focusing on current activities in PDA development and PIV applications in two-phase flows. Applications of LDA and PDA for analyzing flows with combustion were reviewed by Heitor et al. (1993), and a more industrial orientated review on particle sizing methods can be found in Black et al. (1996).

The present review summarizes developments in LDA, PDA, and signal processing relevant for both measurement techniques. Besides describing recent developments, also the basic principles of both methods will be introduced and guidelines for selecting an optimal optical system will be given, especially for PDA measurements in two-phase flows. The article is organized in the following way. After an introduction to the principles of LDA, applications for velocity measurements in two-phase flows and developments in applying LDA to particle size and concentration measurements over the past 20 years will be briefly reviewed. This section includes also recent developments on extended optical systems which allow sizing of non-spherical particles using extended LDA optical systems. The second part is related to PDA. First the principles of PDA will be presented, followed by a section on the optimum layout of PDA for certain applications. This section demonstrates the use of theoretical and numerical tools, such as geometrical optics, Mie-calculations, and GLMT (generalized Lorenz-Mie theory) for designing PDA systems. One of the most important features of PDA, namely the possibility of accurate particle concentration or mass flux measurements, will be considered in a separate section and recent developments will be introduced. For demonstrating the potential of PDA, some recent developments on improved and very specific optical systems will be considered. The last section of this review will be devoted to signal processing and recent developments. An accurate and fast signal processing is essential for reliable measurements and equally important as the appropriate design and application of the optical system.

7.2 LASER-DOPPLER ANEMOMETRY

The dual beam configuration of LDA is most widely used today, where the Doppler difference frequency is directly measured and the receiving optics may be placed at an arbitrary position with respect to the transmitting beams. Laser-Doppler anemometry has been first applied to measurements of mean velocities and turbulence properties in single phase flows. In this case small particles, which follow the flow and the turbulent fluctuations, need to be present in the flow or must be added to it (i.e. seeding the flow with a tracer). The principles of LDA are, for example, described in detail by Durrani and Greated (1977), Durst et al. (1981), and Durst et al. (1987).

7.2.1 Principles of LDA for Two-Phase Flows

The basic ideas for applying LDA to measurements in two-phase flows were put forward, for example, by Farmer (1972, 1974), Durst and Zaré (1975) and Roberts (1977). Durst and Zaré (1975) showed that LDA may be applied to velocity measurements of large reflecting and refracting particles. The light waves produced by the two incident laser beams reflect or refract at large particles, interfere, and produce fringes in space which move across the detector at the Doppler difference frequency. The theoretical derivations of Durst and Zaré (1975) revealed that the relations for the Doppler difference frequency for large reflecting or refracting particles are identical with the universal equation of Laser-Doppler anemometry valid for seeding particles when the intersection angle of the two incident beams is small and the photodetector is placed at a large distance from the measurement volume. The analysis is simplified by considering spherical particles. For a non-deformable, reflecting particle, the rate at which the fringes cross any point in space, i.e. at the photo detector, is the same at all points in the surrounding space and is given by (Figure 7- 2 a)):

$$f_D = \frac{2\left(U_{\perp}\cos\beta \pm U_{II}\sin\beta\right)\sin\theta}{\lambda} \tag{7.1}$$

The angle β is a function of L/R and θ. Since L/R is usually large and θ is small it follows that also that β is small and Equation 7.1 yields the universal equation of laser Doppler anemometry:

$$f_D = \frac{2\,U_\perp \sin\theta}{\lambda} \tag{7.2}$$

When the incident light beams are transmitted through a transparent particle they are refracted twice (Figure 7- 2 b)) and the Doppler difference frequency detected in the forward direction for the simple case of a non-rotating, spherical particle is given by (Durst and Zaré 1975):

$$f_D = \frac{2\,U_\perp (\sin\theta - \sin\beta)}{\lambda} \tag{7.3}$$

Again, for large values of L/R one obtains the classical equation of LDA.

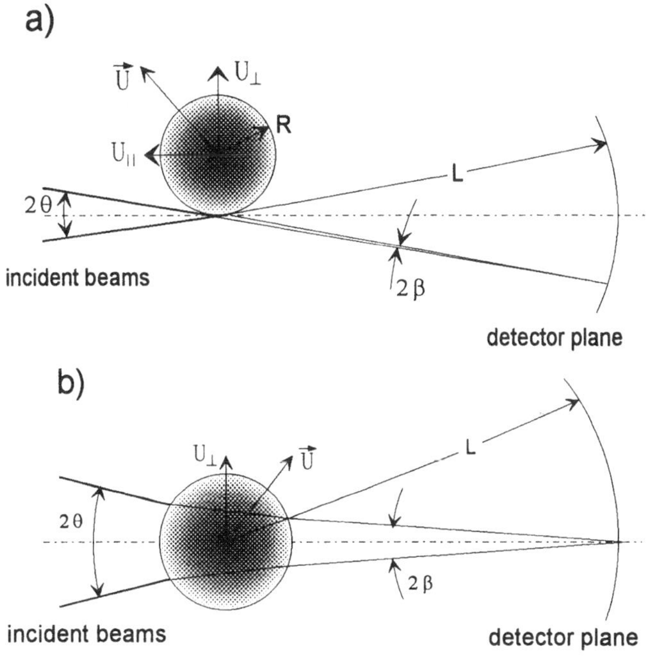

Figure 7-2 Interference of the two laser beams for large (a) reflecting and (b) refracting particles

The findings described above are the basis for the application of LDA for particle velocity measurements in two-phase flows. A pre-requisite for successful applications of LDA for velocity measurements in two-phase flows is an unhindered optical access, putting a constraint on the permissible volume concentration and/or penetration depth. Nevertheless, numerous studies have

been published in the past where LDA was applied to velocity measurements in various types of gas-solid two-phase flows, liquid sprays, and bubbly flows.
There have also been several attempts to apply Laser-Doppler anemometry to the simultaneous measurement of particle velocity, size and concentration (e.g. Farmer 1972, Chigier et al. 1979, Durst 1982, Negus and Drain 1982, Hess 1984, Hess and Espinola 1984, Allano et al. 1984, Grehan and Gouesbet 1986, Maeda et al. 1988). The sizing of particles using LDA may be based on two methods (Figure 7-1):

- the absolute value of the scattering intensity (i.e. pedestal of the Doppler signal),
- or the signal visibility.

The pedestal of the Doppler signal is obtained by applying a low-pass filter unit to the photodetector signal. As shown in Figure 7-3, the intensity of the scattered light depends in a characteristic way on particle size. From this relation three scattering regimes may be identified, where also the Mie-parameter, defined as $\alpha = \pi D_p/\lambda$, is used to non-dimensionalize the particle size.

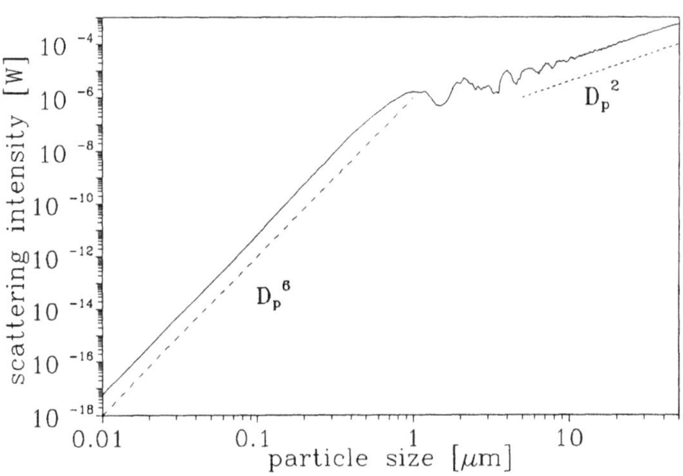

Figure 7-3 Dependence of scattering intensity on particle size obtained from a Mie-calculation (intensity of incident beam: $I_0 = 1.0 \cdot 10^7\ W/m^2$, wavelength: $\lambda = 632.8$ nm, scattering angle: $\varphi = 15°$, aperture angle of receiving optics: $\Delta\delta = 10°$, refractive index of particle: $n = 1.5$)

- The so-called Rayleigh-scattering applies for particles that are small compared with the wavelength of the incident light, i.e. $\alpha << 1$ or $D_p < \lambda/10$. This regime is named after Lord Rayleigh, who first derived the basic scattering theory for such small particles. Characteristic for this regime is a variation of the scattering intensity with the fourth to sixth power of the particle diameter, depending on the particle properties (i.e. refractive index) and the angle of observation.

- For very large particles, i.e. $\alpha >> 1$ or $D_p > 4 \cdot \lambda$, the laws of geometrical optics (also called the Fraunhofer regime) are applicable (van de Hulst 1981). The light scattering intensity varies approximately with the square of the particle diameter.
- The intermediate regime (i.e. $D_p \approx \lambda$) is called the Mie-region (Mie 1908) and is characterized by large oscillations in the scattering intensity, depending on the particle properties, the observation angle, and the receiving aperture. Hence, the scattering intensity cannot be uniquely related to the particle size.

Since the scattering intensity depends additionally on refractive index and particle shape, particle sizing based on intensity measurements generally requires calibration.

For large particles, i.e. $D_p >> \lambda$, the geometrical optics interpretation of the scattered light leads to three components, namely diffracted, externally reflected and internally refracted light as indicated in Figure 7-4. The refracted light may be separated in several modes depending on the number of internal reflections, i.e. P1, P2, P3, ... Pn. Light diffraction is concentrated in the forward scattering direction, i.e. the so-called forward lobe, and is the dominant scattering phenomenon. Therefore, the regime of geometrical optics is also called Fraunhofer diffraction regime. The diffraction pattern and the angular range of this forward lobe is dependent on the wavelength of the light and the particle diameter, more specifically on the Mie-parameter. The angular extent of the first lobe of diffracted light decreases with increasing particle diameter and is given by $\varphi < \pm\varphi_{diff}$ with:

$$\sin\varphi_{diff} = \frac{5}{4}\frac{\lambda}{D_p} \tag{7.4}$$

Moreover, it is important to note that the intensity of the diffracted light is independent of the optical constants of the particle material, which is an advantage in sizing particles of different or unknown refractive index.

Externally reflected light is scattered over the entire angular range, i.e. $0° < \varphi < 180°$ (see Figure 7-20), whereas refracted light of the first order (i.e. P1) does not exceed an upper angular limit φ_{refr} which is given for $n_p/n_m > 1.0$ by geometrical optics:

$$\cos\left(\frac{\varphi_{refr}}{2}\right) = \frac{n_m}{n_p} \tag{7.5}$$

Hence, this upper angular limit is determined by the relative refractive index. For a given fluid and $n_p/n_m > 1.0$, φ_{refr} decreases and more refracted light is concentrated in the forward direction with decreasing refractive index of the particle (see also Figure 7-20). A similar relation can be derived for $n_p/n_m < 1.0$, e.g. for bubbles in a liquid.

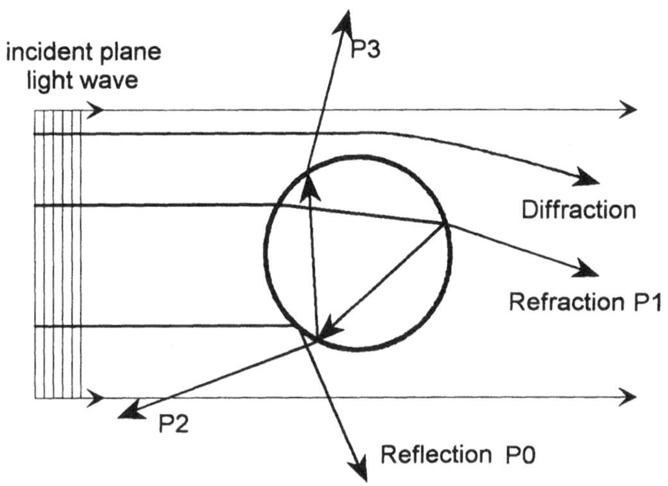

Figure 7-4 Different scattering modes for a spherical particle in the geometrical optics regime

An additional problem for sizing particles by a standard LDA-system is the effect of the non-uniform distribution of intensity within the measurement volume, being the volume of the beam intersection. Laser beams normally have a Gaussian intensity distribution. This results for the same particle in lower scattering intensities when they pass the outer rim of the measurement volume and hence they are detected as smaller particles. An additional consequence of this effect, which is also called trajectory dependent scattering, is that the effective measurement volume size is dependent on particle size (Figure 7-5). A small particle passing through the edge of the measurement volume may not be detected by the data acquisition due to its low scattering intensity, whereas a large particle at the same location still produces a signal which lies above the detection level. Hence, the probability of detecting large particles is higher than for small particles, potentially leading to biased statistical measurements. This effect also has consequences for the determination of the particle concentration, which will be discussed below. Therefore, measurements of particle size and concentration by LDA requires extensions of both the optical system and the data acquisition in order to reduce errors due to the Gaussian beam effect.

The following techniques have been introduced to reduce particle sizing errors when using signal amplitude methods:

- Limitation of the measurement volume size by additional optical systems (i.e. gate photodetector (Chigier et al. 1979) or two-color systems with two measurement volumes of different diameter (Yeoman et al. 1982)).
- Modification of the laser beam to produce a "top-hat" intensity distribution (Grehan and Gouesbet 1986).
- Computational deconvolution of signal intensity distributions (Chigier et al. 1979).

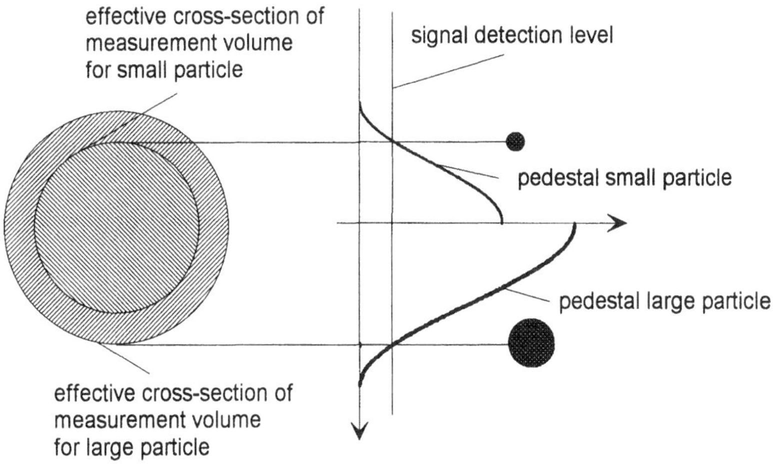

Figure 7-5 Illustration of the Gaussian beam effect on intensity measurements by LDA and its effect on the effective cross-section of the measurement volume.

In the following section some examples of particle size measurements using LDA are given which have been mostly developed some time ago, before the PDA-technique was extensively used for sizing spherical particles. Most of the techniques described below, which are based on intensity and visibility measurements, are inferior to the PDA in the case of spherical particles. There is, however, still a potential for reliable instruments for local size and velocity measurements in two-phase flows with non-spherical particles found in many industrial processes. Some examples of novel LDA systems for sizing non-spherical particles will also be presented.

7.2.2 Special LDA-Systems for Two-Phase Flow Studies

In order to limit the region of the detection volume, Chigier et al. (1979) used an additional receiving optics which was placed at 90° off-axis and was used to trigger the main receiving system mounted in the forward scattering direction. For a further reduction of the trajectory ambiguity an inversion routine to convolute the signal amplitude distributions obtained from many particles was used, by applying an equation relating the signal peak amplitude to both the particle diameter and the particle location in the measurement volume. A comparison of particle size distribution measurements by LDA with results obtained by the slide impaction method gave only fair agreement (Chigier et al. 1979).

By superimposing two measurement volumes of different diameter and color, it is possible to trigger the data acquisition only when the particles pass through the central part of the larger measurement volume, where the intensity is more uniform. Such a coaxial arrangement of two measurement volumes may be realized by using a two-component LDA-system with different waist diameter for the two colors (Yeoman et al. 1982, Modarres and Tan 1983) or by overlapping a large diameter single beam with the LDA measurement volume (Hess 1984). When a particle passes through the LDA measurement volume, the light scattering intensity from the larger diameter single beam is measured to determine the particle size (Figure 7-6). Also a combination of LDA with an independent white light scattering instrument has been used for simultaneous particle size and velocity measurements by Durst (1982).

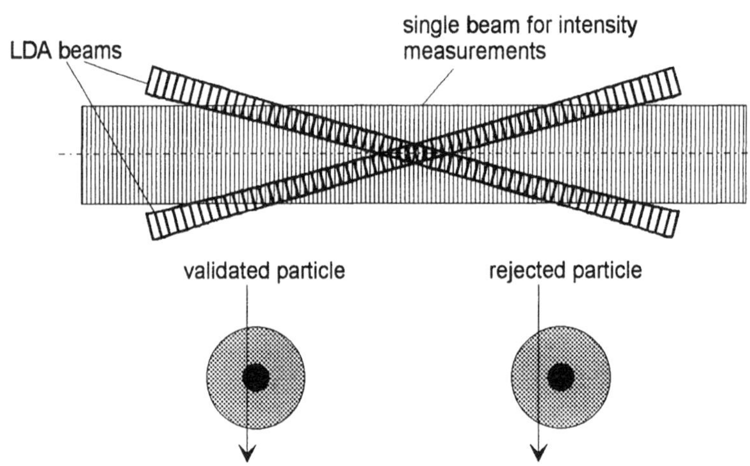

Figure 7-6 Coaxial arrangement of two measurement volumes of different color

For producing laser beams with uniform intensity distribution the so-called top-hat technique may be applied. In order to produce such a top-hat profile, Allano et al. (1984) used a holographic filter and subsequently related the measured scattering intensity to the particle diameter using the Lorenz-Mie theory. Similar to the configuration shown in Figure 7-6, a large sizing beam with top-hat intensity profile and a small LDA measurement volume were used for simultaneous size and velocity measurements. Grehan and Gouesbet (1986) tested this system for simultaneous measurements of droplet size and velocity in mono-dispersed sprays. A four beam, two-color LDA-system was used by

Maeda et al. (1988) to produce two concentric measurement volumes of different color and size. Using a system of pinholes and lenses the larger measurement volume had a top-hat intensity distribution for particle sizing.

Alternative to using signal amplitude, the signal visibility or signal modulation depth may be used for particle sizing (Farmer 1972). Compared to scattering intensity measurements this method has a number of advantages, since visibility does not depend on scattering intensity and hence, is not influenced by laser power and detector sensitivity. The visibility is determined from the maximum and minimum amplitudes of the Doppler signal as indicated in Figure 7-7.

$$V = \frac{I_{max} - I_{min}}{I_{max} + I_{min}} \tag{7.6}$$

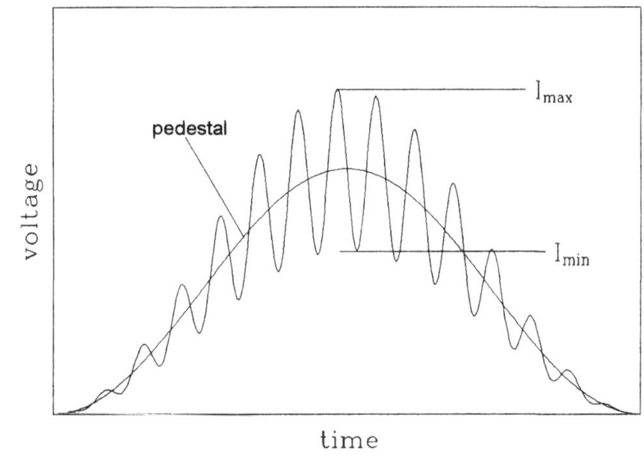

Figure 7-7 Doppler signal and definition of signal visibility or modulation depth

The visibility of the Doppler signal decreases with increasing particle size as illustrated in Figure 7-8. The first lobe in the visibility curve covers the measurable particle size range. With increasing particle size, secondary maxima appear in the visibility curve (Figure 7-8). The visibility curve depends strongly on the optical configuration of the receiving optics, i.e. the off-axis angle and the size and shape of the imaging mask in the receiving optics. The latter effect was evaluated in detail by Negus and Drain (1982). As an example, Mie calculations of the visibility curves for different optical configurations are shown in Figure 7-9. It is obvious, that in direct forward scatter the sizing range is very limited and that the measurable particle size range is considerably influenced by the shape of the imaging mask. Using an off-axis arrangement of the receiving optics the measurable size range can be considerably increased (Figure 7-8).

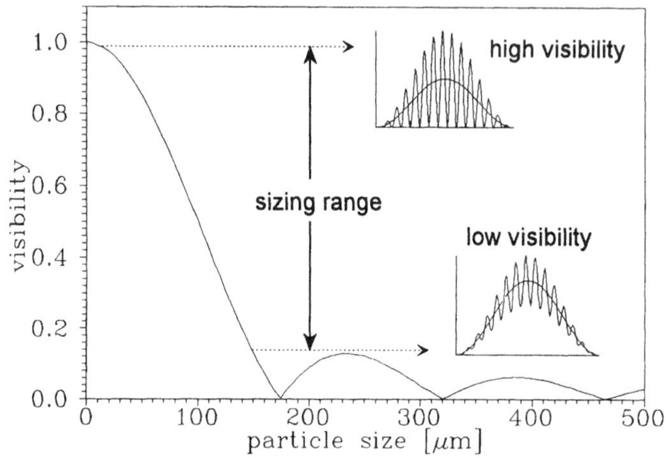

Figure 7-8 Variation of signal visibility with particle size (Mie calculation for an off-axis light collection, $\varphi = 15°$ ($\lambda = 632.8$ nm; $d_f = 6.55$ μm, circular mask, receiving cone angle $\delta = 4°$)

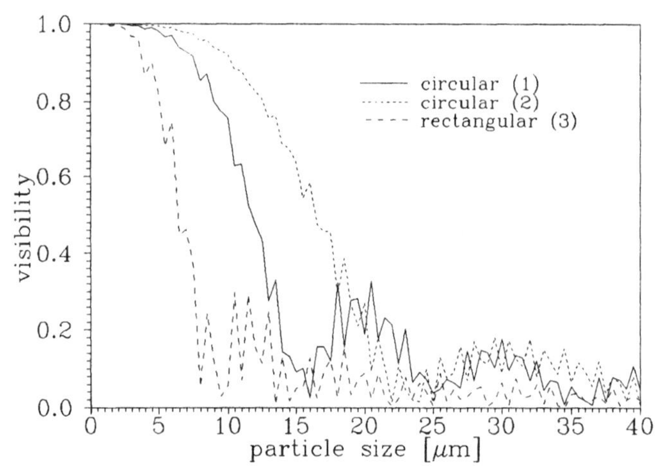

Figure 7-9 Mie calculations of visibility curves for different optical configurations of the receiving optics in direct forward scatter ($\lambda = 632.8$ nm; **1**: fringe spacing $d_f = 10.2$ μm, circular mask, receiving cone angle $\delta = 4°$; **2**: $d_f = 18.0$ μm, circular mask, $\delta = 4°$; **3**: $d_f = 6.55$ μm, rectangular mask, receiving aperture angle in horizontal and vertical direction, $\delta_h = 11°$, $\delta_v = 4°$)

Extensive research has been performed on the suitability of the visibility method for particle sizing. It was found that this method seems to be very sensitive with regard to a careful positioning of the aperture mask, accurate dimensions of the mask and the particle trajectory through the LDA measurement volume. The last effect may be minimized by using a two-color measurement volume with an appropriate validation scheme to insure that only particles passing the center of the measurement volume are validated, as for example suggested by Yeoman et al. (1982). A detailed review about visibility methods was given by Tayali and Bates (1990), where also a number of other LDA-based particle sizing methods are described which are not considered here.

Since many industrial and technical processes involve two-phase flows with non-spherical particles, such as coal combustion or powder production, recently several attempts have been made to extend LDA to such applications. The use of light scattering intensity for sizing non-spherical particles has limitations with regard to the location of the receiving optics. In side scatter, where the scattered light is composed of reflection and refraction, the scattering intensity is strongly affected by particle shape, orientation and surface quality. Therefore, sizing non-spherical particles only seems to be possible in near-forward scatter where diffracted light is dominant. The intensity of diffractively scattered light is related to the projected area of the particle but insensitive to particle shape and refractive index.

An extended LDA for sizing non-spherical particles based on diffracted light was recently developed and tested by Morikita et al. (1994). The optical system is based on a two-color, three-beam system and the use of an Argon-Ion laser. Two beams with a wavelength of λ = 480 nm are used to produce the LDA measurement volume and the third beam with λ = 514.5 nm is directed along the bi-sector of the LDA beams to form two concentric measurement volumes of different diameter (Figure 7-10). In order to reduce the trajectory ambiguity due to the Gaussian intensity distribution in the sizing beam (green beam), only light scattering from the central region of the measurement volume is collected, by using the LDA measurement volume as a detection volume (Figure 7-10). Hence, a signal on the sizing channel is only accepted when at the same time a signal is present on the LDA channel. The diameter of the sizing beam is 375 μm and that of the LDA measurement volume 100 μm.

The receiving optics is positioned in the forward scatter direction on the bi-sector of the LDA system as illustrated in Figure 7-10. The receiving lens collects the scattered light where the central portion (i.e. the incident beam and the central part of the diffraction lobe) is blocked using a circular mask. Hence scattered light is only collected in an annular region around the central lobe of the diffraction pattern. Behind the receiving lens a beam splitter and two color filters are introduced. Then the scattered light from the LDA measurement volume (blue) and the sizing measurement volume (green) is focused by two lenses onto two photodiodes. In order to limit the length of the measurement volume and to allow rejection of de-focused particles, a spatial filter is introduced in the receiving optics. Furthermore, the effective length of the sizing beam is limited through the coincidence constraint (i.e. the simultaneous presence of signals on both channels) by the length of the LDA measurement volume, which is 1 mm.

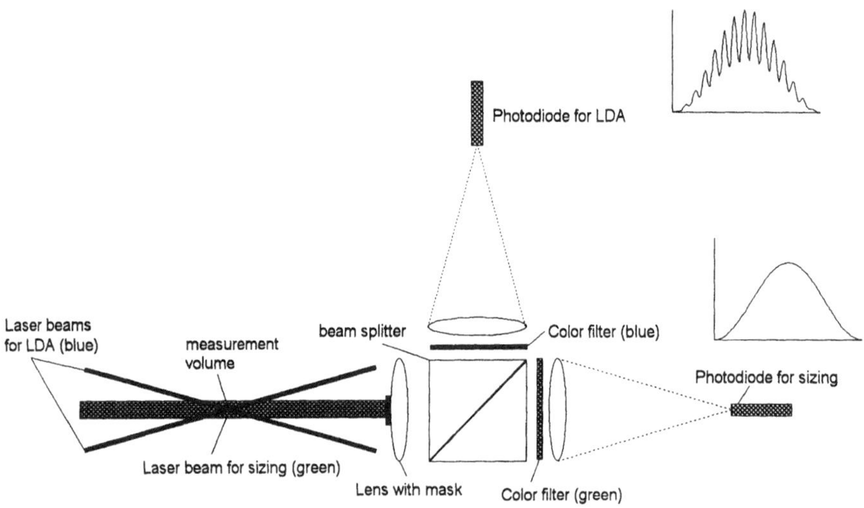

Figure 7-10 Optical arrangement of receiving optics for the extended LDA using diffracted light for sizing as proposed by Morikita et al. (1994)

The information provided by the intensity of the diffractively scattered light is an equivalent diameter D_e obtained from the projected area of the particle S_p:

$$D_e = \sqrt{\frac{4\,S_p}{\pi}} \tag{7.7}$$

The relation between equivalent diameter and scattering intensity can be determined from Fraunhofer diffraction theory. The intensity at a point on the receiving plane diffracted by a circular aperture is determined from (Hecht and Zajac 1982):

$$I(x) = \frac{E^2\,A^2}{2\,R^2}\left[\frac{2\,J_1(x)}{x}\right]^2 \tag{7.8}$$

where E is the beam intensity per unit area, A is the area of the scattering aperture, R is the distance from the center of the aperture to a point on the receiver plane, and J_1 is the first-order spherical Bessel function. The particle size parameter x is obtained from:

$$x = \frac{k\,a\,r}{R} \tag{7.9}$$

with the wave number k, the radius of the aperture a, and the radius r from the center of the receiver plane to any point on the receiver plane. Integrating over the receiving aperture from r_{min} to r_{max} (i.e. the angular region over which the light is collected for sizing) one obtains the total intensity collected by the receiver. The solution of the above equations yields response curves, i.e. intensity versus particle equivalent diameter, for different geometries of the receiver aperture, according to which the system can be optimized. The response curves for different minimum collection angles are shown in Figure 7-11 for a given outer diameter of the receiving aperture of 50 mm and a collection lens with a focal length of f = 300 mm. It is obvious that larger minimum collection angles yield a better linearity of the response curve. Hence, Morikita et al. (1994) selected an angle of 1.43 degree for their optical system.

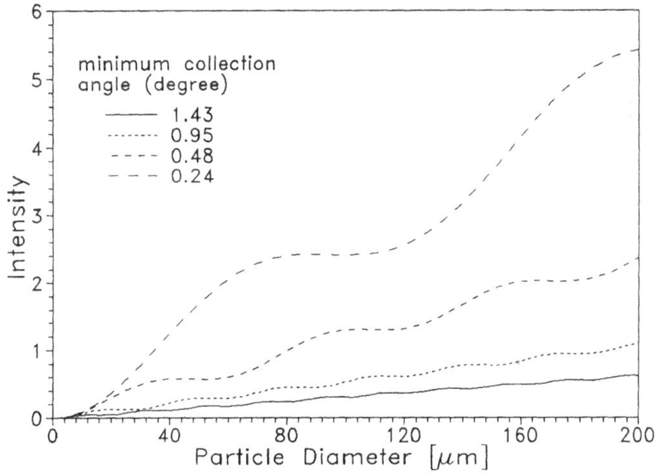

Figure 7-11 Calculated response curves for the intensity of diffracted light based on Fraunhofer approximation for different minimum collection angles (maximum collection angle: 4.76°, laser wave length: 514.5 nm)

For demonstrating the performance of the particle sizing instrument Morikita et al. (1994) performed measurements for various kinds of spherical, non-spherical, transparent, and opaque particles, such as polyethylene particles, glass beads, copper and stainless steel particles, aluminum oxide and morundum particles. The sizing was performed based on a calibration curve obtained by using precision pinholes of different and known size. A comparison of the size measurements with a microscope analysis showed reasonable agreement for the different particles considered. An example of the results is shown in Figure 7-12 for different particles.

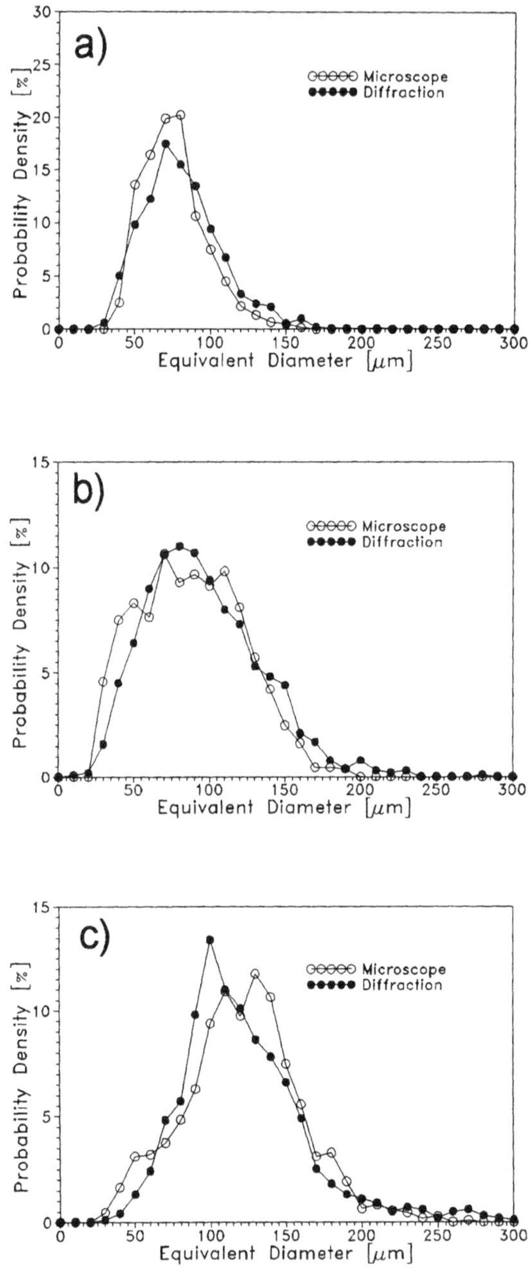

Figure 7-12 Distribution of the equivalent size of spherical and non-spherical particles obtained by laser diffraction measurements and a microscope, a) polystyrene spheres, b) copper particle, c) aluminum oxide

The authors concluded from their results that the size measurement was not very sensitive with respect to particle orientation in the measurement volume, due to the concentric receiving aperture. Also for transparent particles, where refracted light will be also collected, reasonably accurate measurements could be obtained. One should however keep in mind that intensity measurements are very sensitive to variations in laser power, photodetector sensitivity, and contamination of windows. Moreover, variations of the particle concentration within a cross-section of a flow field will result in variations of light absorption for the incident and scattered light, depending on the measurement location (see for example Kliafas et al. (1990) for a detailed analysis of the turbidity). Therefore, the application of this method is again limited to very dilute two-phase flows.

Another recently developed optical technique for sizing non-spherical opaque particles, the so-called shadow-Doppler technique, combines LDA with a direct imaging of the particle (Hardalupas et al. 1994). The optical system consists of a standard LDA transmitting optics, a receiving optics for velocity measurements, and an additional receiving optics which creates a magnified image of the projected area of the particle onto a linear photo-diode array. The collection optics of the shadow-Doppler velocimeter is illustrated in Figure 7-13. It consists of a pair of receiving lenses, a x10 microscope objective and a horizontally-oriented, 35-element, linear photo-diode array (i.e. perpendicular to the plane of the incident LDA beams). The overall magnification of the receiving optics is 200 so that particles between 30 and 140 μm can be measured.

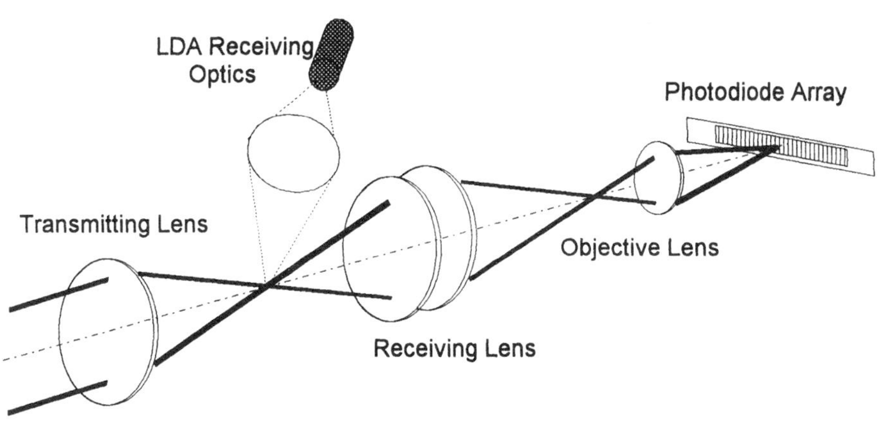

Figure 7-13 Schematics of the receiving optics of the shadow-Doppler velocimeter

As a particle passes through the measurement volume, the magnified image sweeps across the detector in the direction of particle motion. Hence, the output signal of those elements of the linear photo-diode array exposed to the shadow vary in time (Figure 7-14). The width of the particle shadow as a function of time is obtained from the linear dimension of those pixels exposed to the image and the magnification factor. The linear dimensions of the particle in the vertical direction (i.e. perpendicular to the linear array) can be obtained by repeatedly reading the linear array in quick succession and relating the elapsed time to the vertical coordinate through the particle velocity measured using the LDA. Applying an amplitude normalization and a thresholding procedure, the images of the particles can be reconstructed. Due to the Gaussian intensity distribution in the measurement volume the threshold level has to be calibrated in order to get the correct particle size (Hardalupas et al. 1994).

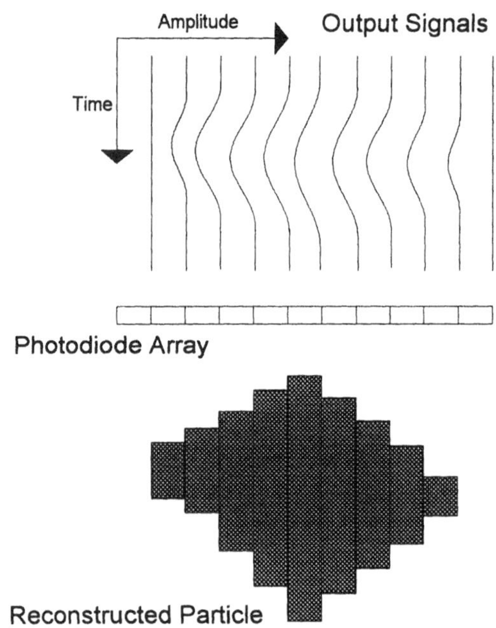

Figure 7-14 Output of photo-diode array and particle image reconstruction

Sizing errors may result for particles moving through the measurement volume outside the central region. This error depends on the depth of field of the receiving system and is illustrated in Figure 7-15. For particles with trajectory A the images of both laser beams fall together and create a dark circular shadow. Particles moving slightly out-of-focus create double images which overlap for some portion, depending on the out-of-focus distance (trajectory B). Hence, the resulting signal on the pixels of the photo-detector array has three levels. Finally, for completely out-of-focus particles, two separate circular shadows are created and the output signal has two separate but lower peaks (trajectory C).

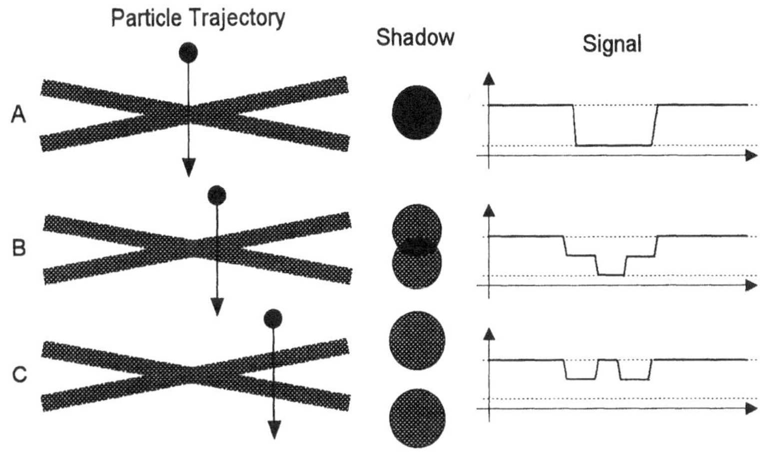

Figure 7-15 Effect of particle trajectory on particle shadow and signal

For ensuring that only in-focus particles are considered for size measurement, the upper and lower threshold levels have to be set appropriately. Extensive studies on the required threshold level have been performed by Hardalupas et al. (1994). With appropriate settings the sizing error can be reduced to about 10%. The shadow-Doppler technique is presently being developed further. Recently, first attempts to measure the particle mass flux were also made (Maeda et al. 1996 a)).

As a result of the particle size-dependent dimensions of the measurement volume and the difficulties associated with particle size measurements using LDA, the measurement of particle concentration is generally based on a calibration procedure using information about the global mass balance. In principle, however, accurate particle concentration measurements using LDA are only possible for simple one-dimensional flows with mono-sized particles. In that case, the measurement volume size may be determined by calibration. This however does not remove the problem related to the spatial distribution of particles in the flow and the associated turbidity effect (Kliafas et al. 1990). This effect causes a dependence of the scattering intensity received by the photodetector on the measurement location as a result of the different optical path lengths through the particle-laden flow and the associated different rates of light absorption.

For a simultaneous determination of fluid and particle velocity by LDA the fluid flow has to be additionally seeded by small tracer particles which are able to follow the turbulent fluctuations. The remaining task is the separation of the Doppler signals resulting from tracer particles and the dispersed phase particles. In most cases this discrimination is based on the scattering intensity combined with some other method in order to reduce the error due to the Gaussian beam effect. The discrimination procedure introduced by Durst (1982) for example, was based on the use of two receiving optical systems and two photodiodes

which detect the blockage of the incident beams by large particles. Together with a sophisticated signal processing it was possible to successfully separate signals produced by large particles and tracers.
An improved amplitude discrimination procedure using two superimposed measurement volumes of different size and color was developed by Modarres and Tan (1983). The smaller or detection measurement volume was only used to trigger the measurements by the larger control volume. Thereby, it was ensured that the sampled signals were only received from the central part of the larger measurement volume, where the spatial intensity distribution does not exhibit strong variations.
A combined amplitude-visibility discrimination method which did not rely on additional optical components was proposed by Börner et al. (1986). After first separating the signals based on the signal amplitude, the visibility of all signals was determined to ensure that no samples from large particles passing the edge of the measurement volume were collected as tracer particles. This method required additional electronic equipment and a sophisticated software signal processing.
A much simpler amplitude discrimination method was introduced by Hishida and Maeda (1990). In order to ensure that only particles traversing the center of the measurement volume are sampled, a minimum number of zero crossings in the Doppler signal was required for validation.
All the above described discrimination procedures can be successfully applied only when the size distribution of the dispersed phase particles is well separated from the size distribution of the tracer particles.

7.3 PHASE-DOPPLER ANEMOMETRY

The principle of phase-Doppler anemometry (PDA) relies on the Doppler difference method used for conventional laser-Doppler anemometry and was first introduced by Durst and Zaré (1975). By using an extended receiving optical system with two or more photodetectors it is possible to measure simultaneously size and velocity of spherical particles. For obtaining the particle size the phase shift of the light scattered by refraction or reflection from the two intersecting laser beams is used.

7.3.1 Principles of PDA

A typical optical set-up of a two detector PDA-system is shown in Figure 7-16. The transmitting optics is a conventional dual beam LDA optics, in this case with two Bragg cells for frequency shifting. The PDA receiver module is positioned at the off-axis collection angle φ in the y-z plane and consists of a collection lens which collimates the scattered light. This parallel light then passes a mask which defines the elevation angles of the two photodetectors (i.e.

the angles out of the y-z plane). In this case the mask has two rectangular slits. The slits are located symmetric with respect to the y-z plane at the elevation angle $\pm\psi$. Then the light is focused onto a spatial filter, i.e. a vertical slit, which defines the effective length of the measurement volume from where the scattered light may be received (see also Figure 7-27). The effective length of the measurement volume results from the width of the spatial filter l_s (typically about 100 μm) and the magnification of the receiving optics, i.e. the ratio of the focal lengths of the collecting lens to the second lens: $L_s = f_1/f_2\ l_s$. Finally, the scattered light passing the two rectangular slits is focused onto the photo-detectors using two additional lenses.

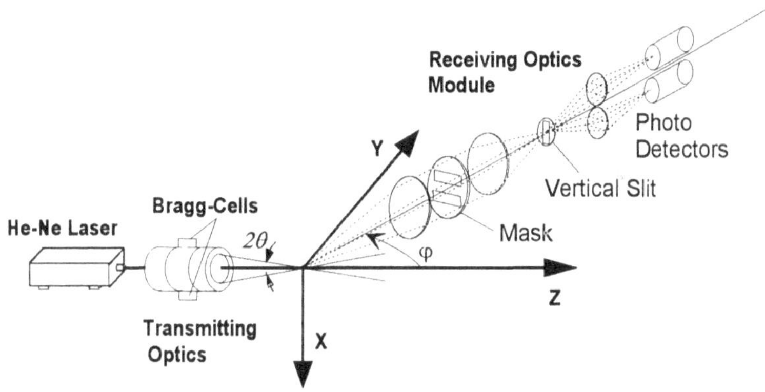

Figure 7-16 Optical configuration of a two-detector phase-Doppler anemometer

For explaining the operational principle of PDA the simple fringe model may be used, by assuming that the interference fringes in the intersection region of the two incident light beams of the LDA are parallel light rays (Saffman 1987 a)). A spherical transparent particle placed into this fringe pattern will act as a kind of lens, whereby the light rays will be projected into space as indicated in Figure 7-17. The separation of the projected fringes at a distance f_r from the particle is approximately given by:

$$\Delta s \approx (f_r - f)\frac{d_f}{f} \tag{7.10}$$

where, d_f is the fringe spacing in the measurement volume. Since small particles are considered and f_r is usually much larger than the particle focal length, one obtains:

$$\Delta s \approx f_r \frac{d_f}{f} \tag{7.11}$$

By introducing the focal length of the particle:

$$f = \frac{m}{m-1}\frac{D_p}{4} \qquad (7.12)$$

with $m = n_p/n_m$ being the relative refractive index of the particle compared with the surrounding medium, the separation of the projected fringes is obtained with:

$$\Delta s \approx \frac{4\, f_r\, d_f}{D_p}\frac{m-1}{m} \qquad (7.13)$$

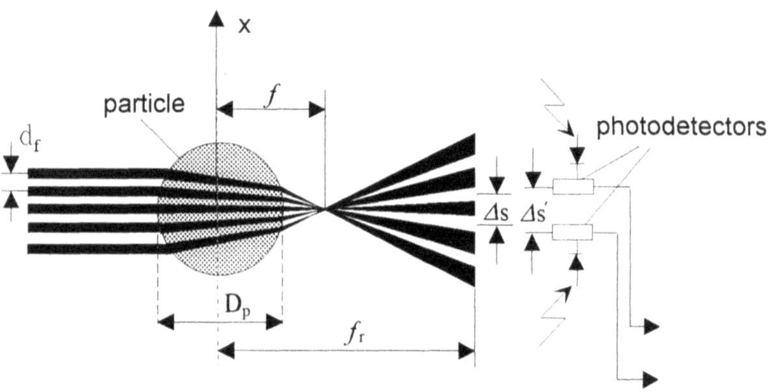

Figure 7-17 Fringe model of the phase-Doppler principle for the case of refraction

Since in general the particles move through the measurement volume it is hardly possible to measure this spatial separation of the fringes. However, if now two photo detectors are symmetrically placed at f_r with a separation Δs' (Figure 7-17) the fringes produced by the moving particle will sweep across the two detectors at the Doppler difference frequency. The signals seen by the two detectors will have a relative phase difference given by:

$$\phi = 2\pi\frac{\Delta s'}{\Delta s} = 2\pi\frac{2\, f_r}{\Delta s}\sin\psi \qquad (7.14)$$

Using Equation 7-13 and introducing the fringe spacing this becomes:

$$\phi = \pi\frac{D_p}{d_f}\frac{m}{m-1}\sin\psi = \frac{2\,\pi\, D_p}{\lambda}\frac{m}{m-1}\sin\theta\,\sin\psi \qquad (7.15)$$

Here ψ is the elevation angle of one photodetector measured from the bisector plane of the two incident beams (i.e. the y-z-plane in Figure 7-16 where also the optical axis of the PDA-receiver is located). It should be emphasized that

Equation 7.15 is an approximation valid only for small scattering angles φ which represents the off-axis angle measured from forward scattering direction, the z-direction defined in Figure 7-16. This equation however is very useful to estimate roughly the measurable particle size range for a given system or to perform an approximate design of the optical configuration for small off-axis angles. For the determination of the particle size from the measured phase difference the required correlations are derived from geometrical optics, which is valid for particles large compared to the wavelength of light (van de Hulst 1981). The phase of the scattered light is given by:

$$\phi = \frac{2\pi\, D_p\, n_m}{\lambda}\left(\sin\tau - p\,\frac{n_p}{n_m}\,\sin\tau'\right) \qquad (7.16)$$

where n_m and n_p are the refractive indices of the medium surrounding the particle and the medium of the particle itself. The parameter p indicates the type of scattering, i.e. p = 0, 1, 2, ... for reflection, first order refraction, second order refraction, etc. Moreover, τ and τ' are the angles between the surface tangent and the incident or refracted ray, respectively. For a dual-beam LDA system the phase difference of the light scattered from each of the two beams is given in a similar way:

$$\Delta\phi = \frac{2\pi\, D_p\, n_m}{\lambda}\left\{\left(\sin\tau_1 - \sin\tau_2\right) - p\,\frac{n_p}{n_m}\left(\sin\tau'_1 - \sin\tau'_2\right)\right\} \qquad (7.17)$$

where the subscripts 1 and 2 are used to indicate the contributions from both incident beams. For two photodetectors placed at a certain off-axis angle φ and placed symmetrically with respect to the bisector plane at the elevation angles $\pm\,\psi$ one obtains the phase difference (see for example Bauckhage 1988):

$$\Delta\phi = \frac{2\pi\, D_p\, n_m}{\lambda}\,\Phi \qquad (7.18)$$

The parameter Φ depends on the scattering mode and is given for reflection and refraction by:
reflection (p = 0):

$$\Phi = \sqrt{2}\left\{\begin{aligned}&\left(1 + \sin\theta\,\sin\psi - \cos\theta\,\cos\psi\,\cos\varphi\right)^{1/2}\\ &-\left(1 - \sin\theta\,\sin\psi - \cos\theta\,\cos\psi\,\cos\varphi\right)^{1/2}\end{aligned}\right\} \qquad (7.19)$$

refraction (p = 1):

$$\Phi = 2\left\{\begin{aligned}&\left[1 + m^2 - \sqrt{2}\,m\left(1 + \sin\theta\,\sin\psi + \cos\theta\,\cos\psi\,\cos\varphi\right)^{1/2}\right]^{1/2}\\ &-\left[1 + m^2 - \sqrt{2}\,m\left(1 - \sin\theta\,\sin\psi + \cos\theta\,\cos\psi\,\cos\varphi\right)^{1/2}\right]^{1/2}\end{aligned}\right\} \qquad (7.20)$$

where $m = n_p/n_m$ has been used for convenience and 2θ represents the angle between the two incident beams. Since the phase difference is a function of p, one expects a linear relation of the correlation between particle size and phase (Equation. 7.17) only for those scattering angles, where one scattering mode is dominant (i.e. reflection or refraction). Therefore, the values for Φ have been given for these two scattering modes only (i.e. Equations 7.19 and 7.20). Other scattering modes, i.e. p = 2, may also be used for phase measurements, especially in the region of back scattering, as will be shown later. Such a back-scatter arrangement might have advantages with regard to optical access, since both incident beams and scattered light may be transmitted through one window.

By recording now the band-pass filtered Doppler signals from the two photodetectors, the phase difference $\Delta\phi$ is determined from the time lag between the two signals as indicated in Figure 7-18.

$$\Delta\phi = 2\pi \frac{\Delta t}{T} \tag{7.21}$$

where T is the time of one cycle in the signal. With Equation 7.18 it is now possible to determine the particle diameter for a given refractive index n_m and wavelength λ:

$$D_p = \frac{\lambda}{2\pi\, n_m} \frac{1}{\Phi} \Delta\phi \tag{7.22}$$

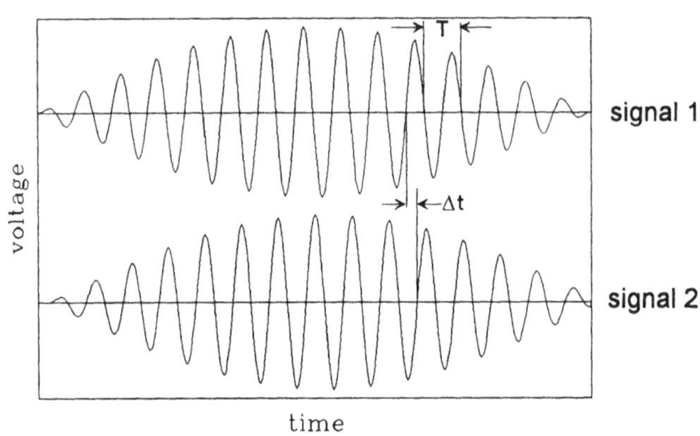

Figure 7-18 Determination of the phase shift from the two band-pass filtered Doppler signals

From Equation 7.21 and Figure 7-18 it is also obvious that only a phase shift between zero and 2π can be distinguished with a two-detector PDA-system, which limits the measurable particle size range for a given optical configuration.

Therefore, also three-detector systems are used, whereby two phase differences are obtained from detector pairs having different spacing (Figure 7-19). This method allows the measurable particle size range to be extended while still maintaining the resolution of the measurement. Moreover, the ratio of the two phase measurements may be used for additional validation, e.g. a sphericity check for deformable particles such as liquid droplets or bubbles.
The interpretation of PDA principles based on geometrical optics is valid only for particles considerably larger than the wavelength and also when only one scattering mode is present on the detector aperture. Extensions can be introduced to account for the Gaussian beam intensity distribution (Sankar and Bachalo 1991).

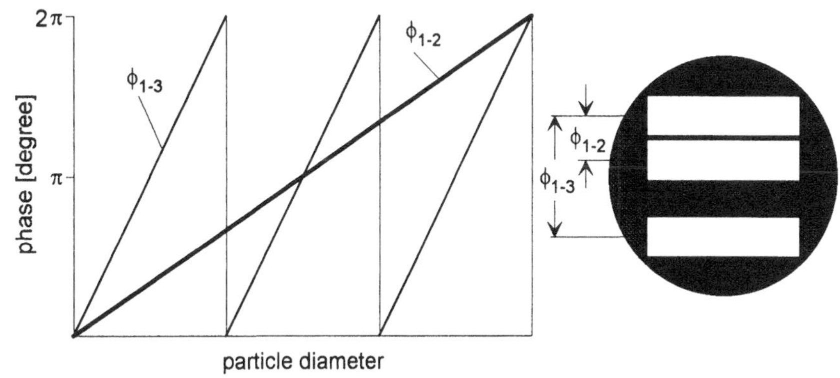

Figure 7-19 Phase-size relations for a three-detector phase-Doppler system

Especially for small particles however, diffraction becomes an important contribution to the light scattering, which may influence and disturb the phase measurement. Therefore, the more general Mie-theory must be applied to determine the scattering characteristics of small size particles and for more precise results for larger particles. The Mie-theory relies on the direct solution of Maxwell's equations for the case of the scattering of a plane light wave by a homogeneous spherical particle of arbitrary size and refractive index. In order to calculate the scattered field of a PDA-system it is necessary to add the contributions of the two incident beams and average over the receiving aperture, taking into account the polarization and phase of each beam. Hence it is possible to determine the intensity, visibility and phase of the detector signal for arbitrary optical configurations. Consideration of the influence of the Gaussian beam has also been made available recently, using for instance the generalized Lorenz-Mie theory (GLMT) (Grehan et al. 1992) or the Fourier-Lorenz-Mie theory (FLMT) (Albrecht et al. 1995). Light scattering programs incorporating such theories are indispensible for the optimization of PDA systems.

7.3.2 Layout of PDA-Systems

In the following, various aspects of the optimum selection of set-up parameters will be discussed for different types of particles (i.e. reflecting and transparent particles) based on calculations by geometrical optics and Mie-theory (DANTEC/Invent 1994). The calculations based on geometrical optics are performed for a point-like aperture while the Mie calculations consider the integration over a rectangular aperture with given half angles in the horizontal (δ_h) and vertical (δ_v) directions with respect to the y-z-plane (Figure 7-16). It should be noted that the integration of the scattered light over the receiving aperture is important for reducing strong oscillations in the phase-size relation.
For totally reflecting or strongly absorbing particles any scattering angle may be used except the near forward scattering range, where diffraction will destroy the linearity of the phase-size relation.
Transparent particles may be distinguished between those having a refractive index larger or smaller than the surrounding medium. Liquid droplets or glass beads in air have a relative refractive index m which is larger than unity, typically in the range 1.3 to 1.5, and water droplets in oil or bubbles in liquid have a relative refractive index below unity.
In this case the selection of the optimum optical configuration should be mainly based on the relative balance of the different scattering modes (i.e. reflection, refraction or second order refraction) with one mode dominating. The linearity of the phase-size relation is the second selection criterion. The relative intensities of the different scattering modes, i.e. reflection, or first and second order refraction, are determined by using geometrical optics calculations, where both parallel (p) and perpendicular (s) polarization are considered (Figure 7-20). Parallel polarization refers to light with polarization in the beam/detector plane, i.e. in the y-z plane of Figure 7.16.
Reflected light covers the entire angular range for refractive index ratios below and above unity. However, a distinct minimum is found for parallel polarized light at the so-called Brewster's angle which is given for a sphere by:

$$\varphi_B = 2\tan^{-1}(1/m) \tag{7.23}$$

The Brewster's angle decreases with increasing refractive index ratio.
First order refraction is concentrated in the forward scattering range and extends up to the critical angle which is given for different relative refractive indices $m = n_p/n_m$ as follows:

$$\varphi_C = 2\cos^{-1}(m) \qquad \text{for: } m < 1 \tag{7.24}$$

$$\varphi_C = 2\cos^{-1}(1/m) \qquad \text{for: } m > 1 \tag{7.25}$$

The critical angle increases with increasing relative refractive index (m > 1) and first order refraction becomes dominant over reflection over a wider angular range. Second order refraction covers again the entire angular range for a relative refractive index below unity. For m larger than unity, second order refraction is concentrated in the backward scattering range and limited by the Rainbow angle (Naqwi and Durst 1991).

$$\varphi_R = \cos^{-1}\left\{\frac{2}{m^4}\left(\frac{4-m^2}{3}\right)^3 - 1\right\} \tag{7.26}$$

With increasing relative refractive index the angular range of second order refraction is reduced and the rainbow angle increases. The characteristic scattering angles given above are summarized in Figure 9-21 as a function of refractive index and scattering angle. Based on the location of these characteristic angles Naqwi and Durst (1991) proposed a map for the presence of the different scattering modes as a function of scattering angle and relative refractive index for supporting the layout of the optical configuration of PDA systems. Recently, Naqwi and Menon (1994) have introduced a more rigorous procedure for the design and optimization of PDA-systems by additionally considering the light absorption characteristics of particles. Scattering mode chart for 15 scattering domains and 5 attenuation levels were introduced by indicating regions where the three scattering modes (i.e. reflection, refraction and internal reflection) are dominant with high and low level of confidence.

In the following, the angular distribution of the scattering intensities resulting from different modes are discussed in more detail for different relative refractive indices which are typical for practical two-phase flow systems (e.g. air bubbles in water, water droplets in air and glass particles in air). Moreover, Mie-calculations are performed for the range of the optimum scattering angle suggested by the relative intensity distributions.

For bubbles in water the optimum scattering angle seems to be rather limited i.e. between 70° and about 85° where reflection is dominant for either polarization (Figure 7-20 a)). The phase-size relations show reasonable linearity in this range, but also a scattering angle of 55° gives a linear response function (Figure 7-22). In forward scattering strong interference with refracted light exists and the phase-size relation becomes nonlinear (i.e. at a scattering angle of 30°). Similar observations are made for water droplets or glass particles in oil.

For two-phase systems with relative refractive indices larger than unity refraction is dominant for parallel polarization in the forward scattering range up to about 70 to 80° depending on the value of the refractive index ratio (Figure 7-20). Since below about 30° diffraction interferes with the refracted light, especially for small particles, the lower limit of the optimum scattering angle is limited by this value. This is also obvious from the angular distribution of the phase for different particle diameters shown in Figure 7-23. The phase-size relations for water droplets in air show that a reasonable linearity is obtained in the range between 30 and 80° (Figure 7-24).

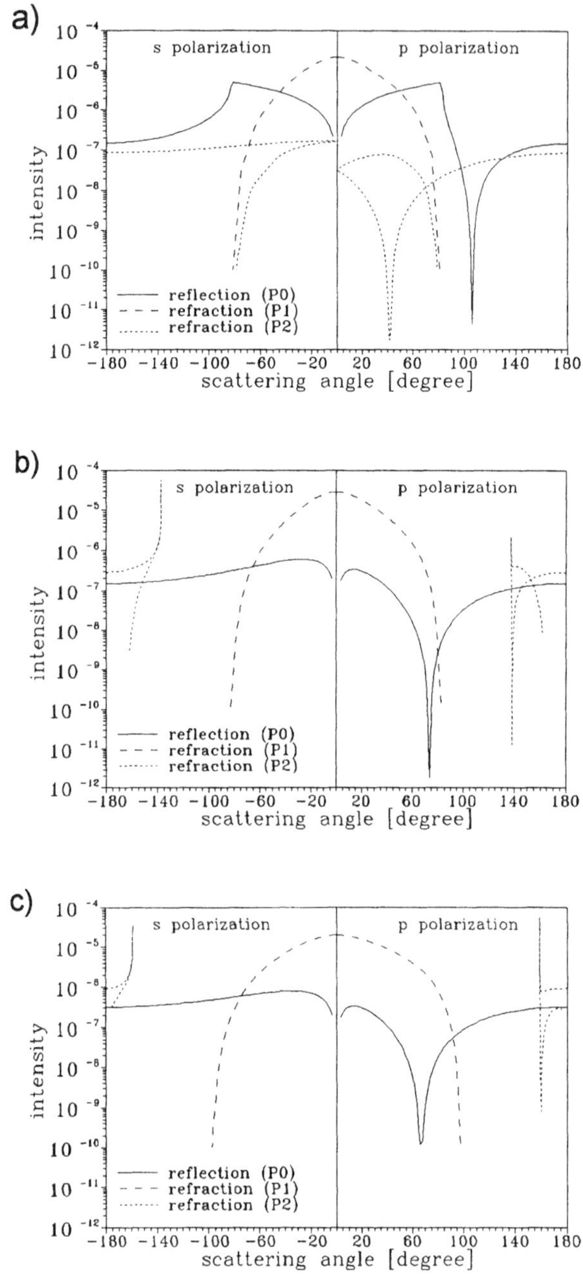

Figure 7-20 Angular intensity distribution of the different scattering modes obtained by geometrical optics for a point receiving aperture (λ = 0.6328 μm, θ = 2.77°, D_p = 30 μm; a) m = 0.75, b) m = 1.33, c) m = 1.52)

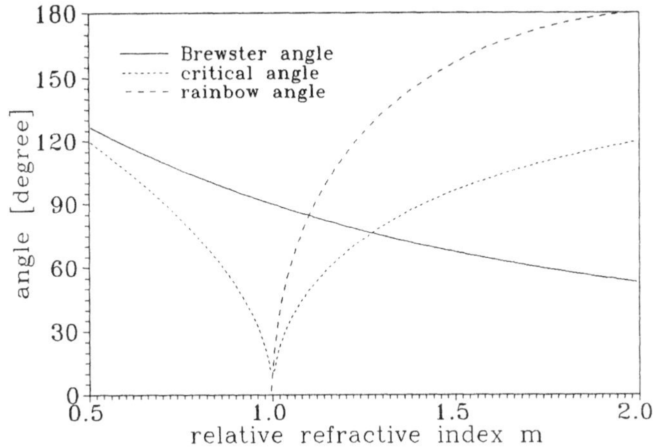

Figure 7-21 Location of characteristic scattering angles as a function of relative refractive index (Naqwi and Durst 1991)

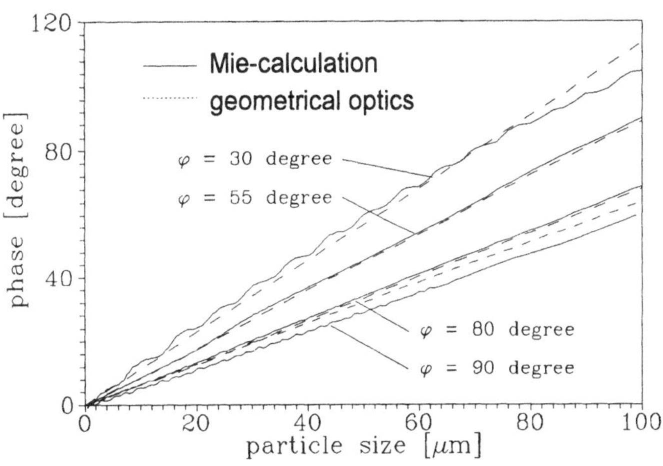

Figure 7-22 Mie-calculation of phase-size relations for different scattering angles between 30 and 80° (λ = 0.6328 μm, p polarization, m = 0.75 (i.e. air bubble in water), $\theta = 2.77°$, $\psi = 1.85°$, $\delta_h = 5.53°$, $\delta_v = 1.85°$)

The relative intensity distributions in Figure 7-20 also suggest that for m > 1 reflected light is dominant between the critical angle and the Rainbow angle. However, here interference with third order refraction exists (not shown in Figure 7-20) and this angular range can be only recommended for perpendicular polarization where a reasonable linearity of the phase response curve is obtained for water droplets in air only around 100° (Figure 7-25). At 120° the Mie-calculations do not correspond to the geometrical optics result and therefore this result is not shown in Figure 7-25.

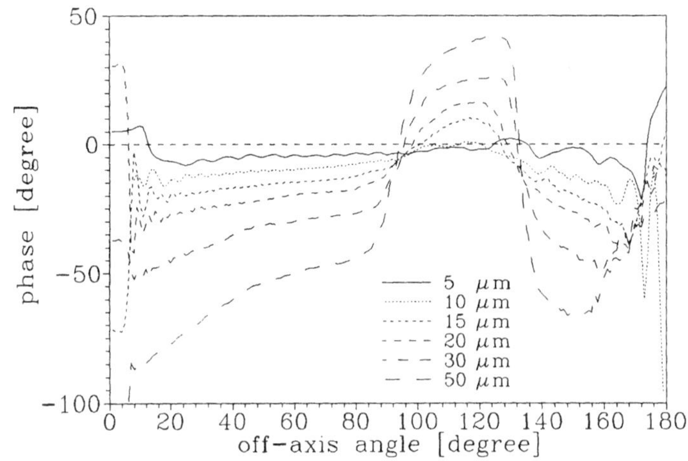

Figure 7-23 Angular distribution of phase for different particle diameters (λ = 0.6328 μm, p polarization, m = 1.33 (i.e. water droplet in air), θ = 2.77°, ψ = 1.85°, δ_h = 5.53°, δ_v = 1.85°)

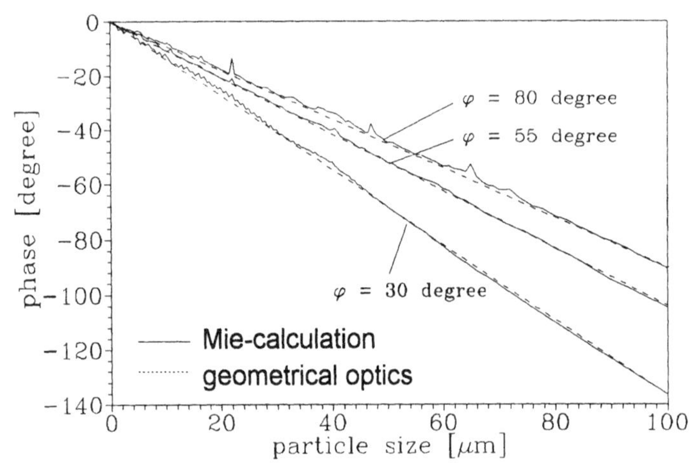

Figure 7-24 Mie-calculation of phase-size relations for different scattering angles between 30 and 80° (λ = 0.6328 μm, p polarization, m = 1.33 (i.e. water droplet in air), θ = 2.77°, ψ = 1.85°, δ_h = 5.53°, δ_v = 1.85°)

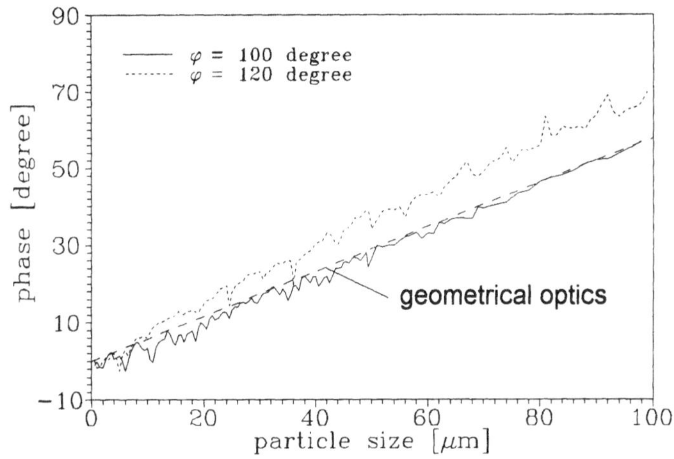

Figure 7-25 Mie-calculation of phase-size relations for different scattering angles of 100° and 120° (λ = 0.6328 μm, s polarization, m = 1.33 (i.e. water droplet in air), θ = 2.77°, ψ = 1.85°, δ_h = 5.53°, δ_v = 1.85°)

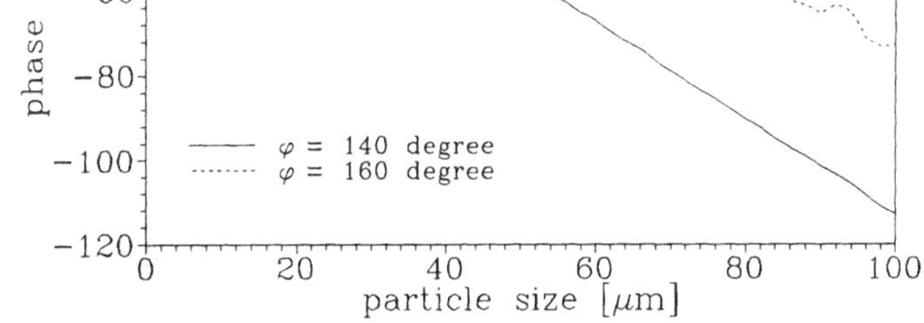

Figure 7-26 Mie-calculation of phase-size relations for scattering angles of 140° and 160° (λ = 0.6328 μm, s polarization, m = 1.33 (i.e. water droplet in air), θ = 2.77°, ψ = 1.85°, δ_h = 5.53°, δ_v = 1.85°)

In the region of backscatter the intensity of secondary refraction is only dominant in a narrow range above the Rainbow angle for perpendicular polarization. The optimum location of the receiving optics however strongly

depends on the value of the relative refractive index (Figure 7-26) which is critical for applications in fuel sprays where the refractive index varies with droplet temperature and hence the location of the rainbow angle is not constant. From Figure 7-26 it becomes obvious that just above the rainbow angle, i.e. for $\varphi = 140°$, a linear phase-size relation is also obtained.
As described above, the proper application of PDA requires that one scattering mode is dominant and the appropriate correlation (i.e. Equation 7.18 and Equation 7.17 or Equation 7.20) has to be used to determine the size of the particle from the measured phase. However, on certain trajectories of the particle through the focused Gaussian beam the wrong scattering mechanism might become dominant and lead to erroneous size measurements (Sankar and Bachalo 1991, Grehan et al. 1992). This error is called Gaussian beam effect or trajectory ambiguity and is illustrated in Figure 7-27, where a transparent particle moving in air is considered, with the desired scattering mode being refraction, which is dominant for collection angles between 30 and 80°. When the particle passes through the part of the measurement volume located away from the detector (i.e. on the negative y-axis), it is illuminated non-homogeneously. Thus refracted light is coming from the outer portion of the measurement volume where the light intensity is relatively low, while reflected light comes from a region closer to the center of the measurement volume where the illuminating light intensity is considerably higher due to the Gaussian intensity profile. In this situation the reflected light becomes dominant, resulting in wrong size measurement since the particle diameter is determined from the correlation for refraction. It is obvious from Figure 7-27 that the trajectory ambiguity is potentially of great importance for large particles whose size is comparable to the dimensions of the measurement volume.

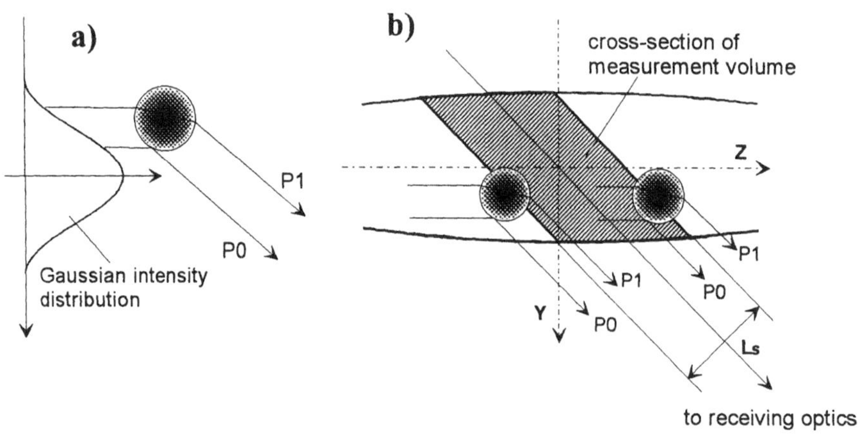

Figure 7-27 Illustration of Gaussian beam effect (a) and slit effect (b)

The phase error as a function of measurement volume diameter to particle diameter is illustrated in Figure 7-28 for a particle moving along the y-axis through the measurement volume. The phase and amplitude were calculated using GLMT (Grehan et al. 1992). The largest phase error is found on the negative y-axis and it may become negative or positive depending on the diameter ratio. The smallest errors are however observed for small particles which leads to the recommendation that the measurement volume diameter should be about 5-times larger than the largest particles in the size spectrum considered. This requirement however has restrictions for applications in dense particle-laden flows, where the measurement volume must be small enough to ensure that the probability that two particles are simultaneously in the measurement volume is small.

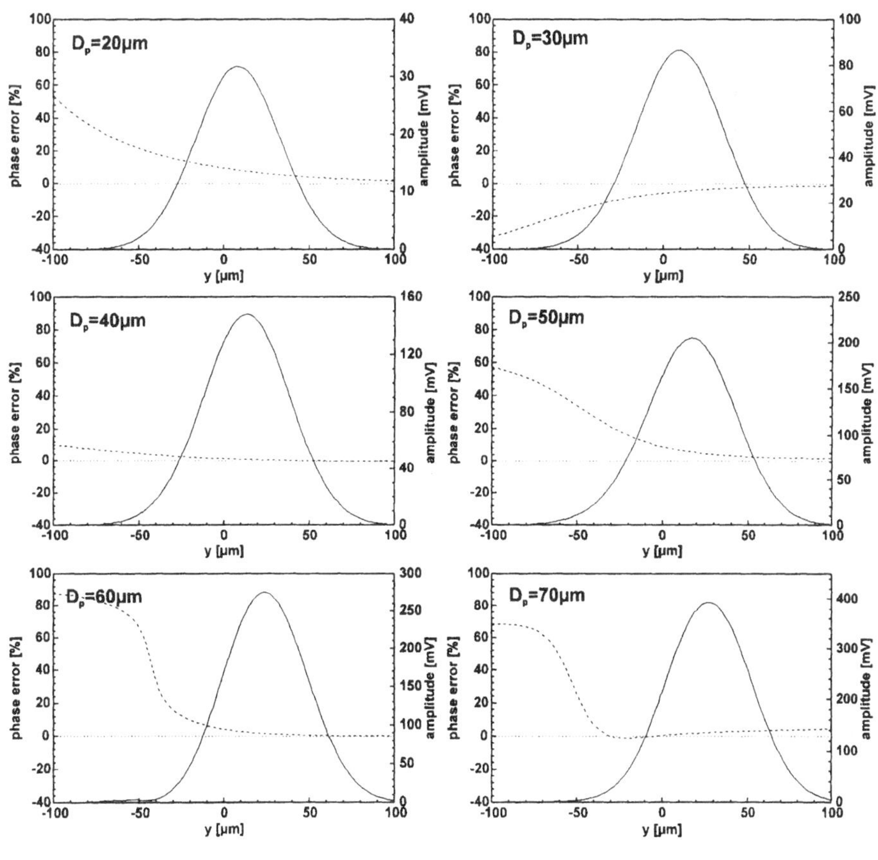

Figure 7-28 Phase error (----) and scattering amplitude (——) along the y-axis for different particle diameters and a measurement volume diameter of 100 μm

From the profiles of the signal amplitude one may recognize that the maximum is shifted towards the positive y-values. This implies, that the effective location of the measurement volume (i.e. the region from where the signals are detected) is not identical with the geometric location of the laser beams. This shift depends on the ratio of particle size to measurement volume size and on the specific optical configuration. With increasing particle size the effective measurement volume is shifted in the positive y-direction and the negative z-direction as illustrated in Figure 7-29. Especially for particles with a diameter comparable to or larger than the measurement volume diameter, the effective measurement volume may be located completely outside the geometric measurement volume (Qiu and Hsu 1996, Panidis and Sommerfeld 1996).

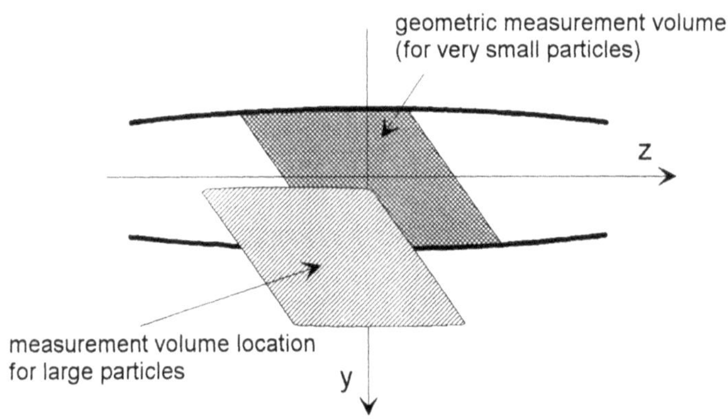

Figure 7-29 Shift of measurement volume cross-section in the y-z-plane with increasing particle size

Similarly the so-called slit effect may result in erroneous particle size measurements as reported by Durst et al. (1994). As described previously, only a portion of the measurement volume is imaged onto the photodetector using a slit aperture in the receiving optics (see Figure 7-16 and Figure 7-27 b)). Due to the finite size of the particles, scattered light will reach the detector even when the center of the particle is outside the slit aperture image. No problems result for particles passing the edge of the measurement volume on the negative z-axis (Figure 7-27 b)). However, when particles pass the edge of the measurement volume located opposite the transmitting optics, the refracted light is blocked by the spatial filter to a large extent while reflected light may still reach the photodetector. When considering particles in air, where the collection angle is typically between 30° and 80°, the intensity of reflected light is much lower than that of refracted light (see Figure 7-20). Therefore, particles passing the right edge of the measurement volume will not be detected by the data acquisition, i.e. the scattering amplitude of reflected light is lower than the trigger level.

Hence the slit effect will not be a major source of sizing errors. This was also demonstrated by the theoretical analysis of Qiu and Hsu (1996) and the experimental studies of Maeda et al. (1996 b)).

In order to reduce sizing errors due to the trajectory ambiguity, additional validation criteria have been proposed recently which are summarized below:

- burst centering, whereby only the central portion of the Doppler signal is used for estimating signal phase and frequency (Qiu et al. 1991, Qiu and Sommerfeld 1992); generally the phase is correct at the point of maximum amplitude,
- a correlation between particle size and scattering amplitude is used, in order to reject signals from particles having a scattering amplitude which is too low for the corresponding size (Qiu and Sommerfeld 1993, Sankar et al. 1995, Sommerfeld and Qiu 1995); note that this requires a measurement of the signal amplitude,
- additional validation based on the phase ratio obtained by using a three detector system (Hardalupas and Taylor 1994, Maeda et al. 1996 b)).

Moreover, the use of extended optical systems combined with additional validation criteria, such as the dual-mode phase-Doppler anemometer (Tropea et al. 1995) may effectively reduce errors resulting from the trajectory ambiguity. An overview about extended PDA-systems will be given in section 7.3.4.

7.3.3 Particle Concentration and Mass Flux Measurements by PDA

Since PDA allows the measurement of particle size and velocity, it is also possible to estimate the particle number or mass concentration and the particle mass flux. The particle number concentration is defined as the number of particles per unit volume. This quantity however, cannot be measured directly, since PDA is a single particle counting instrument and therefore requires that at most one particle is in the measurement volume at a time. The particle concentration has to be derived from the number of particles moving through the measurement volume during a given measurement period. For each particle one has to determine the volume which is sweeping together with the particle across the measurement volume cross-section during the measurement time Δt_s. This volume depends on the instantaneous particle velocity $\vec{U}_p$ and the measurement volume cross-section perpendicular to the velocity vector, i.e. $\mathrm{Vol} = A' \left|\vec{U}_p\right| \Delta t_s$ (Figure 7-30). Additionally, the effective cross-section of the measurement volume is a function of the particle size and therefore, $A' = A'(\alpha_k, D_i)$, where α_k is the particle trajectory angle for each individual sample k and D_i is the particle diameter for size class i. Hence, the concentration associated with one particle is given by:

$$C_n = \frac{1}{Vol} = \frac{1}{\left|\vec{U}_p\right| A'(\alpha_k, D_i)\, \Delta t_s} \tag{7.27}$$

This implies that for accurate particle concentration measurements one has to know the instantaneous particle velocity and the effective measurement volume cross-section. Hence accurate particle concentration measurements require the following:

- Correct particle size measurements, especially for large particles which have the highest contribution to the local mass concentration. Therefore, sizing errors due to the Gaussian beam effect are a potential source of erroneous concentration measurements.
- The measurement of the instantaneous particle velocity in complex flows.
- An on-line determination of the effective particle size-dependent cross-section of the measurement volume.
- That all particles are detected by the data acquisition system. Since mainly the detection of small particles is a problem, this error is usually quite small for concentration measurements.
- A high validation rate.

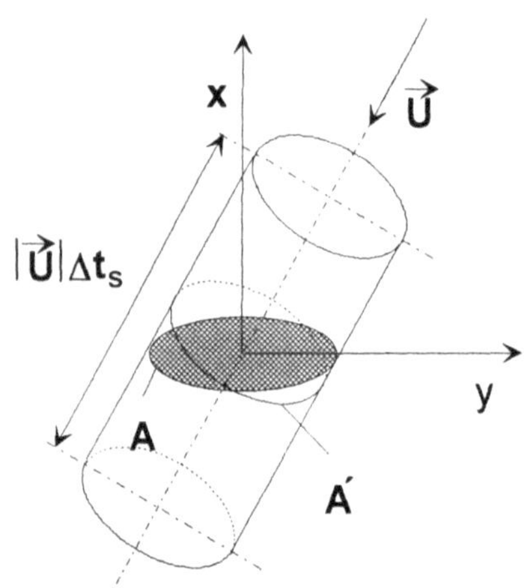

Figure 7-30 Measurement volume associated with one particle moving across the detection region during the measurement time Δt_s

The validation rate is the number of validated samples normalized by the total number of analyzed samples. The validation criteria are mainly applied to insure that the signal information, such as signal frequency and phase, received from two or more channels or two pairs of channels are within certain limits in order to ensure that the signals come from the same particle or that the particle is spherical. Especially in complex flows, e.g. for high particle concentration, the validation rate may decrease considerably. For concentration measurements, however, the rejected particle should be considered in some way, since they have passed the measurement volume, although they have not provided acceptable signals for accurate phase and frequency measurements. In order to account for these missing particles, usually the measured concentration is corrected by multiplying with the inverse of the validation rate (Sommerfeld and Qiu 1995). This procedure relies on the assumption that the rejected particles have the same size distribution as the validated particles and only yields acceptable results if the validation rate is rather high, typically larger than about 80%. Otherwise this approach may completely fail since actually no information on the size of the rejected particles is available.

The dependence of the measurement volume cross-section on particle size is a result of the Gaussian intensity distribution in the measurement volume and the finite signal noise. As illustrated in Figure 7-7, a large particle passing the edge of the measurement volume will scatter enough light to produce a signal above the detection level. A small particle will produce such a scattering intensity only for a smaller displacement from the measurement volume center (Figure 7-7). Therefore, the measurement volume cross-section decreases with particle size and approaches zero for $D_p \rightarrow 0$ as shown in Figure 7-31.

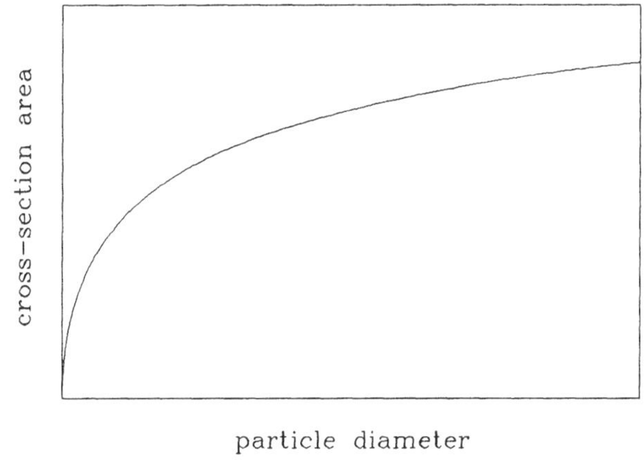

Figure 7-31 Cross-section of effective measurement volume as a function of particle size

Additionally, the effective measurement volume cross-section is determined by the off-axis position of the receiving optics and the width of the spatial filter used to limit the length of the measurement volume imaged onto the photodetectors. For a one-dimensional flow along the x-axis (Figure 7-16) the effective size-dependent cross-section of the measuring volume for a given off-axis angle of the receiving optics, φ, is determined from (Figure 7-27):

$$A(D_i)_x = \frac{2\, r(D_i)\, L_s}{\sin\varphi} \tag{7.28}$$

Here, L_s is the width of the image of the spatial filter in the receiving optics which depends on the slit width and the magnification of the optics, D_i is the diameter of the considered particle size class and $r(D_i)$ is the particle size-dependent radius of the measurement volume (Figure 7-32). For any other trajectory of the particle through the measurement volume the effective cross-section perpendicular to the particle trajectory is obtained with the particle trajectory angle α_k:

$$A'(\alpha_k, D_i) = \frac{2\, r(D_i)\, L_s}{\sin\varphi} \cos\alpha_k \tag{7.29}$$

The particle trajectory angle can be determined from the different instantaneous particle velocity components with:

$$|\cos\alpha_k| = \frac{A'(\alpha_k, D_i)}{A(D_i)} = \left|\frac{u}{|\vec{U}|}\right|_k = \frac{1}{\sqrt{1+\left(\frac{w}{u}\right)^2+\left(\frac{v}{u}\right)^2}} \tag{7.30}$$

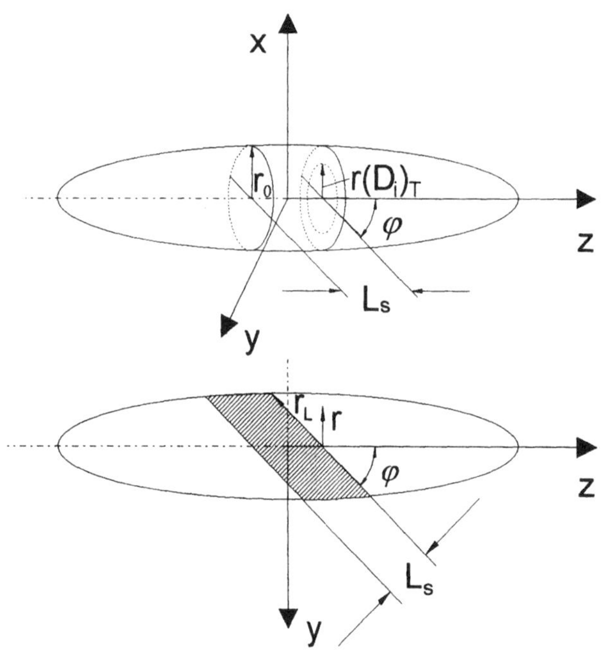

Figure 7-32 Geometry of PDA measurement volume

The particle size-dependent radius of the measurement volume $r(D_i)$ may be determined in-situ by using the burst length method (Saffman 1987 b)) or the so-called logarithmic mean amplitude method (Qiu and Sommerfeld 1992). The latter is more reliable for noisy signals, i.e. low signal to noise ratio as demonstrated by Qiu and Sommerfeld (1992).

The above discussion reveals that in complex two-phase flows with random particle trajectories through the measurement volume, a three-component PDA-system is required for accurate concentration measurements. For a spectrum of particle sizes the local particle number concentration is then evaluated from:

$$C_n = \frac{1}{\Delta t_s} \sum_{i=1}^{N_i} \left(\sum_{k=1}^{N_k} \frac{1}{A'(\alpha_k, D_i)} \sum_{j=1}^{N_j} \frac{1}{|\vec{U}_{kj}|} \right) \quad (7.31)$$

The sums in Equation 7.31 involve the summation over the individual realizations of particle velocities (index j) in a pre-defined directional class (index k) and size class (index i). The summation over the particle size classes (index i) include the appropriate particle size-dependent cross-section of the measurement volume for each directional class (Equation 7.29). It should be stated that the use of a mean velocity either in a directional class or a size class is not appropriate to determine the particle concentration, since the mean velocity may become zero or close to zero resulting in an infinite concentration as pointed out by Hardalupas and Taylor (1989).

The particle mass concentration can be obtained by multiplying Equation 7.31 with the mass of the particles. Quite often the particle mass flux in a considered flow direction is a useful quantity to characterize a two-phase flow. The mass flux in direction n is obtained from:

$$f_n = \frac{\pi \rho_p}{6 \, \Delta t_s} \sum_{i=1}^{N_i} \left(\sum_{k=1}^{N_k} \frac{D_i^3}{A'(\alpha_k, D_i)} \sum_{j=1}^{N_j} \frac{u_n}{|\vec{U}_{kj}|} \right) \quad (7.32)$$

Here u_n is the particle velocity component in the direction for which the flux shall be determined.

For a directed two-phase flow, i.e. when the temporal variation of the particle trajectory through the measurement volume is relatively small, for example in a spray, the mean particle trajectory angle may be determined from independent measurements of the individual velocity components as shown by Qiu and Sommerfeld (1992). In complex turbulent two-phase flows, generally a three-component PDA-system is required for accurate concentration and mass flux measurements.

An alternative method for determining the particle number concentration is based on the averaged residence time of the particles in the measurement volume (Hardalupas and Taylor 1989).

$$C_n = \frac{1}{\Delta t_s} \sum_{i=1}^{I} \left(\frac{\sum_{N_i} t_{ri}}{Vol(D_i)} \right) \tag{7.33}$$

Here t_{ri} is the particle residence time in the measurement volume, $Vol(D_i)$ is the particle size-dependent volume and N_i is the number of samples in one particle size class (index i). As demonstrated by Qiu and Sommerfeld (1992), the particle residence time or burst length cannot be accurately determined for noisy Doppler signals. Hence this alternative method is not very reliable and yields considerable errors in particle concentration measurements.

Recently a novel method was introduced which allows accurate particle concentration or mass flux measurements even in complex flows with a one-component PDA-system (Sommerfeld and Qiu 1995), using the integral value under the envelope of the band-pass filtered Doppler signal. Since this value depends on both the particle trajectory through the measurement volume and the particle velocity it can be used to estimate the instantaneous particle velocity and the direction of motion when only one velocity component is measured (Figure 7-33). The integral of the envelope of the band-pass filtered Doppler signal can be written for a particle of given size and velocity as a function of time or using $ds = |\vec{U}_{ikj}|\,dt$ as a function of particle travel distance in the following way:

$$Int_{ikj} = \int_0^{t_j} V(D_i, t)\,dt = \int_0^{s_j} V(D_i, x, y, z)\,ds \tag{7.34}$$

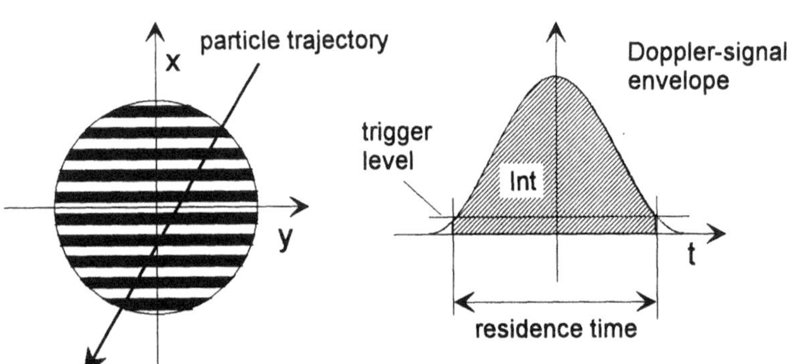

Figure 7-33 Determination of the integral value under the envelope of the band-pass filtered Doppler signal

The indices i, k, and j again refer to particle size classes, directional classes, and individual samples in each directional class in order to account for velocity variations in one directional class. Assuming that the probability of a particle passing the measurement volume cross-section at any location is constant, the summation of all integral values in one directional class is obtained by integrating also over the cross-section of the measurement volume:

$$\begin{aligned} INT_{ik} &= \sum_{j=1}^{N_j} Int_{ikj} = \frac{1}{A'(\alpha_k, D_i)} \left[\sum_{j=1}^{N_j} \frac{1}{\left|\vec{U}_{ikj}\right|} \right] \iint_{A_k} \left[\int_0^{s_j} V(D_i, x, y, z)\, ds \right] dA \\ &= \frac{1}{A'(\alpha_k, D_i)} \left[\sum_{j=1}^{N_j} \frac{1}{\left|\vec{U}_{ikj}\right|} \right] \iiint_{Vol(D_i)} V(D_i, x, y, z)\, dv \end{aligned} \tag{7.35}$$

For each of the directional classes the number concentration can be introduced into this equation.

$$C_n(D_i)_k = \frac{1}{A'(\alpha_k, D_i)\, \Delta t_s} \left[\sum_{j=1}^{N_j} \frac{1}{\left|\vec{U}(D_i)_{kj}\right|} \right] \tag{7.36}$$

$$INT_{ik} = C_n(D_i)_k\ \Delta t_s \iiint_{Vol(D_i)} V(D_i, x, y, z)\, dv \tag{7.37}$$

Now the sum over all the integral values for the individual directional classes for a given size class i is evaluated.

$$INT(D_i) = \Delta t_s \left[\sum_{k=1}^{N_k} C_n(D_i)_k \right] \iiint_{Vol(D_i)} V(D_i, x, y, z)\, dv \tag{7.38}$$

Finally, the total particle number concentration can be obtained by summation over all the size classes which yields the following equation:

$$C_n = \frac{1}{\Delta t_s} \sum_{i=1}^{N_i} \frac{INT(D_i)}{\iiint_{Vol(D_i)} V(D_i, x, y, z)\, dv} \tag{7.39}$$

Similarly, the particle mass concentration is obtained as:

$$C_m = \frac{\pi\, \rho_p}{6\, \Delta t_s} \sum_{i=1}^{N_i} \frac{D_i^3\ INT(D_i)}{\iiint_{Vol(D_i)} V(D_i, x, y, z)\, dv} \tag{7.40}$$

The particle mass flux in any direction n can be obtained from the following equation:

$$F_n = \frac{\pi\ \rho_p}{6\ \Delta t_s} \sum_{i=1}^{N_i} \frac{D_i^3 \sum_{k=1}^{N_k} \sum_{j=1}^{N_j} Int_{ikj}\ u_n}{\iiint_{Vol(D_i)} V(D_i, x, y, z) dv} \tag{7.41}$$

Note that the mass flux is a vector quantity and F_n stands for the flux in the direction of velocity component u_n. Since this velocity component is connected with the individual realizations, the integral value according to Equation 7.34 is introduced in Equation 7.41 for convenience. The volume integral of the Doppler signal envelope in the denominator is obtained from the logarithmic mean amplitude (LMA) method (Qiu and Sommerfeld 1992). A detailed derivation for the determination of this integral is given by Sommerfeld and Qiu (1995).

In order to demonstrate the performance of the Doppler-burst envelope integral value method for the estimation of the instantaneous particle velocity vector and the particle mass flux or concentration, measurements were performed in a liquid spray issuing from a hollow cone pressure atomizer (cone angle 60°) and a swirling flow which exhibits complex particle trajectories (Sommerfeld and Qiu 1993). All the measurements were conducted using the one-component phase-Doppler anemometer. The integration of the mass flux profiles provided the dispersed phase mass flow rate which agreed to ± 10 % with independent measurements of the mass flow rate (Sommerfeld and Qiu 1995).

The methodology and the fundamentals for measurements of the instantaneous local particle density in pneumatic conveying using phase-Doppler anemometry were recently explored by van de Wall and Soo (1994) and Bao and Soo (1995). The concept for their approach was based on associating with each validated particle signal the appropriate measurement volume

$$Vol_i = u_{p,i}\ A(D_i)\ \Delta t_i \tag{7.42}$$

as illustrated in Figure 7-34. Here $u_{p,i}$ is the particle velocity, $A(D_i)$ the particle size-dependent cross-section of the measurement volume and Δt_i the time between the arrival of successive particles in the measurement volume.

Local instantaneous time averaging was performed over a period Δt which was typically $10^2 \cdot \Delta t_i$ by considering the scale relation:

$$\sum_{i=1}^{k} u_{p,i}\ \Delta t_i < L_{syst} \tag{7.43}$$

where L_{syst} is a characteristic dimension of the flow field; in this case the pipe diameter. The instantaneous particle mass concentration is then found from:

$$c_m = \overline{c_m} + c'_m = \frac{1}{\Delta t} \sum_{i=1}^{k} \frac{m_{p,i}}{u_{p,i}\ A(D_i)} \tag{7.44}$$

The instantaneous values averaged over the time slot Δt are then used to determine long-time averages and rms.-values of the particle mass concentration.

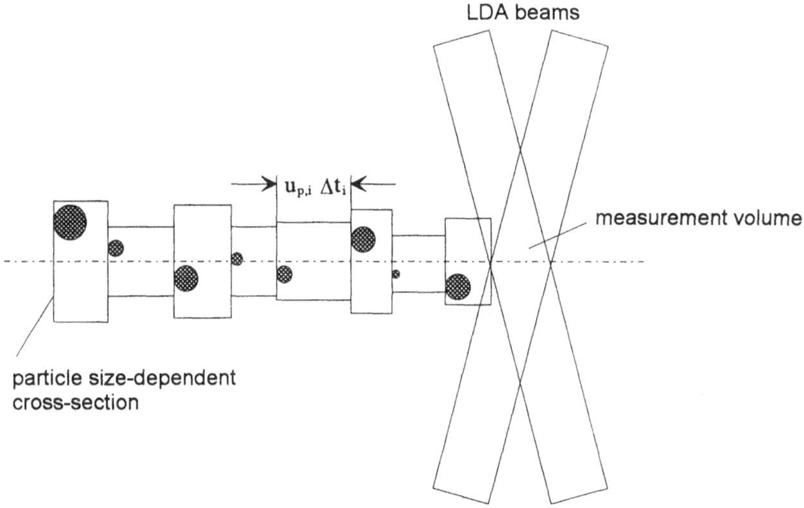

FIGURE 7-34 Approach for determination of instantaneous particle concentration (van de Wall and Soo 1994)

7.3.4 Novel PDA-Systems

Leaving the realm of conventional PDA instruments, there exist a large number of novel concepts for either improving the measurement accuracy of PDA and/or extending measurement capabilities to the determination of further particle properties. This discussion begins with the introduction of the planar PDA and its integration into conventional PDA systems, resulting in the dual-mode PDA, as mentioned briefly above. This is followed by a discussion of the so-called dual burst technique, which allows under certain conditions the refractive index of a particle and the concentration of a solid suspension in a droplet to be determined. These examples illustrate some of the many possibilities still remaining with the use of elastic light scattering.

The planar PDA is shown schematically in Figure 7-35, in which the laser beams, their polarization direction and the photodetectors, all lie in the same plane (i.e. y-z plane). As with conventional PDA arrangements, the position of the detectors, i.e. their elevation angles, must be chosen to yield a linear relationship between measured phase and particle diameter. This is possible for most liquids, resulting in general in a substantially lower slope in the phase/diameter dependence. On the other hand the typical oscillations of phase at low particle diameter may be higher in amplitude because the detectors of the planar PDA are situated at different scattering angles. The planar PDA has been discussed for various applications in the past, for example for the measurement of very small particles (Naqwi et al. 1992) and for the measurement of cylindrical particles (Mignon et al. 1996) or for the elimination of the Gaussian beam effect (Aizu et al. 1993). It is the latter context in which the planar PDA is used in the dual-mode PDA.

Figure 7-35 illustrates also the combination of a conventional PDA with a planar PDA to form a dual-mode PDA, inherently able to measure two velocity components. The corresponding phase/diameter relations for the two PDA systems are shown in Figure 7-36. Similar to the use of three detectors in a conventional PDA system, the two phase measurements in a dual-mode system can be used to resolve any 2π ambiguity. The real value of combining the two systems lies however, in the fact that each system responds quite differently to the Gaussian beam effect (trajectory ambiguity). Particles passing through regions of the measurement volume in which reflective rather than refractive scattering dominates, will lead to improper measurements in each system. However the independent measurements will no longer be in agreement and thus, this can be used as a validation criterion to omit erroneous measurements due to the Gaussian beam effect (Tropea et al. 1995, Tropea et al. 1996). Also the so-called slit effect can be eliminated using the dual-mode PDA approach (Durst et al. 1994).

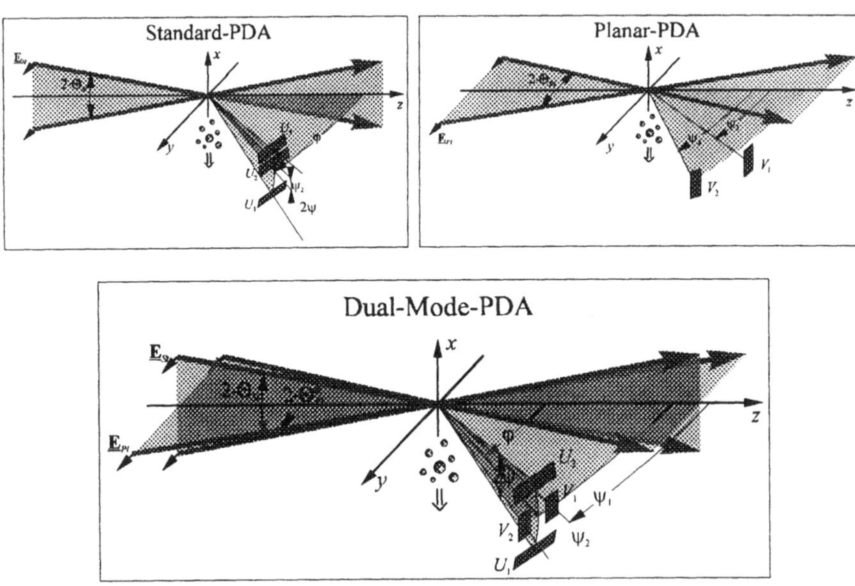

Figure 7-35 Optical arrangement of the standard PDA, the planar PDA and the dual-mode PDA

The dual-mode PDA therefore allows the measurement volume to be made much smaller without danger of compromising measurement accuracy. This in turn leads to the possibility of measuring in flows of higher densities of the dispersed phase, for example near injection spray nozzles. Due to the increased reliability of the particle diameter measurement and the availability of two velocity components, the dual-mode PDA results in improved estimations of the mass flux (Dullenkopf et al. 1996).

The mass flux of the dispersed phase is in fact an essential measurement quantity in many experimental investigations. The accuracy of mass flux measurements will depend not only on the instrumentation, but also on the flow field and the size distribution of the dispersed phase, so that a general accuracy estimate is not feasible. In simple spray flows however, an accuracy of ±10% on the local mass flux can be expected (Sommerfeld and Qiu 1995, Mundo 1996).

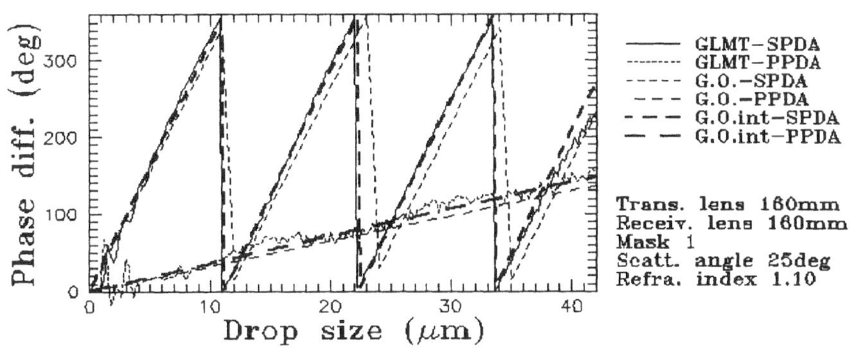

Figure 7-36 Computed phase/diameter relations for conventional and planar PDA in a dual-mode PDA (Tropea et al. 1996), (GLMT: generalized Lorenz-Mie theory, G.O.: geometrical optics, SPDA: standard PDA, PPDA: planar PDA)

One pre-requisite for such estimates is that all measured particles are spherical. This may be obtainable in modeled flows with selected particles, but is certainly not the rule in practical situations. The instrumentation problem is therefore two-fold. If the PDA system can detect non-sphericity, as indicated above using a three-detector receiver, then as a minimum the mass contained in all rejected non-spherical particles will be missed. If on the other hand, many non-spherical particles are in fact accepted as spherical particles, their computed size may differ from the volume equivalent diameter of a spherical droplet, thus also falsifying the measured mass flux.

Presently most commercial PDA systems assume that the non-sphericity validation is reliable and the measured mass flux is adjusted according to the percentage of rejected particles. Here a word of caution is necessary, since there are indications that a conventional three-detector PDA system is not so sensitive to non-sphericity. This is demonstrated in Figure 7-37, in which even for highly non-spherical droplets, good agreement is found between the sizes measured with the two sets of detectors 1-2 and 1-3 (see Figure 7-19), thus leading to a validation. The phase distortion due to non-sphericity appears to effect all three detectors about equally. In this sense the dual-mode arrangement, using two pairs of detectors arranged orthogonal to one another, is much more sensitive to non-sphericity (Damaschke at al. 1997). Nevertheless, the estimation of mass flux under such circumstances remains an unsolved measurement problem.

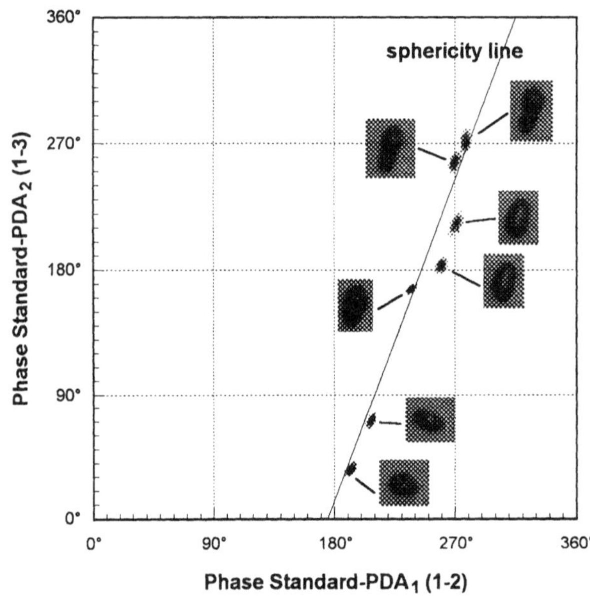

Figure 7-37 Comparison of phase differences (1-2) and (1-3) in a conventional, three detector PDA for various non-spherical droplets.

Still a more recent innovation is the dual-burst technique (DBT), which in fact uses the previously discussed Gaussian beam effect to its advantage (Onofri et al. 1996). An operating premise of the DBT is that the laser beams in the measurement volume are focused to a much smaller size than the particle. Otherwise, the optical arrangement is similar to a conventional, two or three detector PDA system as pictured in Figure 7-38. Furthermore, detector positions are chosen, such that both reflective and refractive components of scattered light can be expected. However, due to the relatively large size of the particles, these components appear one after the other and not mixed. This is illustrated in Figure 7-39 for a water droplet and for a 16% ink solution.

From the burst arising from reflected light, it is possible to determine the particle size according to Equations 7.18 and 7.19. The bursts from refractive light also yield the particle size, according to Equations 7.18 and 7.20 or, if the refractive index is unknown, it can be estimated (Onofri et al. 1996).

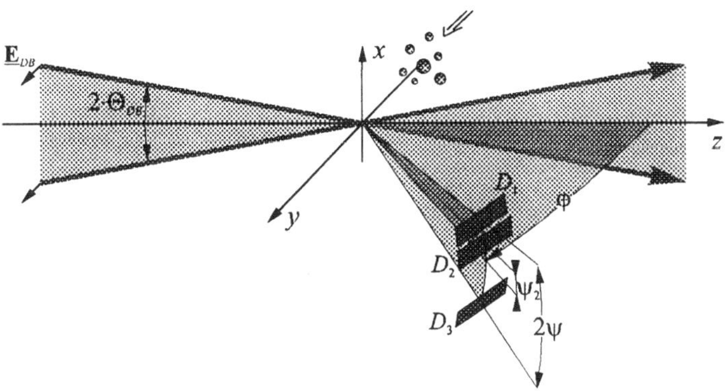

Figure 7-38 Optical arrangement for the dual burst technique

Thus, the DBT yields an estimate of refractive index, which in multiphase flows opens the possibility of distinguishing among different dispersed particles. Secondary properties of the particle, for instance the temperature, may also be estimated; however, this hinges on the accuracy and resolution of the refractive index measurements. For liquid droplets an accuracy of ±0.02 in the determination of the refractive index is typical, at least for droplets larger than 30-40 μm. For smaller droplets the technique is no longer suitable, primarily because the phase/diameter fluctuations at small diameters, together with inherent estimator variability, become larger than the measurement value.

Attention can now be turned to the amplitude of the refractive bursts shown in Figure 7-39. In the case of the 16% ink solution, the amplitude is considerably smaller due to the light absorption in the particle. The difference in amplitude can therefore be used to estimate the ink concentration, or more directly, the absorption coefficient can be given in terms of light intensity into the droplet and intensity leaving the droplet. The difficulty in implementing this is that the incident intensity, or alternatively the signal amplitude with pure water, is not known beforehand, only the signal amplitude of the reflected burst is known. To overcome this difficulty, the theoretical relation between the amplitude of reflective to refractive contributions for the particular optical arrangement is used, as computed by a light scattering program. Therefore, the ratio of the measured amplitude ratios to the theoretical amplitude ratios for pure liquid are used to estimate the absorption coefficient. Figure 7-40 illustrates some example measurements of absorption coefficient, compared with measurements taken with a refractometer.

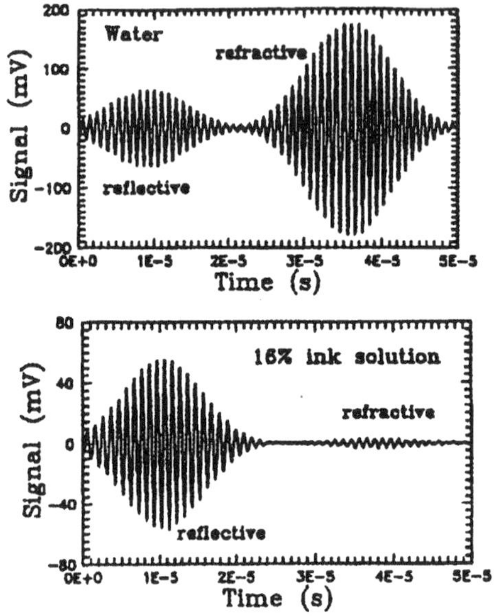

Figure 7-39 Signal received for a water droplet (upper figure) and a droplet with a 16% ink solution (lower figure)

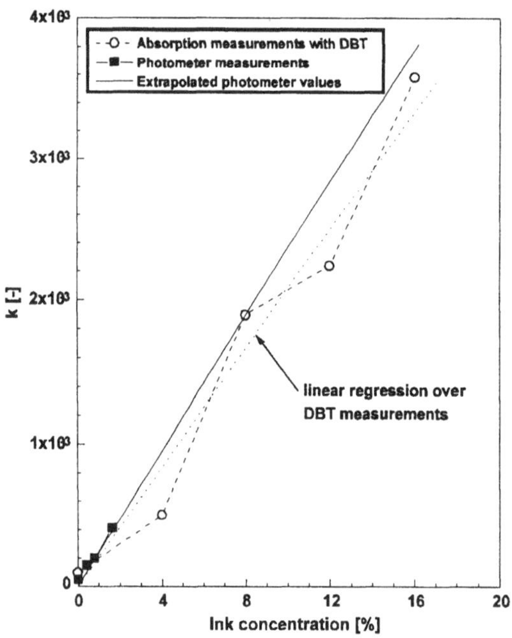

Figure 7-40 Absorption coefficient: comparison between DBT and photometer.

Clearly the DBT requires additional information from the received signals over what is currently available in commercial instruments. To date therefore, solutions based on transient recorders and software signal processing have been implemented. Furthermore the DBT does not measure the velocity component of the main particle flow, which incidentally must be more or less aligned with the *y* axis. Therefore a second LDA channel, with appropriate color separation in the receiving optics, must be added. Despite these restrictions and limitations, the DBT is a good example of novel light scattering techniques related to the PDA, which can be used for specific laboratory studies.

Finally some brief remarks will be directed to the use of rainbow refractometry, as this technique, although not yet mature, is intimately related to the light scattering involved in PDA and has also been combined with PDA instruments to extend measurement capabilities to refractive index. The monochromatic rainbow for spherical, homogeneous particles is a well documented scattering phenomenon, which results in very characteristic scattering intensity patterns in the far field (Bohren and Huffman 1983). Such an intensity distribution, the primary rainbow, is pictured in Figure 7-41 for a water droplet and a wavelength of λ=632.8*nm* (van Beeck and Reithmuller 1996 a)). The position of the first maximum is primarily a function of the refractive index of the particle. Further maxima, Airy fringes, also exhibit a characteristic scattering angle frequency, as does the ripple structure superimposed on these fringes. The spatial frquency of the Airy fringes is a strong function of particle size.

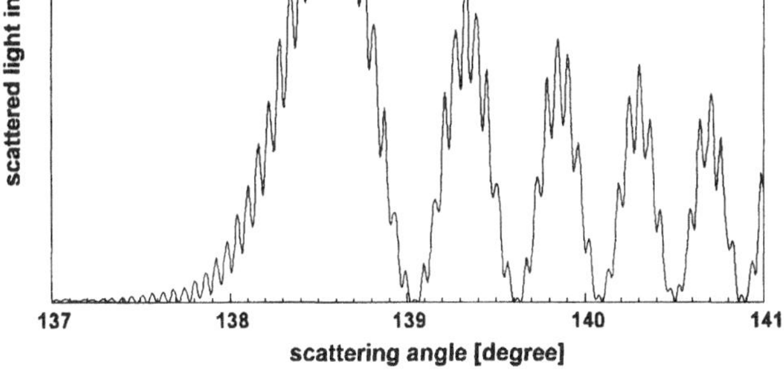

Figure 7-41 Far-field Lorenz-Mie scattered light intensity distribution, characterizing the primary, monochromatic rainbow (van Beeck and Riethmuller 1996 a)).

The properties of this scattering pattern have been studied extensively in an effort to extract size, sphericity, velocity and refractive index information about the particle (Roth et al. 1992, Roth et al. 1996, Marston 1980). Two concepts for implementation have been successfully demonstrated. One uses a line detector to capture the intensity pattern directly, and has been combined with a conventional PDA for size and velocity information (Sankar et al. 1996). The second uses a single photomultiplier and yields also particle velocity directly (van Beeck and Riethmuller 1996 b)). The main application to date is for the in-situ determination of fuel droplet size and temperature, however further development work is necessary before this technique can be routinely used.

7.4 SIGNAL PROCESSING

Attention is now turned to the signal processing and data processing tasks involved in LDA and PDA. For each scattering center passing the measurement volume, a signal of the form shown in Figure 7-42 is obtained, whereby the amplitude, the duration, the noise level, etc., depend on the particular optical set-up, the flow and the properties of the scattering center. It is the task of the signal processing to detect when a signal is present and then to estimate from the signal several primary measurement quantities, including the frequency (which yields the velocity), the arrival time of the particle, possibly the duration of the signal and in PDA the phase of the signal with respect to another signal. Sometimes amplitude information is also required from the signal, depending on the type of processor and data processing used.

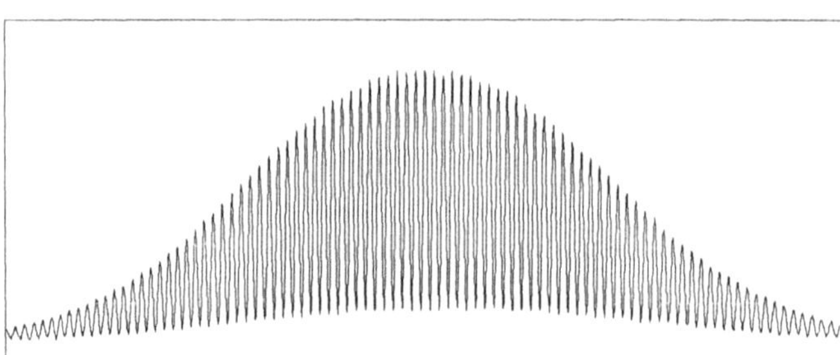

Figure 7-42 Signal from photodetector when a scattering particle passes the LDA measurement volume.

The data processing then estimates from the primary data the desired fluid mechanic properties, such as the mean velocity, the turbulence level, spectral densities, or in the case of PDA, particle distributions, concentrations, mass flux, etc. Furthermore, the data processing usually has some validation checks about whether the individual measurements were within acceptable bounds. Generally a purpose-built device is used for the signal processing, whereas the data processing is performed on a PC.

The first step of the signal processing is therefore to detect when a signal is present, which refers to distinguishing a signal from the noise background. The noise background arises from several sources, including stochastic noise coming from the photodetector and electronics (shot noise and Johnson noise) as well as from the physical processes themselves (scattering, laser). Noise can also arise from unwanted reflections or stray light associated with the flow rig. The noise in LDA/PDA signals is usually considered to be white in spectral content. The signal strength on the other hand is dependent on a variety of factors, as outlined in section 7.2.1.

The particle size is of particular importance, due to the squared dependence of scattered intensity, so that this represents an important optimization step in laying out an experiment with LDA. Large particles scatter more light and increase the signal strength, however they also respond less to flow velocity fluctuations. Neutrally buoyant particles are therefore particularly attractive as seeding particles for the continuous phase in dispersed, two-phase flows. The response of particles in a given flow-field can be estimated using a simplified equation of motion containing only the drag force (C_D = 24/Re) and the acceleration force.

$$\frac{d}{dt}U_p = -18\frac{\nu}{d_p^2}\frac{U_p - U_f}{(\rho_p / \rho_f)} \tag{7.45}$$

U_p - particle velocity $\quad$ U_f - fluid velocity

ρ_p - particle density $\quad$ ρ_f - fluid density

d_p - particle diameter $\quad$ ν - fluid kinematic viscosity

Obviously, U_p - U_f represents the slip velocity. Table 7-1 gives allowable seed particle sizes for a 99% amplitude response to sinusoidal fluctuations at 1 kHz and 10 kHz.

Finally, the choice of detector, either a photomultiplier, an avalanche-photodiode (APD) or a PIN diode, can greatly influence the final signal-to-noise ratio (SNR) of the signal, depending on the frequency of the signal involved and the wavelength of the light. Further details on choice of detectors and their influence on signal quality can be found in Durst and Heiber (1977) and Dopheide et al. (1987). Typically noise contributions are reduced by using band-pass filters prior to the signal processing, however great care must be taken in choosing cut-off frequencies, to avoid suppressing particle signal information.

Particle	Medium	Density ratio	Allowable diameter	
			f = 1 *kHz*	f = 10 *kHz*
silicone oil	air	900	2.6	0.8
TiO_2	air	3.5×10^3	1.3	0.4
MgO	methane-air flame (1800°C)	1.8×10^4	2.6	0.8

Table 7-1 Summary of allowable particle diameters for 99% amplitude response to sinusoidal fluctuations at the given frequencies.

The signal detection can be performed either in time domain or frequency domain. A simple time domain detection involves an amplitude level as an indicator for signals, as shown schematically in Figure 7-43. This method, although widespread in commercial processors, has many drawbacks and is rapidly being replaced by more advanced spectral techniques, in which the SNR of the signal is continually monitored and submitted to an acceptance level (Qiu et al. 1994; Ibrahim and Bachalo 1992). The SNR ratio can be derived either from the spectrum or from the autocovariance function, and both techniques are used in commercial processors. In the particular case of dispersed, two-phase or multi-phase flows, the signal detection may have the further task of distinguishing between the phases and this aspect is discussed in more detail below because most practical schemes employ combinations of signal detection, signal processing and optical techniques to make this distinction.

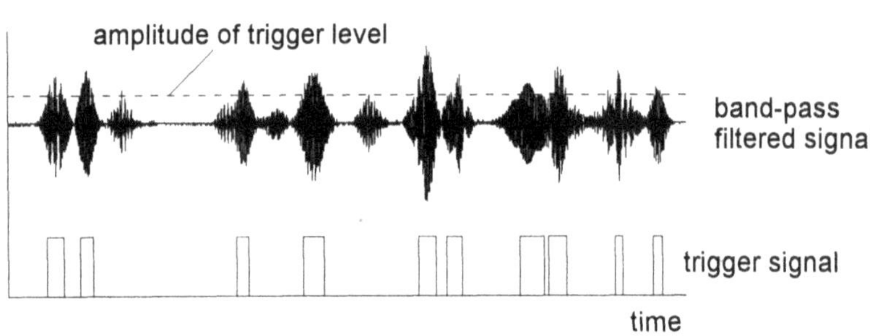

Figure 7-43 Signal detection using an amplitude level

The estimation of signal frequency is the main task of the signal processing and there have been a large number of techniques used in the past to accomplish this. Whereas time domain methods such as zero or level crossing detectors (counters) were used in the first generation of instruments, spectral methods are used almost exclusively now. Most processors are also realized in digital electronics. There are three basic spectral approaches used, as outlined in Figure 7-44, in which appropriate references are also given. Not uncommon are customized processing systems based on a fast digitizer, for instance a transient recorder or digital oscilloscope, and software programs for implementing the spectral analysis.

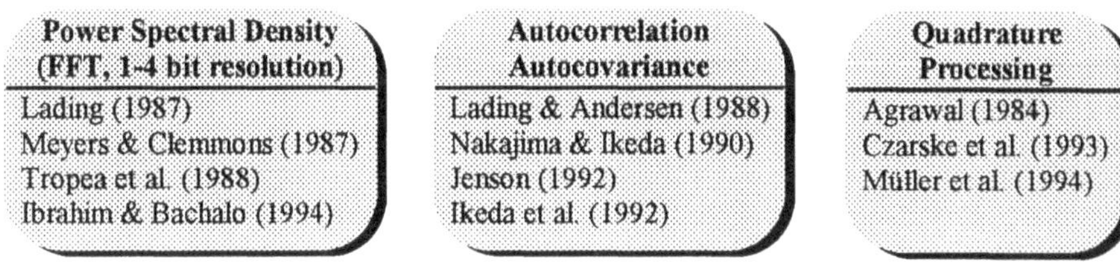

Figure 7-44 Overview of modern signal processing techniques in LDA and PDA.

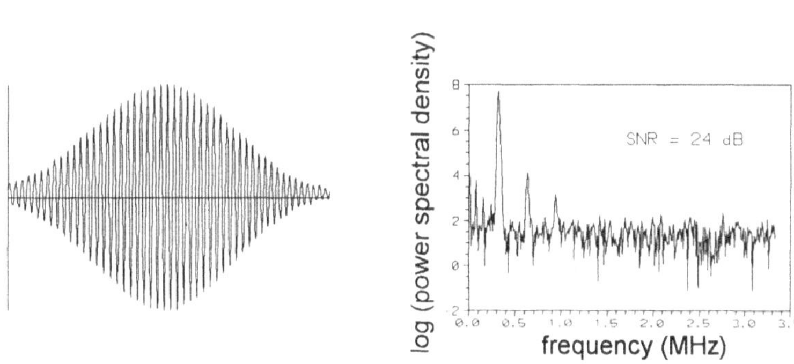

Figure 7-45 An example LDA signal after high-pass filtering (The power spectral density separates effectively the signal content from the noise content.)

The performance of any given processor will depend on a large number of parameters and can be very specific to a given application, as discussed by Tropea (1989). The main issues concerned are robustness, i.e. an insensitivity to front panel settings, ability to detect and estimate frequency at low SNR, typically below 0dB, and processing speed, which determines maximum achievable data rates. The advantages of spectral processing are manifold, the most decisive being the clear distinction between signal and noise. In the power spectral density the white noise appears as a constant value over all frequencies, in the autocorrelation as a peak at lag time zero and in the quadrature method as an amplitude variation of the rotating phasor. This is illustrated in Figure 7- 45, showing a high-pass filtered Doppler signal and its corresponding power spectral density. The techniques used for actually estimating the frequency from either the power spectral density or the autoccorelation are quite advanced, typically incorporating validation criteria to increase the reliability. For instance, acceptance may be dependent on a minimum SNR being exceeded and/or a maximum signal duration, derived from the system optical parameters and estimated velocity.

A remark concerning the interplay between optical parameters and signal processing performance is appropriate here, especially for two-phase flows. Clearly the signal-to-noise ratio should be maximized. A good approximation for SNR is given by Stieglmeier and Tropea (1992).

$$SNR \propto \frac{\eta P_0}{\Delta f}\left(\frac{D_a d_{in}}{f_t f_r}\right)^2 d_p^2 G V^2 \tag{7.46}$$

η	- quantum efficiency of detector	P_0	- incident light power
Δf	- bandwidth of system	d_p	- particle diameter
G	- scattering coefficient	V	- visibility
η	- quantum efficiency of detector	P_0	- incident light power
D_a	- receiving aperture diameter	d_{in}	- beam diameter
f_t	- focal length, tranmitting lens	f_r	- focal length, receiving lens

The most readily varied parameters here are the focal lengths and the scattering coefficient *G*, through the choice of the scattering angle. Focal lengths however are often dictated by the flow rig, as is the scattering angle. For PDA the scattering angle must also yield an appropriate phase/diameter response, see section 7.3.2. Another method of increasing SNR is through beam expansion, i.e. the laser beams before the front lens are expanded in order to achieve a smaller measurement volume, hence a higher light intensity. A smaller measurement volume will also, in principle, allow measurements in a more dense two-phase flow, without violating the fundamental pre-requisite for LDA/PDA, namely that only one particle resides in the volume at any one time.

However there are also drawbacks to reducing the measurement volume size indefinitely. In particular, the variance of frequency estimation is inversely proportional to the duration of the signal. Smaller measurement volumes result in short signals and hence more statistical scatter in the velocity values. This is none other than the Heisenberg uncertainty principle taking effect. This can be seen by examining the minimum possible estimator variance, the so-called Cramer-Rao-Lower Bound (CRLB), which in fact is closely achieved in many processors and is given by Rife and Boorstyn (1974) as:

$$\sigma_f^2 = \frac{3}{\pi^2 SNR \bullet N(N^2 - 1)} f_s^2 \tag{7.47}$$

Here f_s is the sample frequency and N represents the number of samples. The signal duration is given by N/f_s and thus the variance decreases with increasing signal duration.
Just to complete the discussion on measurement volume size, smaller volumes will cause systematic errors in particle sizing (Gaussian beam effect), as discussed in the following section. Also in multi-component LDA/PDA systems, very small measurement volumes, say below 50µm, become difficult to align on top of one another. Therefore the conclusion is that some optimization is necessary in choosing the measurement volume size.
The determination of arrival time is important to reconstruct the time series of velocity in the data processing stage. The duration of the signal, termed the residence or transit time of the particle, is also important for the data processing as described below. The accuracy requirement on these two quantities lies considerably below that of the signal frequency estimation. In PDA however, the transit time is often used to indirectly estimate the measurement volume size (Saffman 1987 b)) and for this a higher accuracy is required. Several refined techniques have therefore been proposed, also based on spectral analysis (Qiu and Sommerfeld 1992).
The phase of the signal is unimportant in LDA, however in PDA the phase difference between two simultaneous signals is the primary measurement quantity which corresponds to the particle size. Again, spectral domain estimates are most widely used either through the covariance function (Lading and Andersen 1988) or from the cross spectral density (Domnick et al. 1988). The latter is a complex quantity obtained after the Fourier transform of the two input signals *x* and *y*. The ratio of the imaginary to the real part gives the phase relation between the two signals at the chosen frequency. The frequency is chosen as the peak of the spectral magnitude and corresponds to the fundamental signal frequency. This is illustrated in Figure 7- 46, showing the function G_{xy} and θ_{xy} for a pair of PDA signals with SNR=25dB.

$$\left|G_{xy}(f_k)\right|^2 = \frac{2\Delta t}{N}\left|X_k^* Y_k\right| \qquad k = 0,1\ldots,N/2-1$$

$$G_{xy}(f) = \underbrace{C_{xy}(f)}_{\textit{Coherence}} - \underbrace{jQ_{xy}(f)}_{\textit{Quadrature}} \tag{7.48}$$

$$\theta_{xy}(f) = \arctan\left[\frac{Q_{xy}(f)}{C_{xy}(f)}\right] \tag{7.49}$$

Generally the peak position of the cross-spectral density magnitude is chosen by interpolating between two or more of the coefficients, yielding a frequency resolution at least one order of magnitude better than the coefficient spacing. A number of interpolation procedures have been proposed, usually employing either a parabolic fit on the logarithmic amplitude (Domnick et al. 1988) or a

Gaussian fit on the linear amplitude (Hishida et al. 1989). The appropriateness of the Gaussian curve form for interpolation can be improved by windowing the input data in time domain, typically with a Hanning or $\cos^2$ window (Matovic and Tropea 1991). Further improvements are achievable by using more points around the peak (Qiu et al. 1991) or by strategically spacing the points according to the spectral peak width (Matovic and Tropea 1991). Performance tests indicate, that optimized routines for spectral peak interpolation can be made reliable even for signals with a SNR as low as -10 dB (Qiu et al. 1991).

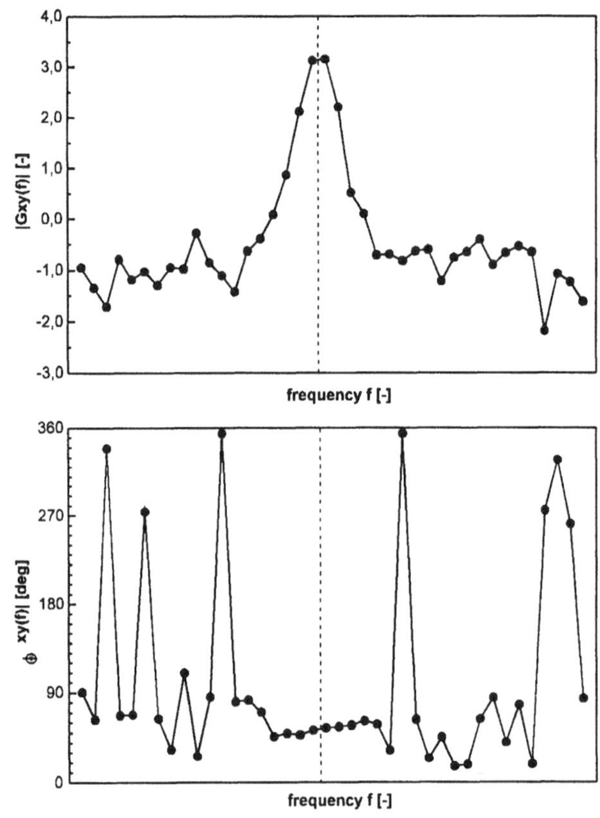

Figure 7-46 Cross spectral density function and phase of a PDA signal with SNR=25dB.

Whereas the method of signal processing in LDA and PDA is fixed by the choice of processor, the data processing task, performed with software, often requires considerable input from the user and must be matched carefully to the flow situation. There are basically two reasons for this.

- The flow velocity or particle velocity information from an LDA or PDA system is available only at irregular (and almost random) time intervals.
- The rate of velocity information is usually correlated to the quantity being measured, i.e. the flow velocity.

The first condition means that in principal, the constraint imposed for equidistant sampled data by the sampling theorem can be circumvented. Thus spectral content of velocity fluctuations can be estimated beyond half of the mean data rate, however this is generally achieved at the expense of estimator variability. There exists a large body of literature on the estimation of spectra from randomly sampled velocity data, such as with LDA, and this remains an active area of research (Adrian and Yao 1987, Gaster and Roberts 1977, Roberts and Ajmani 1986, Nobach et al. 1996). One aspect of these developments which is of particular interest when measuring in two-phase flows, is that of signal reconstruction, i.e. the estimation of fluid velocity between particles (Müller et al. 1994 a) b), Veynante and Candel 1988). In this way the velocity of the continuous phase can be approximated at the instance when the dispersed phase is measured and can thus lead to improved estimators of the slip velocity (Prévost et al. 1996).

The second feature of LDA data, the correlation between sample rate and sampled quantity, demands consideration when formulating estimators, even for the simplest quantities such as the mean velocity. This is illustrated in Figure 7-47 in which the measured particles are superimposed on a hypothetical velocity time trace. A simple arithmetic average of all particle velocities will result in an overestimation of the mean velocity, because more particles are seen at high velocities than at low velocities. Thus, the arithmetic average is a biased estimator, as first discussed by McLaughlin and Tiederman (1973).

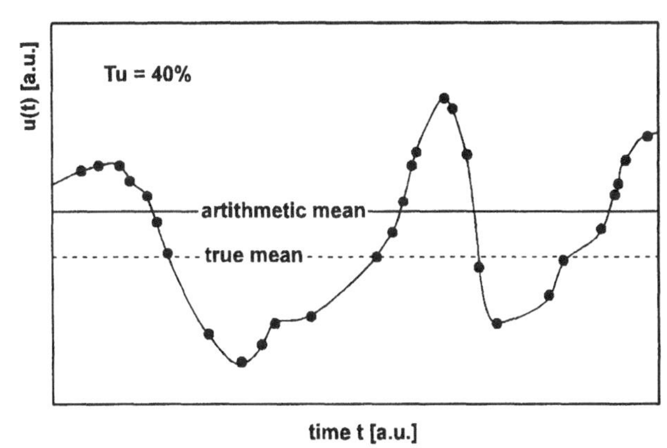

Figure 7-47 The deviation of an arithmetic average velocity over all measured particles compared to the true mean velocity.

It is apparent that the magnitude of the error on the mean velocity will increase with increasing turbulence level. For moderate levels of turbulence, up to about 40%, the maximum error is given by Erdmann and Tropea (1982) as:

$$\beta = \frac{\langle U \rangle - U}{U} = Tu^2 \tag{7.50}$$

where <U> is the arithmetic mean, U is the true mean and Tu is the turbulence intensity.

To avoid such difficulties with moment estimators, it is sufficient to weigh the individual samples with a factor inversely proportional to the velocity vector magnitude at the time of the sample. The velocity vector magnitude itself can be used if all three velocity components are measured, or if at least the dominating components are measured. Alternatively the duration of the signal $\Delta\tau$, representing the transit or residence time of the particle in the measurement volume, can be used, since this will decrease linearly proportional with flow velocity. Thus a reliable estimator for the first and higher moments can be given as:

$$\langle U \rangle = \frac{\sum_1^N U_i \Delta\tau_i}{\sum_1^N \Delta\tau_i} \qquad \langle u^n \rangle = \frac{\sum_1^N (U_i - \langle U \rangle)^n \Delta\tau_i}{\sum_1^N \Delta\tau_i} \tag{7.51}$$

For this purpose, most signal processing electronics also measure the residence time of particles.

7.4 RECAP AND FUTURE DIRECTIONS

This chapter has concentrated on single-point measurements for two-phase flows using elastic light scattering, with a focus on the laser-Doppler and phase-Doppler techniques. Despite many recent improvements in both techniques, the limitations of these methods, especially when applied to two-phase flows, must be recognized and understood. The phase-Doppler anemometer for instance, is restricted to spherical particles and, although non-sphericity can be reliably detected, there is presently no means to estimate the mass flux contained in the non-spherical particles which are usually excluded from further processing by validation criteria. For this and other reasons, the mass flux and concentration measurements in dispersed two-phase flows using PDA can be of widely varying accuracy and often below acceptable limits for use as verification data of numerical simulations. The consequence of these limitations is that in many studies a tailoring of the experiment can be recommended, for instance the dispersed phase can be modelled using spherical particles. Furthermore, independent and/or consistency checks of the measurement quantities should be planned as an integral part of the experiment. The most obvious example is that the total mass flux is measured also on the feed or collection line and compared to measured values integrated across given flow planes. This of course, is only possible in selected experiments and is seldom applicable in flows with evaporation.

Nevertheless the outlook for further improvements of both laser-Doppler and phase-Doppler instruments is very encouraging. Some recent development trends in LDA-systems have been summarized in Tropea (1995). These include miniaturization of the optical systems using semiconductor or solid state light sources, integrated optics, fiber devices and holographic elements. This miniaturization is generally also associated with increased robustness and performance, in terms either of signal-to-noise ratio and of measurable quantities. Rapid improvements have also been made in the field of data processing, in particular spectral analysis, which meets the rising need to resolve small scale turbulent motions.

The field of phase-Doppler anemometry or PDA-like instruments appears to have even larger potential for new developments. Some of these novel systems have been indicated in this chapter, however, further extensions can be expected in the measurement of non-spherical, oscillating or inhomogeneous particles. Unavoidably, these extensions will only be possible by detecting scattered light at additional spatial orientations and the challenge is to minimize the number of detectors, while gaining maximum significant information about the particles. One major constraint, both for laboratory and commerical systems, is that it is very difficult to focus physically separated detectors independently onto the same detection volume. This dilemma requires innovative technical solutions. One pre-requisite for these developments consists in readily accessible methods to compute the scattered light field from arbitrary particle positions in arbitrary beams. This goal is equally challenging as the design of appropriate optical systems and has been achieved to date only for very restricted classes of beams and particles. This field has recently been reviewed in a series of articles dedicated to the measurement of non-spherical and non-homogeneous particles (see the special issue of Measurement Science and Technology to appear in Vol. 9, Feb. 1998).

To summarize, the field of single-point measurements continues to undergo rapid developments in all of its aspects and applications and will undoubtly continue to play an important role in the study of dispersed two-phase flow. Such developments are presented at regular conferences, such as the International Symposium on Application of Laser Techniques to Fluid Mechanics (Lisbon) and the Conference on Optical Particle Sizing. Moreover, international journals, such as Measurement Science and Technology, Particle and Particle Systems Characterization, and Experiments in Fluids are devoted to recent developments in measurement techniques for two-phase flows.

7.5 REFERENCES

Adrian, R.J. and Yao, C.S., Power spectra of fluid velocities measured by laser Doppler velocimetry. Exper. in Fluids, 5, 17-28 (1987)

Agrawal, Y., Quadrature demodulation in laser Doppler velocimetry. Applied Optics, 23, 1685-1686 (1984)

Aizu, Y., Durst, F., Gréhan, G., Onofri, F. and Xu, T.H., PDA-system without Gaussian beam defects. 3rd Conference on Optical Particle Sizing, Yokohama, Japan, 461-470 (1993)

Albrecht, H.-E., Borys, M. and Hübner, K., Generalized Theory for the simultaneous measurement of particle size and velocity using laser Doppler and lase two-focus methods. Part. Part.. Syst. Charact. 10, 138-145 (1993)

Albrecht, H.-E., Bech, H., Damaschke, N. und Feleke, M., Berechnung der Streulichtintensität eines beliebig im Laserstrahl positionierten Teilchens mit Hilfe der zweidimenstionalen Fouriertransformation, Optik, 100, 118-124 (1995)

Allano, D., Gouesbet, G., Grehan, G. and Lisiecki, D., Droplet sizing using a top-hat laser beam technique. J. of Physics D: Applied Physics, 17, 43-58 (1984)

Bao, J. and Soo, S.L. Measurement of particle flow properties in a suspension by a laser system. Powder Technology, 85, 261-268 (1995)

Bauckhage, K., The phase-Doppler-difference-method, a new laser-Doppler technique for simultaneous size and velocity measuremets. Part. Part. Syst. Charact., 5, 16-22 (1988)

Black, D.L., McQuay, M.Q. and Bonin, M.P., Laser-based techniques for particle size measurements: Areview of sizing methods and their industrial applications. Prog. Energy Comb. Sci., 22, 267-306 (1996)

Bohren, C.F. and Huffman, D.R., Absorption and Scattering of Light by Small Particles. Wiley: New York, Chap. 4, 1983

Börner, Th., Durst, F. and Manero, E., LDV measurements of gas-particle confined jet flow and digital data processing, Proc. 3rd Int. Symp. on Applications of Laser Anemometry to Fluid Mechanics, Paper 4.5. (1986)

Chigier, N. A., Ungut, A. and Yule, A. J., Particle size and velocity measurements in planes by laser anemometer, Proc. 17th Symp. (Int.) on Combustion, 315-324 (1979)

Czarske, J., Hock, F. and Müller, H., Quadrature demodulation - A new LDV burst signal frequency estimator. Proc. SPIE 2052, 79-86 (1993)

Damaschke, N. Gouesbet, G., Grehan, G., Mignon, H. and Tropea, C., Response of PDA systems to non-spherical droplets. 13th Ann. Conf. Liquid Atomization and Spray Systems, Florence, Italy (1997)

DANTEC/Invent, STREU: A computational code for the light scattering properties of spherical particles. Instruction Manual (1994)

Domnick, J., Ertl, H., Tropea, C. Processing of phase/Doppler signals using the cross-spectral density function. 4th Int. Symp. on Appl. of Laser Anemometry to Fluid Mechanics, Lisbon, July 11-14 (1988)

Dopheide, D., Faber, M., Reim, G., Taux, G., Laser- und Avalanche-Dioden für die Geschwindigkeitsmessung mit Laser-Doppler-Anemometrie, Technisches Messen, 54, 291-303 (1987)

Dullenkopf, K., Willmann, M., Schöne, F.,. Stieglmeier, M., Tropea, C. and Mundo, Chr. Comparative mass flux measurements in sprays using patternator and phase-Doppler anemometers. 8th Int. Symp. on Appl. of Laser Techniques to Fluid Mechanics, Lisbon, 8.-11.7. (1996)

Durrani, T.S. and Greated, C.A., Laser Systems in Flow Measurement. Plenum Press, New York (1977)

Durst, F. and Zaré, M., Laser-Doppler measurements in two-phase flows. Proceedings of the LDA-Symposium, University of Denmark (1975)

Durst, F. and Heiber, K.F., Signal-Rausch-Verhältnisse von Laser-Doppler-Signalen, Optica Acta, 24, 43-67 (1977)

Durst, F., Melling, A. and Whitelaw, J.H. Principles and Practice of Laser-Doppler Anemometry. 2nd Edition, Academic Press, London (1981)

Durst, F., Review-combined measurements of particle velocities, size distribution and concentration. Transactions of the ASME, J. of Fluids Engineering, 104, 284-296 (1982)

Durst, F., Melling, A. and Whitelaw, J.H. Theorie und Praxis der Laser-Doppler Anemometrie. G. Braun, Karlsruhe (1987)

Durst, F., Tropea, C. and Xu, T.-H. The Slit Effect in Phase Doppler Anemometry. 2nd Int. Conf. on Fluid Dynamic Measurements and its Application, Beijing, China (1994)

Erdmann, J.C. and Tropea, C., Statistical bias of the velocity distribution function in laser anemometry. Proc. Int. Symp. on Application of LDA to Fluid Mechanics, Lisbon, Portugal, Paper 16.2 (1982)

Farmer, W. M., Measurement of particle size, number density and velocity using a laser interferometer. Applied Optics, 11, 2603-2612 (1972)

Farmer, W. M., Observation of large particles with a laser interferometer. Applied Optics, 13, 610-622 (1974)

Gaster, M. and Roberts, J.B. The spectral analysis of randomly sampled records by a direct transform. Proc. R. Soc. Lond. A 354, 27-58 (1977)

Grehan, G. and Gouesbet, G., Simultaneous measurements of velocities and size of particles in flows using a combined system incorporating a top-hat beam technique. App. Opt. 25, 3527-3538 (1986)

Grehan, G., Gouesbet, G., Nagwi, A. and Durst, F., On elimination of the trajectory effects in phase-Doppler systems. Proc. 5th European Symp. Particle Characterization (PARTEC 92), pp. 309-318 (1992)

Hardalupas, Y. and Taylor, A.M.K.P., On the measurement of particle concentration near a stagnation point. Exper. in Fluids, 8, 113-118 (1989)

Hardalupas, Y. and Taylor, A.M.K.P., Phase validation criteria of size measurements for the phase Doppler technique. Exper. in Fluids, 17, 253-258 (1994)

Hardalupas, Y., Hishida, K., Maeda, M., Morikita, H., Taylor, A.M.K.P. and Whitelaw, J.H., Shadow Doppler technique for sizing particles of arbitrary shape. Applied Optics, 33, 8417-8426 (1994)

Hecht, E. and Zajac, A., Optics, Addison-Wesley Publishing Company, Inc., New York (1982)

Heitor, M.V., Starner, S.H., Taylor, A.M.K.P. and Whitelaw, J.H., Velocity, size and turbulent flux measurements by laser Doppler velocimetry. in: Instrumentation for flows with combustion (Ed. A.M.K.P. Taylor), Academic Press, London, 113-250 (1993)

Hess, C. F., Non-intrusive optical single-particle counter for measuring the size and velocity of droplets in a spray, Applied Optics, 23, 4375-4382 (1984)

Hess, C. F. and Espinosa, V. E., Spray characterization with a nonitrusive technique using absolute scattered light. Optical Engineering, 23, 604-609 (1984)

Hishida, K., Kobashi, K. and Maeda, M., Improvement of LDA/PDA using a digital signal processor (DSP). Proc. 4th Int. Symp. on Appl. of Laser Anemometry to Fluid Mechanics, Swansea, UK (1989)

Hishida, K. and Maeda, M., Application of laser/phase Doppler anemometry to dispersed two-phase flow. Part. Part. Syst. Charact., 7, 152-159 (1990)

Ibrahim, K. and Bachalo, W., The significance of the Fourier Analysis in Signal Detection and Processing in Laser Doppler and Phase Doppler Applications. Proc. of the 6th Int. Symp. on Appl. of Laser Techn. to Fluid Mechanics, Lisbon, Portugal, paper 21.5 (1992)

Ibrahim, K. and Bachalo, W., Time-Frequency Analysis and Measurement Accuracy in Laser Doppler and Phase Doppler Signal Processing Applications.

Proc. of the 7th Int. Symp. on Appl. of Laser Techn. to Fluid Mechanics, Lisbon, Portugal, paper 8.3 (late) (1994)

Ikeda, Y., Shimazu, M., Yoshida, N., Nagayama, H., Kurihara, N. and Nakajima, T., Burst Digital Correlator for Wide-Band and Low SNR LDV Measurements. Proc. of the 6th Int. Symp. on Appl. of Laser Techn. to Fluid Mechanics, Lisbon, Portugal, paper 21.3 (1992)

Jenson, L.M., LDV Digital signal processor based on autocorrelation. Proc. of the 6th Int. Symp. on Appl. of Laser Techn. to Fluid Mechanics, Lisbon, Portugal, paper 21.4 (1992)

Kliafas, Y., Taylor, A.M.K.P. and Whitelaw, J.H., Errors due to turbidity in particle sizing using laser-Doppler anemometry. Trans. of the ASME,J. Fluid Engineering, 112, 142-148 (1990)

Lading, L. Spectrum Analysis of LDA Signals. Proc. of The Use of Computers in Laser Velocimetry, ISL, France, paper 20 (1987)

Lading, L. and Andersen, K., A Covariance processor for velocity and size measurement. 4th Int. Symp. on Appl. of Laser Anemom. to Fluid Mech., Lisbon, July 11-14, paper 4.8 (1988)

Maeda, M., Hishida, K., Sekine, M., and Watanabe, N., Measurements of spray jet using LDV system with particle size discrimination. Laser Anemometry in Fluid Mechanics-III (Eds. R.J. Adrian et al., Selected Papers from the 3rd Int. Symp. on Appl. of Laser Anemometry to Fluid Mechanics, 375-386 (1988)

Maeda, M., Morikita, H., Prassas, I.,Taylor, A.M.K.P. and Whitelaw, J.H., Accuracy of particle flux and concentration measurement by shadow-Doppler velocimetry. Proc. 8th Int. Symp. on Appl. of Laser Techn. to Fluid Mech., Paper 6.4 (1996 a))

Maeda, T., Morikita, H., Hishida, K. and Maeda, M. Determination of effective measuring area in a conventional phase-Doppler anemometer. Proceedings of the Eigth International Symposium on Applications of Laser Techniques to Fluid Mechanics, Lisbon, Portugal, Vol. 1, Paper 2.5 (1996 b))

Marston, P.L. Rainbow phenomena and the detection of non-sphericity of drops. Applied Optics 19, 680-685 (1980)

Matovic, D. and Tropea, C., Spectral peak interpolation with application to LDA signal processing. Measurement Science and Technology, 2, 1100-1106 (1991)

McLaughlin, D.K. and Tiederman, W.G., Biasing correction for individual realization of laser anemometer measurements in turbulent flows. Physics of Fluids, 16, 2082-2088 (1973)

Meyers, J.F. and Clemmons, J.I. Jr., Frequency domain laser velocimeter signal processor. NASA Techn. Paper 2735 (1987)

Mie, G., Beiträge zur Optik trüber Medien, speziell kolloidaler Metallösungen. Ann. der Physik, 25, 377-422 (1908)

Mignon, H., Gréhan, G., Gouesbet, G., Xu, T.-H. and Tropea, C. Measurement of cylindrical particles using phase-Doppler anemometer. Applied Optics, , (1996)

Modarress, D. and Tan, H., LDA signal discrimination in two-phase flows, Experiments in Fluids, 1, 129-134 (1983)

Morikita, H., Hishida, K. and Maeda, M., Simultaneous measurement of velocity and equivalent diameter of non-spherical particles. Part. Part. Syst. Charact., 11, 227-234 (1994)

Müller, E., Nobach, H. and Tropea, C., LDA signal reconstruction: Application to moment and spectral estimation. 7th Int. Symp. on Appl. of Laser Techniques to Fluid Mechanics, Lisbon, Portugal, Paper 23.2 (1994)

Müller, H., Czarske, J., Kramer, R., Többen, H., Arndt, V., Wang, H. and Dopheide, D., Heterodyning and quadrature signal generation: Advantageous techniques for applying new frequency shift mechanisms in the laser Doppler velocimetry. 7th Int. Symp. on Appl. of Laser Techniques to Fluid Mechanics, Lisbon, Portugal, Paper 23.3 (1994)

Mundo, Chr. Zur Sekundärzersäubung newtonscher Fluide an Oberflächen. Dissertation University of Erlangen (1996)

Nakajima, T. and Ikeda, Y., Theroetical evaluation of burst digital correlation method for LDV signal processing. Meas. Sci. and Techn., 1, 767-774 (1990)

Naqwi, A.A. and Durst, F. Light scattering applied to LDA and PDA measurements. Part 1: Theory and numerical treatments. Part. Part. Syst. Charact., 8, 245-258 (1991)

Naqwi, A., Durst, F. and Liu, X., Two methods for simultaneous measurement of particle size, velocity, and refractive index. Applied Optics, 30, 4949-4959 (1991)

Naqwi, A., Ziema, M., Liu, X., Hohmann, S. and Durst, F. Droplet and particle sizing using the dual cylindrical wave and the planar phase Doppler optical systems combined with a transputer based signal processor, 6th Int. Symp. on Appl. of Laser Techniques to Fluid Mechanics, Lisbon, 15.3 (1992)

Naqwi, A. and Menon, R., A rigorous procedure for design and response determination of phase Doppler systems. Proc. 7th Int. Symp. on Appl. of Laser Techniques to Fluid Mechanics, Lisbon, Portugal, Paper 24.1 (1994)

Negus, C. R. and Drain, L. E., Mie calculations of scattered light from a spherical particle traversing a fringe pattern produced by two intersecting laser beams, J. Phys. D: Applied Physics, 15, 375-402 (1982)

Nobach, H., Müller, E. and Tropea, C., Refined reconstruction techniques for LDA analysis. 8th Int. Symp. On Appl. of Laser Techn. to Fluid Mechanics, Lisbon, Portugal, Paper 36.2 (1996)

Onofri, F., Girasole, T., Gréhan, G., Gouesbet, G., Brenn, G., Domnick, J., Xu, T.H. and Tropea C., Phase-Doppler anemometry with the dual burst technique for measurement of refractive index and absorption coefficient simultaneously with size and velocity. Part. Part. Syst. Charact., 13, 112-124 (1996)

Panidis, Th. and Sommerfeld, M., The locus of centres method for LDA and PDA measurements. Proc. of the Eigth Int. Symp. on Applications of Laser Techniques to Fluid Mechanics, Lisbon, Portugal, Vol. 1, Paper 12.1 (1996)

Prévost, f., Boree, J., Nuglisch, H.J., Charnay, G. Measurements of fluid/particle correlated motion in the far field of an axisymmetric jet. Int. J. Multiphase Flow, 22, 685-701 (1996)

Qiu, H.-H., Sommerfeld, M. and Durst, F., High resolution data processing for phase-Doppler measurements in a complex two-phase flow. Measurement Science and Technology, Vol. 2, 455-463 (1991)

Qiu, H.-H. and Sommerfeld, M., A reliable method for determining the measurement volume size and particle mass fluxes using phase-Doppler anemometry. Experiments in Fluids, Vol. 13, 393-404 (1992)

Qiu, H.-H. and Sommerfeld, M., The impact of signal processing on the accuracy of phase-Doppler measurements. Proc. 6th Workshop on Two-Phase Flow Predictions, Erlangen 1992, (Ed. Sommerfeld, M.), Bilateral Seminars of the International Bureau Forschungszentrum Jülich, 421-430 (1993)

Qiu, H.-H., Sommerfeld, M. and Durst, F., Two novel Doppler signal detection methods for laser-Doppler and phase-Doppler anemometry. Meas. Sci. Techn., 5, 769-778 (1994)

Qiu, H.-H. and Hsu, C.T., A Fourier optics method for the simulation of measurement volume effect by the slit constraint. Proceedings of the Eigth International Symposium on Applications of Laser Techniques to Fluid Mechanics, Lisbon, Portugal, Vol. 1, Paper 12.6 (1996)

Rife, D.C. and Boorstyn, R.R., Single-tone Parameter Estimation from Discrete-time Observations. IEEE Trans. on Information Theory, 20, 591-598 (1974)

Roberts, D.W., Particle sizing using laser interferometry. Applied Optics, 16, 1861-1868 (1977)

Roberts, J.B. and Ajmani, D.B.S., Spectral Analysis of Randomly Sampled Signals Using a Correlation-based Slotting Technique. IEEE Proc., 133, 153-162 (1986)

Roth, N., Anders, K. and Frohn, A. Simultaneous determination of refractive index and droplet size using Mie theory. Proc. 6th Int. Symp. on Appl. of Laser Techniques to Fluid Mechanics, Lisbon, 15.5 (1992)

Roth, N., Anders, K. and Frohn, A. Size insensitive rainbow refractometry: Theoretical aspects. Proc. 8th Int. Symp. on Appl. of Laser Techniques to Fluid Mechanics, Lisbon, Paper 9.2 (1996)

Saffman, M., Optical particle sizing using the phase of LDA signals. Dantec Information, No. 05, 8-13 (1987 a))

Saffman, M., Automatic calibration of LDA measurement volume size., Appl. Optics, 26, 2592-2597 (1987 b))

Sankar, S.V. and Bachalo, W.D., Response characteristics of the phase-Doppler particle analyzer for sizing spherical particles larger than the wavelength. Applied Optics, Vol. 30, 1487-1496 (1991)

Sankar, S.V., Bachalo, W.D. and Robart, D.A., An adaptive intensity validation technique for minimizing trajectory dependent scattering errors in phase Doppler interferometry. 4th Internationsl Congress on Optical Particle Sizing, Nürnberg, Germany (1995)

Sankar, S.V., Buermann, D.H. and Bachalo, W.D. An advanced rainbow signal processor for improved accuracy in droplet temperature measurements. 8th Int. Symp. on Appl. of Laser Techniques to Fluid Mechanics, Lisbon, 9.3 (1996)

Sommerfeld, M. and Qiu, H.-H., Characterization of particle-laden confined swirling flows by phase-Doppler anemometry and numerical calculation. Int. J. Multiphase Flows, 19, 1093-1127 (1993)

Sommerfeld, M. and Qiu, H.-H., Particle concentration measurements by phase-Doppler anemometry in complex dispersed two-phase flows. Exper. in Fluids, 18, 187-198 (1995)

Stieglmeier, M. and Tropea, C., Mobile fiber-optic laser doppler anemometer. Appl. Optics, 31, 4096-4105, (1992)

Tayali, N. E. and Bates, C. J., Particle sizing techniques in multiphase flows: A review. Flow Meas. Instrum., 1, 77-105 (1990)

Taylor, A.M.K.P., Two phase flow measurements. Optical Diagnostics for Flow Processes (Eds. L. Lading et al.), Plenum Press, New York, 205-228 (1994)

Tropea, C., Dimaczek, G., Kristensen, J., Caspersen, Chr., Evaluation of the Burst Spectrum Analyser LDA Signal Processor. 4th Int. Symp. on Appl. of Laser Anemom. to Fluid Mech., Lisbon, July 11-14, paper 2.22 (1988)

Tropea, C. Performance Testing of LDA/PDA Signal Processing Systems. 3rd Int. Conf. Laser Appl.- Advances and Applications, Sept. 26-28, Swansea, Wales (1989)

Tropea, C. Laser Doppler anemometry: Recent developments and future challenges. Meas. Sci. Techn. 6, 605-619 (1995)

Tropea, C., Xu, T.-H., Onofri, F., Grehan, G., Haugen, P. and Stieglmeier, M., Dual mode phase Doppler anemometer. PARTEC 95, Preprints 4th Int. Congress Optical Particle Sizing, 287-296 (1995)

Tropea, C., Xu, T.H., Onofri, F., Gréhan, G., Haugen, P. and Stieglmeier, M. Dual mode phase Doppler anemometer, Part. Part. Syst. Charact., 13, 165-170 (1996)

van Beeck, J.P.A.J. and Riethmuller, M.L., Rainbow Phenomena Applied to the Measurement of Droplet Size and Velocity and to the Detection of Nonsphericity, Appl. Optics, 35, 2259-2266 (1996 a))

van Beeck, J.P.A.J. and Riethmuller, M.L., A Single-Beam Velocimeter Based on Rainbow-Interferometry. Proc. 8th Int. Symp. on Appl. of Laser Techniques to Fluid Mechanics, Lisbon, 9.1 (1996 b))

Van de Hulst, H.C., Light Scattering by Small Particles. Dover Publications, Inc. New York (1981)

van de Wall, R.E. and Soo, S.L. Measurement of particle cloud density and velocity using laser devices. Powder Technology, 81, 269-278 (1994)

Veynante, D. and Candel, S.M., Application of Non Linear Spectral Analysis and Signal Reconstruction to Laser Doppler Velocimetry. Exp. Fluids, 6, 534-540 (1988)

Yeoman, M. L., Azzopardi, B. J., White, H. J., Bates, C. J. and Roberts, P. J., Optical development and application of a two-colour LDA system for the simultaneous measurement of particle size and particle velocity. ASME Winter Annual Meeting, Arizona, 127-135 (1982)

8

Full Field, Time-Resolved, Vector Measurements

Yang Zhao and Robert S. Brodkey

In the history of turbulence there are a number of markers in time. Such points occur when researchers take stock of their efforts and attempt to evaluate its worth to engineering science. The area of coherent structures in turbulent shear flows has attracted the attention of many researchers from the mid-1960's to the present time. These people have made major gains and have established a rudimentary picture of the dynamic details of turbulence: sweeps, ejections, interactions, etc. But we do not have a complete picture. We would like to have full-field, time-resolved, vector velocity data with high enough resolution so that we could determine the stress and vorticity fields. Two approaches show promise: experiments that provide full-field measurements and direct numerical simulation (DNS). Neither approach is fully satisfactory today for highly resolved, practical flows in the time domain! However, we are at a threshold of being able to accomplish this. In this review, some of the experimental approaches that seem fruitful and might be amenable to further development to give us the information we need to progress to the next step in our understanding of turbulence will be outlined and discussed.

8.1 INTRODUCTION

Turbulent flow still is one of the most important and challenging problems for scientists and engineers. There is little doubt that to correctly understand, describe and control turbulence will provide great benefits to the design of processes. Where possible, experimental observations are a first step to provide guidance for theoretical studies and computational simulations. All of these allow the necessary checks of the accuracy of established engineering models. There is a wide range of algorithms available to model steady-state flows. However, many of the flow fields of current interest, such as coherent structures in shear flows, are unsteady. There is a comparative lack of experimental data and models for such unsteady flow fields. Hot-wire or laser Doppler anemometer (LDA) data of such flows are difficult to interpret as both spatial and temporal information of the entire flow field are required and these methods are commonly limited to simultaneous measurements at only one or at most a few spatial locations. They are in reality, single point measurement techniques. Because of new light sources, such as powerful lasers, as well as the rapid development of digital computers, there is now available commercial two-dimensional (2-D) measurement techniques. The most common 2-D measurement techniques, particle image velocimetry (PIV), consist of two steps. First, the flow field is seeded with small particles. A laser light sheet illuminates a selected plane of the flow and the flow pattern in this plane is recorded. Secondly, the recorded flow patterns are processed and analyzed to obtain the desired 2-D information. This method can provide the simultaneous measurement in time in the illuminated plane even for high unsteady flow fields by extending the present commercial techniques to high speed video cameras and advanced computational techniques. However, most turbulent flows are not only highly unsteady but also strongly three-dimensional. Therefore, the development of 3-D measurement techniques is an essential further step for further progress.

We would like to be able to model practical turbulent systems that are of commercial importance, for example, the flow within the cylinder of an internal combustion engine. We also want to understand complex reactor flow phenomena which can be helpful to improve the efficiency and the quality of commercial products. For example, consider the flow in mixing vessels common to the biotechnology industry, the formation and removal of voids in manufacturing polymer composites, etc. We can make rough estimates today by experimental measurements and by numerical simulations. But these are rough because our experiments are not refined enough, we do not have the depth of understanding needed to accurately model the field, or our computers are not adequate. The hope is that a well-founded middle of the road approach will work, but this will involve some degree of approximation. For example, large eddy simulations (LES) might be the path to provide the answers we need. However, to make this approach work we need to understand more about the smaller scale mechanism of turbulent flow, especially when chemical reactions are involved. Such understanding has been the goal of fluid dynamics research for many decades. The researchers of the past started with the simplest of ideas. As each step or approach proved not to be the definitive model, the efforts became more and more complex. Often along the time-line, some researchers developed a frustration that their approach was not going to generate the desired results. They had arrived at a point of reflection about their work. The recent development of digital computers (like the personal computer - PC) has brought a revolutionary change to our approach to flow measurements. In recent years, microprocessors, the PC's central brains, traditionally doubles in speed and halved in price every 18 months (according to a report in USAToday). New subsystems - video, sound, discs and control boards - are built to make best use of that power. This provides a possibility for processing huge image data files in relative short time and can be a new stepping stone for the next thrust to "understand turbulence."

The history of frustration of fluid researchers is not new. For example, early researchers realized that laminar flow models could not describe turbulent flow, so turbulent had to be studied. G.I. Taylor realized that the phenomenological approach of eddy viscosity and mixing lengths could not describe turbulent flow adequately. This was a point of reflection that led Taylor to introduce the statistical approach to turbulence. G.K. Batchelor realized that the statistical approach, in turn, was not giving the answers he desired and was not sure what should be done. He chose to change his research field. Some researchers turned to what we now call the coherent structures approach. They wanted to obtain a picture of the flow in terms of coherent structures and establish some idea of the dynamical interactions that occur in the flow. However, the frustration is back and these researchers now realize that although we have made major gains and have established a rudimentary picture of the dynamical details of turbulence, we *do not* have the picture needed to allow us to model the flow with the degree of reliability we want. Is it time once again for reflection? If so, whence turbulence? What do we need? Where must we go? What are we missing?

Current investigations are directed toward full-field measurement techniques and direct numerical simulation (DNS). The numerical approaches are limited by the need for much bigger and better computers. Previously, visual observations were used for qualitative assessment. Hot-wire/film and LDA measurements were used to provide the hard numbers for a few points in space in the time domain. Today, the visual-based techniques are being extended to allow full-field, time-resolved velocity vector information to be obtained. However, the need for full-field and time-resolved measurements put vast restrictions on what can be accomplished. To get time-resolved results, often today, we must sacrifice resolution. To get resolution, we must sacrifice the dynamics. Ultimately we want both.

Let us think about what might be ideal for the final attack on the turbulence reaction problem. In the most ideal of all worlds, we would like to have full-field, time-resolved, vector velocity measurements with high enough resolution that we could determine the stress and vorticity fields. This requires that we either measure the instantaneous velocity vectors in the entire space-time domain or extend DNS. These efforts must be applied to meaningful turbulent shear flows, not just to idealizations. What do we need for the next attack on the turbulence problem? For the experimental work, we must develop and utilize full-field, time-resolved, scalar and velocity vector information. We need the dependent stress and vorticity fields to aid in solving our problems that involve chemical reactions in turbulent flows. In this brief review, some of the experimental approaches that seem fruitful and that might be amenable to still further development will be outlined and discussed. Where it will be helpful, we will put the work into context of applied fluid flow problems and theoretical approaches. Hopefully, this combination will give us the information we need to progress to the next step in our understanding of turbulence.

What then is the next step? All the single point, time-resolved techniques are to be considered as techniques of the past for this review. All the two-dimensional, time-resolved or not, are also considered a technique of the past, unless there is a hope that it can be extended to full-field, time resolved measurements. We will fold in those possible techniques that are unique and can provide the simultaneous measurement of both the velocity and scalar fields. It might also be well at this point to cite some of the applications that are being addressed for such measurements: flow in an internal combustion engine, mixing in biotechnical reactors, mixing in conventional mixers, mixing and reaction in opposed jet reactors, mixing and reaction in pipe flow reactors with internal elements, flow in the cooling regions of an internal combustion engine, catalyst surface analysis (non-flow), and slow motion flow in polymer composite structures (void formation and removal).

In summary, commercially available flow measurements have been developed from single point techniques to non-time-resolved, two-dimensional (2-D) methods. Also today low resolution, three-dimensional (3-D) measurements can be made. It is the intent of this Chapter to discuss and outline some of the experimental approaches and their applications, including, but not limited to, particle tracking velocimetry (PTV), scanning particle image velocimetry (SPIV), holographic particle image velocimetry (HPIV), laser induced photochemical anemometer (LIPA), laser induced fluorescence (LIF) and scattering methods (Lorenz-Mie, Rayleigh, Raman). These are techniques that might be amenable to further development to give us the full-field, time-resolved, vector information we need to progress to the next step. Let us emphasize, at this point, the development of these techniques is still in progress, no one can currently provide high-resolution, full-field (3-D), time-resolved velocity vector measurements and dependent measures like vorticity. They can provide parts of these, but not all at once. It is also clear, however, that the goal can be reached in time. Finally, it should be pointed out here that with the advent of fast, efficient imaging hardware, the use of image-based measurement has increased tremendously, and the number of annual publications on imaging velocimetry has grown exponentially. It is almost impossible to include every aspect of image-based measurements in this short review. For further background information, the reader is referred to Adrian (1991, 1993) and Grant (1994). A new PIV bibliography edited by Adrian (1996) contains references from early studies done from 1917 to the latest research in 1995.

8.2 PARTICLE TRACKING VELOCIMETRY (PTV)

The simple goal is to obtain a fully automated, computer-based technique that can track a sequence of particle motions in two or more views to allow extraction of the full three-dimensional flow field measurements in the time domain. Such a procedure would involve the fully integrated image processing of the raw images, position location of the particles and their tracks in two or more views, stereo matching to establish the three-dimensional nature of the flow and a final evaluation for consistency of the measurements. Derivative properties such as vorticity, stress and strain rates are calculated from these instantaneous velocity vectors. For turbulent flows, mean flow vector properties are obtained by ensemble averaging the velocity vectors over a large enough number of realizations. Subtracting the mean component from each vector in the instantaneous fields provides the instantaneous fluctuating vectors over the whole flow region. Because particles were tracked over a time sequence (from one frame to the next), there are no velocity direction ambiguity problems such as in PIV. The velocity vectors in the time domain are obtained. The method must be carefully tested on known synthetic data and then validated for a number of real and meaningful flows.

Particle tracking is probably the most popular technique for full-field measurements. This technique usually (but not always) uses two cameras. An image taken by one camera is a projection of flow markers in the three-dimensional (3-D) space onto a two-dimensional (2-D) image plane. Hence, the single image does not contain enough information to establish the third dimension. That is, distances from the objects to the camera are lost unless the 2-D image information is supplemented in some manner. One means of providing more information is two or more images taken from different camera positions. In this manner the three-dimensional structure of the markers can be extracted directly. This is the approach used at The Ohio State University and thus, our stereo, multiframe PTV technique is first briefly described in this section. The major steps involved in the process are calibration, preprocessing, tracking, establishing correspondence and recovering the 3-D information.

1) Camera calibration

In principle, with two views, either taken from two cameras or one camera with a stereo mirror or a prism, the position of an object in the 3-D world can be determined from the analysis of intersecting image rays. Unfortunately, in practice, there often is no exact intersection between the two rays. This occurs because the recorded image will almost certainly have errors due to the imperfection of physical imaging systems and the particular physical limitations that occur in every application in which image data are recorded. For instance, errors in locating the centroid arise from camera lens distortions, finite resolution due to recording of the image on film because of finite grain size or on a video detector array of finite pixel size. Obviously, a calibration is a prerequisite for reliable data. In addition to eliminating errors, a camera calibration also is used to establish the relationship between the 3-D world coordinates and the corresponding 2-D image coordinates on film or a detector array. Once this relationship is established, 3-D information can be inferred from the 2-D images and vice versa. The important issues in camera calibration are

i) ability to deal with matching errors,
ii) compensation for image distortions,
iii) knowledge of camera positions and parameters and
iv) knowledge of the locations of selected markers in the 3-D world.

The camera calibration, in our case, is accomplished by a least-squares method to determine the relative position and orientation of two cameras from a set of matched points. For more detail of camera calibration see Slama (1980).

2) Preprocessing

Preprocessing of images is an important component for any image-based measurement. The purpose of image preprocessing for PTV measurements is to help in the identification of the particles from the background and in the location of their centroids. During preprocessing, the particles must satisfy certain well-defined characteristics, such as threshold value, minimum size, maximum size and maximum aspect ratio. Some standard operations involved in this stage are background subtraction, contrast enhancement, filtering, etc. Special attention must be given to particle overlap. The resolution of a particle overlay into individual particle positions is necessary for good results.

3) Tracking and establishing correspondence

Perhaps, the matching of particles is the hardest and the most important step in using the stereo approach. Once the stereo images are brought into point-to-point correspondence, the 3-D reconstruction process is relatively straightforward. Given two views of a measured field, correspondence needs to be established among the particles identified. Matching strategies can be differentiated according to the primitives used for matching as well as the image geometry. Ideally, we would like to find the correspondences (i.e., the matched locations) of every individual particle in both images of a stereo pair. However, the information content in a single particle is too low for unambiguous matching. It is impossible to stereoscopically match all the individual points in the two views when the concentration of particles is usefully high. Instead, we establish the tracks by a tree type search from frame to frame in both views and then the tracks are stereoscopically matched. In the tracking part of the analysis, use is made of the continuity of position, velocity and acceleration; that is, as much of the physics as possible is brought into play to help in the analysis. Once matched, the vector velocity information can be extracted as described in the next stage.

4) 3-D positions determination

The 3-D reconstruction process can be considered as an inverse procedure of the calibration. Since closed form solutions may not exist for all cases, a more general approach is required for this process. The lines joining the center of projection and the 2-D image point in each of the stereo images are projected backwards into 3-D space. Then the point in space that minimizes the sum of its distance from each of the back-projected lines is chosen as the estimated 3-D position of the matched point. As mentioned above, instead of using a single point, the identified tracks are used as the matching features. The midpoints of the matched tracks are projected backwards in 3-D space and the 3-D positions of the tracks are determined using a similar minimization criterion. When compared with a single frame, multi-exposure PTV techniques, the main advantages of our multiframe, single exposure PTV are

i) the results are time-resolved as particles can be tracked for long times,

ii) this allows a higher number of seed particles, as a consequence, there is higher spatial resolution, and

iii) there are no ambiguous problems about the motion direction.

The entire process is detailed in Guezennec *et al.* (1994) and won't be repeated here. It is important to note that image preprocessing is a necessary step as well as calibration of the geometry because of index of refraction mismatching and lens distortion. It cannot be stressed too much, that accurate calibration and, if possible, as close an index of refraction matching are essential for accurate results.

In the first evaluation of the technique, we used a swirling and tumbling flow field that is something like that experienced during the intake stroke in an internal combustion engine. The flow was generated by simple modeling on the computer, the tracks for a large number of fluid particles were established and were then translated to what would be viewed by two orthogonal views. These views were then evaluated by the PTV techniques and the results compared to the exactly known locations and velocities. There are, of course, many detailed steps along the path to obtain satisfactory results.

The technique works quite well and has now been applied to a variety of important and practical flow fields. Several such results for 3-D measurements are shown in Figures 8-1 and 8-2. All the vectors found by the PTV technique were located randomly. They have been interpolated and projected onto regular grid points in the three orthogonal directions. Figure 8-1a shows one example from a cut through the water simulation of an internal combustion engine during the intake stroke when the piston is at bottom dead center. An example of the flow at the midplane of two opposing jets that are sometimes used in jet reactors is shown in Figure 8-1b. Figure 8-2a shows a mixing vessel common to the biotechnology industry. Vortical structures observed in this mixing vessel is represented in Figure 8-2b and the turbulent dissipation field for the mixing system is shown in Figure 8-2c. This has been hypothesized to correlate with cell destruction during mixing.

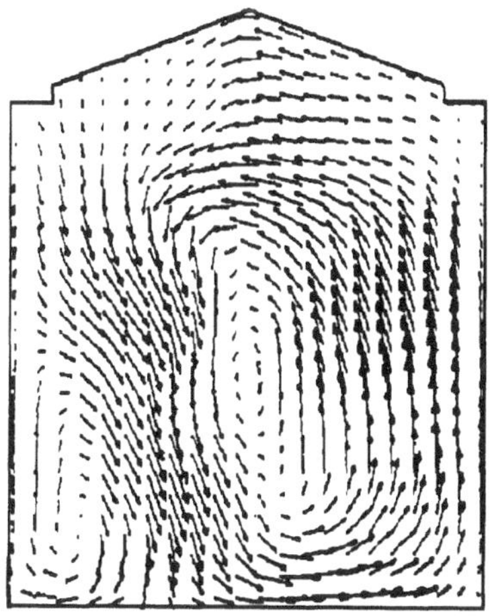

FIGURE 8-1a *One 2-D plane of 19 through the water simulation of an internal combustion engine during the intake stroke when the piston is at bottom dead center (Reprinted by permission of Guezennec et al.)*

However, there are limitations in the number of particles that can be observed in each view. In the analysis, two stereoscopic views are being used to establish the three-dimensional position of the particles. These views are the projections

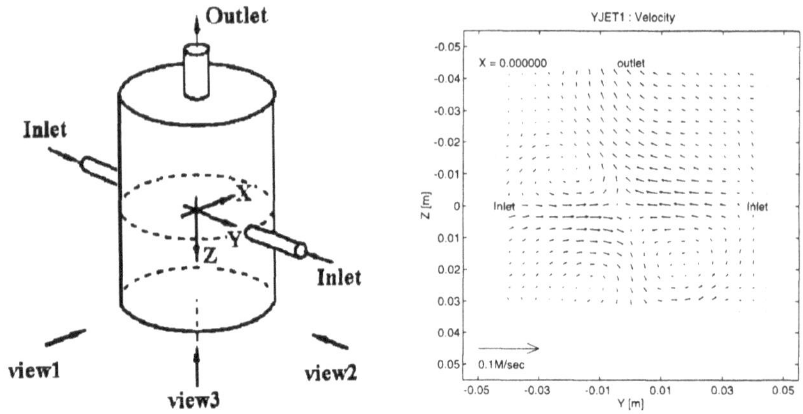

FIGURE 8-1b *One 2-D plane of 19 through the center of the two interacting jets.*

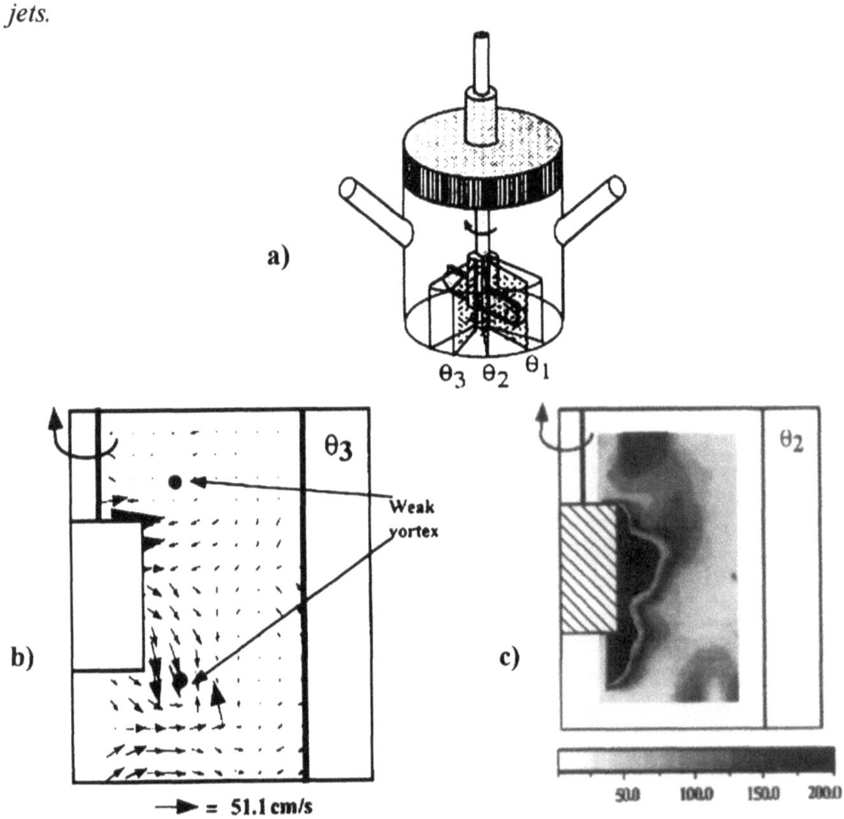

FIGURE 8-2 a) *Mixing vessel typical in the biotechnology industry,* **b)** *vortical structures observed in this vessel, and* **c)** *the turbulent dissipation field for this system (Venkat 1995). (Reprinted by permission of Venkat.)*

on two image planes of all the particles in the flow field. PIV techniques only study a slice of the flow field, while the PTV approach uses all the particles in the flow field at each instant of time. Instead of using two cameras, Racca and Dewey (1988) have used a single camera and a split-view mirror system for 3-D measurements by PTV. This is the same as for the mixing vessel study previously cited. Followed Racca and Dewey (1988), Reese and Chen (1995) have successfully applied the stereo-mirror PTV technique in studying the local, transient flow phenomena in multiphase systems. For a single camera with a split-view system, the main advantage is there are no synchronization problems between cameras. On the other hand, the measurement fields are smaller than using two cameras because two views are now projected onto the single camera's image plane. Clearly, it would not take too high a density of particles to supersaturate the projected view; thus, the particle density is limited so that the analysis can cope with the particle density in each projected view. The analysis is thus limited to modest particle concentrations (several hundred in the volume).

The particle concentration can be increased by increasing the number of views. Although hard numbers are not available, the concentration might be increased by a factor of five or so by this means and not increase the computational time very much. Dracos and his coworkers (Maas *et al.*, 1993; Makik *et al.*, 1993) have used up to four cameras. They investigated the influences of geometry parameters and physical factors on the establishment of a 3-D mathematical model, the system calibration and the particle tracking procedure. They tested and validated the method against simulated turbulent flow trajectories. In addition, a simple experiment was carried out to verify the procedure. A complex approach has been examined by Kasagi and Matsunaga (1995) using three video cameras. In addition to measuring mean and turbulence velocities, three Reynolds stress values and other third-order correlations were obtained for flow over a backward-facing step. One of their 3-D measured results is shown in Figure 8-3. In these examples, instantaneous data were obtained, but the results were ensemble averaged to establish an adequate sample for the curves shown. As point out above, it is in essence, phase averaged information where the local fluctuations in velocity might be composed of a true turbulence contribution and also a cycle to cycle (or sample to sample) contribution. It should be emphasized, that although the particle concentration can be increased, it is often more convenient to look at a series of realizations and then use an ensemble average to gain a picture of the flow field. Of course, this is averaged and cannot provide information about the dynamics of the flow, which requires time-resolved data.

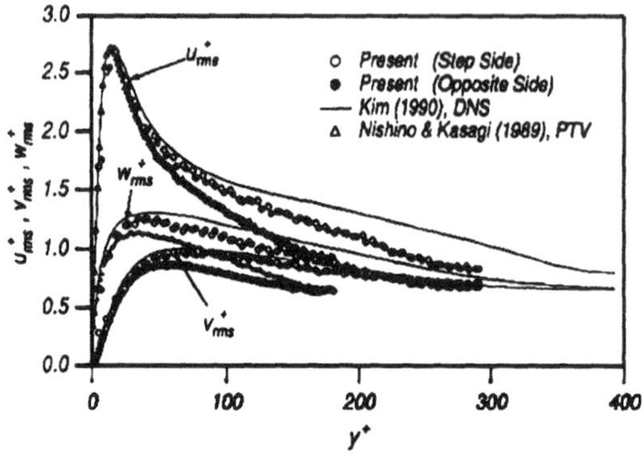

FIGURE 8-3 *Three-dimensional distribution of root-mean-square turbulent velocity fluctuations upstream of a step (Kasagi and Matsunaga 1995). (Reprinted by permission of Butterworth/Heinemann.)*

8.3 OTHER TECHNIQUES

In the remaining part of this paper, possible application of other techniques being investigated by a number of researchers will be discussed. The emphasis will be on the three dimensional and time dependent nature of the measurements. Most of these techniques are being used for idealized systems and the current efforts are directed to the development of the system of measurement itself.

8.3.1 Scanning Particle Image Velocimetry (SPIV)

The commercially available and more conventional PIV approach provides a two-dimensional slice of the flow field. Good starting places for the previous work are the reviews cited at the end of the introduction section. Most systems are not designed to provide time resolution, but rather ensemble averaged results are obtained to establish the statistical two-dimensional average and fluctuating flow field. This limitation can be overcome. The present video-based systems are limited to the video framing rate, which in turn is limited by the band-width of the imaging boards used. Film-based techniques or high band-width and/or high on board storage video systems could be used to extend the results into the time domain. Because the two-dimensional PIV view is an illuminated plane that is normal to the viewing, there is no question as to the location in the third dimension. The particles being observed are only in the lighted field of view. The density of particles (on a volume basis) for the PIV case can be much higher than for the 3-D PTV case. It is simply the difference in looking at a single 2-D plane versus the full 3-D volume. However, the limitation of 2-D rules the technique out for the type of measurements being discussed here. Having ruled such measurements out, one needs to cite at least, the use of stereoscopic viewing to provide a measure for the velocity in the third dimension within the plane as done by Prasad and Adrian (1993). Since, for conventional PIV, the particle images resulting from two or more exposures are stored on the same recording, there is no directional information about the particles' movement. In order to avoid directional ambiguity problems, image-shifting devices are often used. Raffel *et al.* (1995) adopted a very different way for obtaining information about the out-of-plane velocity components. In contrast with the auto-correlation technique normally used in PIV, they estimated out-of-plane velocities by analyzing images of particles within two adjacent laser sheets by spatial cross-correlation. Their imaging arrangement is shown in Figure 8-4. The scanner was only used to alternate the laser sheet location after each second during imaging.

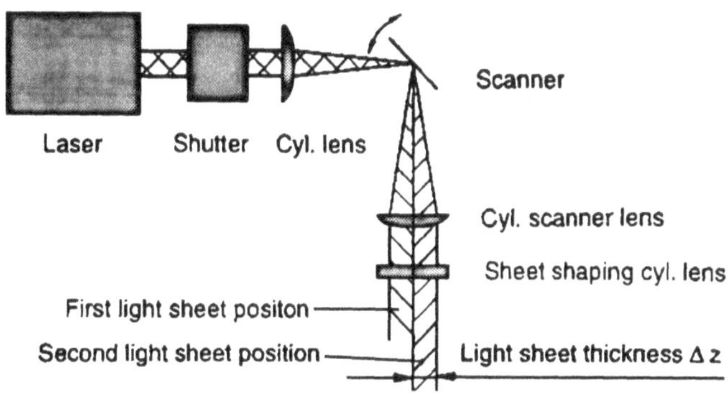

FIGURE 8-4 *Optical components for two parallel light sheet planes used to establish out of plane velocity component (Raffel et al. 1995). (Reprinted by permission of Springer-Verlag.)*

As an alternate, the 2-D plane could be scanned across the volume rapidly and, if fast enough, the full 3-D flow field could be reconstructed and at much higher particle densities than is possible with the PTV approach. Brücker and Althaus (1992) described a scanning apparatus shown in Figure 8-5. The function of the rotating polygonal mirror was to deflect the laser beam rapidly to form a "laser sheet" and to provide multi-exposure. The generation of the scanning laser sheets was then performed by the oscillating mirror. Using this method, a set of 2-D velocity fields was obtained from these scanning sheets by the normal PTV methods. As a test case, the 3-D shape of a vortex ring was reconstructed from these laser sheets. Reviews of the reconstruction of 3-D flow structures from such scanning light sheets can be found in articles by Gad-el-Hak (1989).

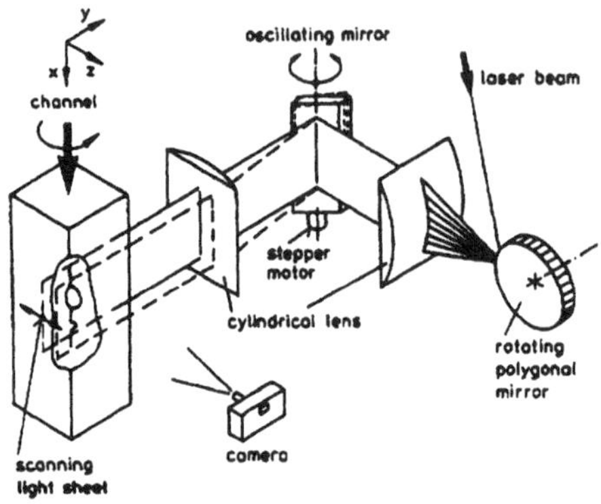

FIGURE 8-5 *Optical components used for scanning a single laser beam to obtain 10 cross sections of the flow (Brücker and Althaus 1992). (Reprinted by permission of Springer-Verlag.)*

More recently, a new approach in scanning systems has been made at The Ohio State University. Figure 8-6 shows the scanning system put forward by Guezennec *et al.* (1994), where the scanning beam is maintained exactly parallel. This avoids a correction introduced by using a rotating mirror, where the planes are not parallel. If the beam is wide (a few mm or so, not for the full field) then 3-D information can be obtained for each plane by using a stereoscopic viewing approach (Figure 8-7). The use of scanning requires obtaining the series of scanned images fast enough so there is little change in the flow field during the cycle time to scan across the volume. This requires high-speed recording. Film-based techniques are faster and of higher resolution than video based techniques. However, modest resolution, but high speed (or high-resolution, modest speed) video based systems are becoming available at a high price. The choice in the work cited here is to use film with a Cu-vapor laser that can record individual pictures at high resolution at 10,000 frames per second (fps). The laser and rotating drum are synchronized to the film speed, since this later is not necessarily constant during the entire run. There can be an appreciable acceleration period at the beginning to get the film up to the final speed. Indeed, some of the older cameras at the higher speeds, operated entirely in the acceleration period. You often ran out of film before getting to the final high speed.

The imaging techniques for individual frames are similar to the present processing by PIV methods. These methods will not be discussed further here. However, the post-processing suggested by Guezennec and Kiritsis (1990) needs to be mentioned further. Figure 8-8 shows the conventional 2-D PIV processing by using cross-correlation in limited sub-volumes between successive scans.

Figure 8-9 illustrates the 2-D "Guided" tracking between successive scans and possibly adjacent laser sheets. Figure 8-10 shows a reconstructed synthetic field, where a) is the synthetic test, b) the conventional 2-D PIV cross-correlation result, c) the actual particle locations and d) the refinement using the subsequent 2-D tracking to add information for each particle trace. These latter two illustrations are a close-up of the vortex structure that is at the lower center of illustration a). Clearly, the loss of detail as a result of the finite area needed for the cross-correlation is recovered by the subsequent guided tracking step.

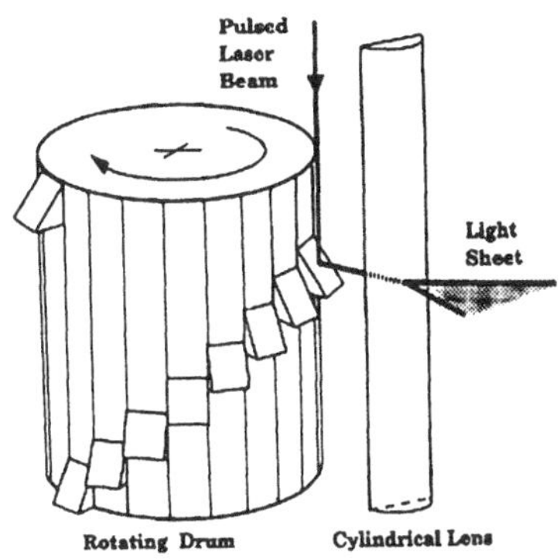

FIGURE 8-6 *Parallel beam scanning system (Guezennec et al. 1994). (Reprinted by permission of Guezennec et al.)*

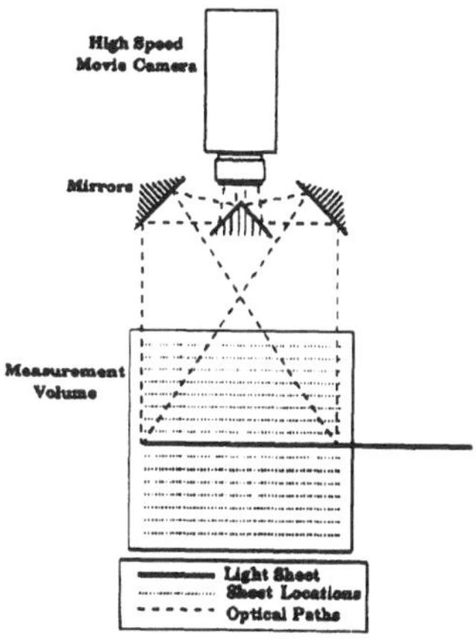

FIGURE 8-7 *Schematic of slit-image arrangement for stereoscopic viewing using a movie camera (Guezennec et al. 1994). (Reprinted by permission of Guezennec et al.)*

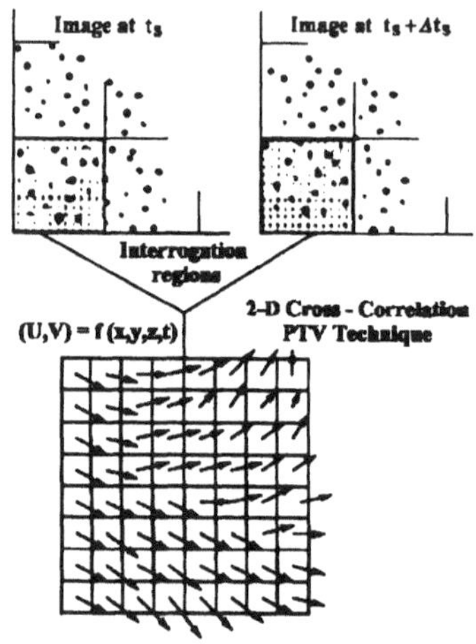

FIGURE 8-8 *Schematic of the 2-D cross-correlation processing between successive scans (Guezennec et al. 1994). (Reprinted by permission of Guezennec et al.)*

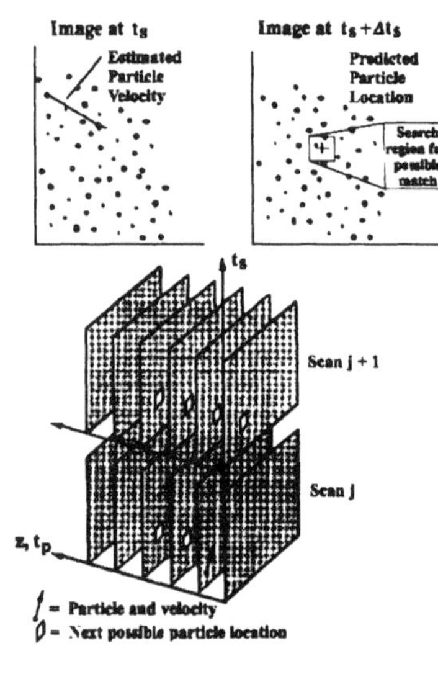

FIGURE 8-9 *Schematic illustration of the 2-D "guided" tracking between successive scans (and possibly adjacent laser sheets) (Guezennec et al. 1994). (Reprinted by permission of Guezennec et al.)*

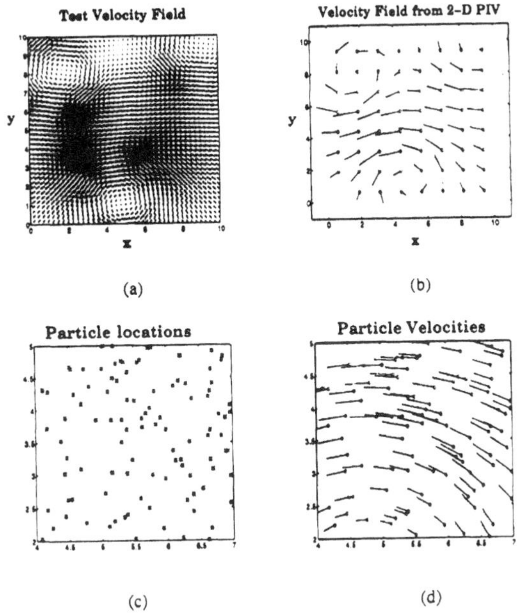

FIGURE 8-10 *Example of 2-D cross-correlation processing and subsequent 2-D tracking (Guezennec et al. 1994). (Reprinted by permission of Guezennec et al.)*

8.3.2 Holographic Particle Image Velocimetry (HPIV)

HPIV methods require a complex and very stable optical system. There are several possible techniques: in-line, off-axis, stereo-viewing, phase conjugate reconstruction and combinations of these, etc. All of the details will not be shown here, but can be found in the references by Adrian and coworkers (Barnhart *et al.*, 1994) and Hussain and coworkers (Meng and Hussain, 1991; Simmons *et al.*, 1993; Hussain *et al.*, 1994). These techniques are truly three-dimensional viewing methods. However, they are currently limited by being very complex and presently are limited to one instant in time. Effective time-based recording has not as yet been accomplished for HPIV. Research to extend HPIV methods to examine complex flows with full space and time resolution is underway in a number of laboratories; however, progress is slow and the efforts are either for limited scales and low time resolution. For example, Weinstein and Beeler (1987) made flow measurements behind a cylinder by using dual-view holographic movies. Two orthogonal in-line holographic cameras were used with the experimental arrangement shown in Figure 8-11. Dual view holographic movies were made at 15 frames/sec for an U_∞ of 2 cm/sec. However, there are efforts to extend PIV measurements by using higher speed digital cameras (Meinhart *et al.*, 1993; Hassan, 1994) and these could then be used for HPIV. Of course, ensemble averages can be obtained by looking at a large number of the individual realizations. This later is the current mode of operation for PIV and will no doubt be extended to HPIV. The computational time involved for high density HPIV images is large and thus has limited what has been accomplished. Figure 8-12 shows two recent examples.

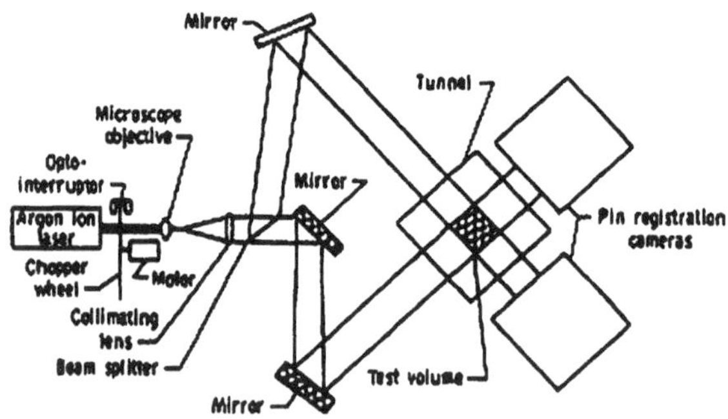

FIGURE 8-11 *Dual view holographic camera system for holocinematographic velocimeter (Weinstein and Beeler 1987). (Reprinted by permission of NASA.)*

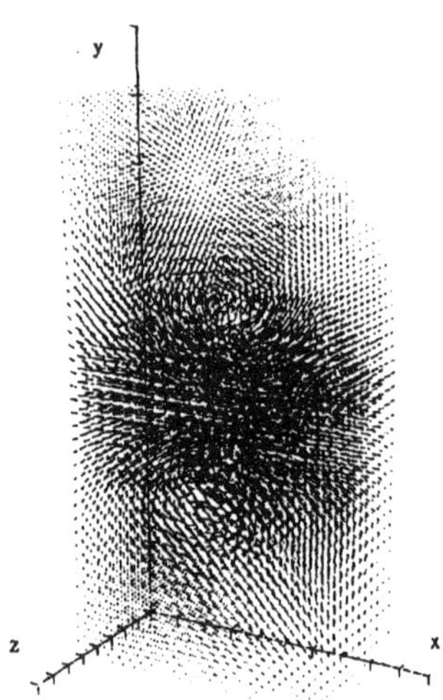

FIGURE 8-12a *Velocity vector field in a 3-D space (21x40x11 mm^3) of half the vortex ring. Grid interval is 1 mm; total number of vectors is 10,824 (Meng and Hussain 1995). (Reprinted by permission of American Institute of Physics.)*

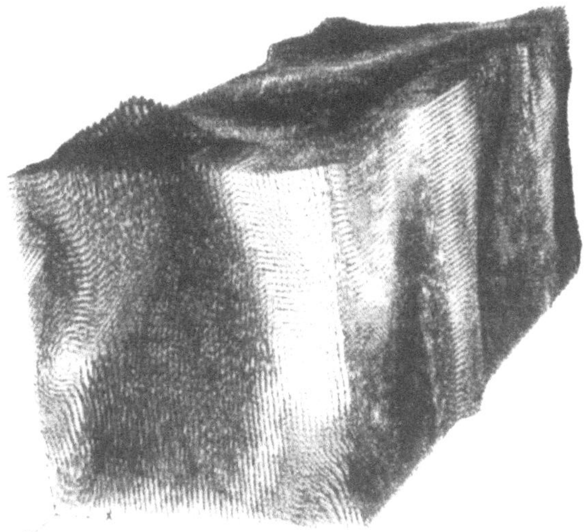

FIGURE 8-12b *Complete 3-D vector field volume, from a pipe channel flow, showing the measured vectors along the sides (24.5x24.5x60 mm^3) of the measurement volume, based on cross-correlation of particle images. More than 400,000 3-D velocity vectors have been extracted. The mean velocity of 0.8 m/s has been subtracted (Barnhart and Adrian 1994). (Reprinted by permission of Optical Society of America.)*

8.3.3 Laser Induced Photochemical Anemometer (LIPA)

LIPA is based on placing a non-invasive grid system in the flow. This is done by marking the fluid with a laser beam and long term fluorescence of the fluid from a dissolved photochromic or photoluminescent chemical. It is easiest to implement on a two-dimensional basis; however, thought has been given to using stereoscopic viewing and a more complex three-dimensional grid. The first efforts to measure velocity date back to the single line work of Hummel and his coworkers (Popovich and Hummel, 1967; Seeley *et al.*, 1975). The grid development technique is due to Falco and his coworkers (Falco and Chu, 1987; Chu *et al.*, 1992), and the most recent improvements are by Hill and Klewicki (1996). This latter work has an excellent literature review.

The starting grid is displaced by the flow and the changes in location of the gird allow the velocity, vorticity, etc. to be determined. The technique is limited in the time domain because of the diffusion or disappearance of the dye marker with time. However, a series of time realizations should be possible, although there may be gaps in the time line. Figure 8-13 from Hill and Klewicki (1996) illustrates the scanning beam technique to generate the grid, a sample grid and it representation, and an example of the grid some time later when it has deformed. These are 2-D results. Results for the extension to 3-D have not been reported yet.

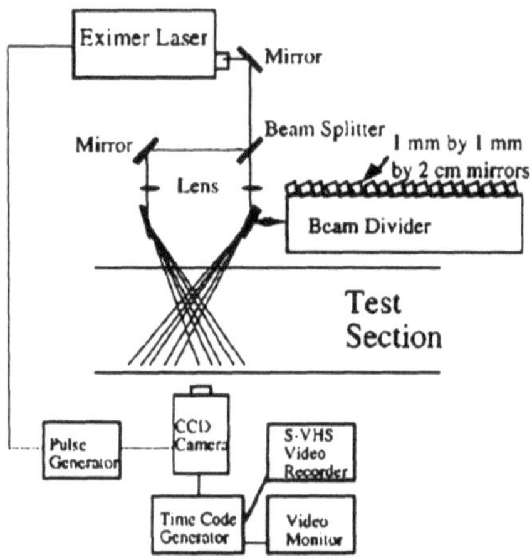

a)

FIGURE 8-13a *LIPA optical configuration for flow tagging velocity measurements (Hill and Klewicki 1996). (Reprinted by permission of Springer-Verlag.)*

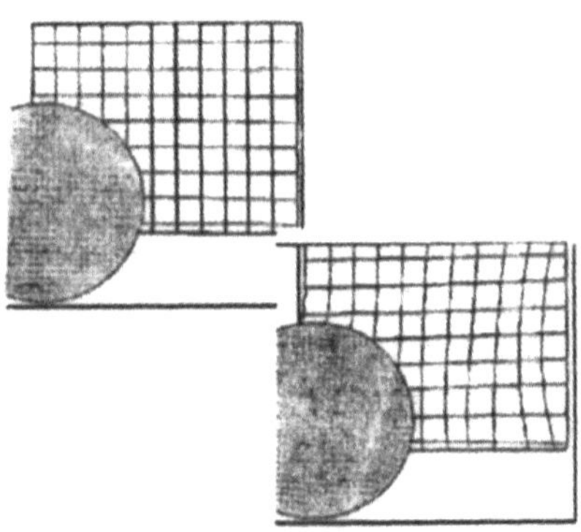

b)

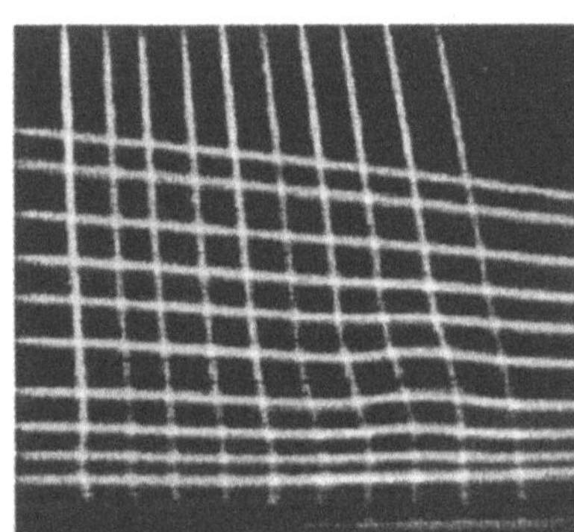

c)

FIGURE 8-13b, c **b)** *Sketch of the original and displaced grids for LPIA,* **c)** *A typical deformed experimental LIPA grid image (Hill and Klewicki 1996). (Reprinted by permission of Hill and Klewicki and Springer-Verlag.)*

8.3.4 Laser Induced Fluorescence (LIF) and Scattering Methods (Lorenz-Mie, Rayleigh, Raman)

LIF and scattering methods are often lumped together because the configuration is essentially the same as that used for PIV. Details of the principles of LIF as well as examples of LIF measurements, were presented by Miles and Nosenchuck (1989). The attraction for this approach is that it only requires the presence of suitable tracer species and avoids the use of particles. Most of the LIF image activity is concerned with species concentration. The species-specific nature of LIF provides direct information on the possible coupling between chemistry and fluid motions. Clearly, LIF techniques have potential to contribute to our understanding of mixing and reacting flows. Most of the LIF techniques are currently found in 2-D configurations. Again, there is no need to provide details about the experimental facilities, since they are parallel to what is currently used for PIV measurements. Just as for PIV, each of these could, have, or are being extended by using a scanning system to provide full 3-D viewing. Where PIV measures the velocity, the LIF and scattering procedures are used for concentration, density and temperature measurements. Combinations of these have also been used.

An important development is underway by Dahm and his coworkers (Dahm *et al.*, 1991, 1992, 1996). In this effort, full field velocity vector information is obtained from scalar measurements. The formal facilities to do the measurements are again very similar to a scanning PIV setup. The data storage requirements are massive as they obtain 200 images in the time domain, each image being a volume of 256 elements on a side. In their first article, the technique has been adopted to analyze the fine scale structure of molecular mixing in turbulent flows. The means of making space-time, fine scale scalar measurements are addressed. Although the velocity fields are not obtained in this article, the vector fields of the scalar dissipation rate are presented. In their second article, the analysis of such data to extract the fully resolved space-time vector velocity field is considered, which they call scalar imaging velocimetry (SIV). In this work, the scalar transport equation has been applied to recover the velocity fields. The final article of the three is an excellent review of their work and the work of others in this area. A number of examples of what can be accomplished are offered. One example from their work is given in Figure 8-14 that shows the scalar field and the corresponding vector field along the local gradient vector.

This approach, although complex in its implementation, is relatively straightforward and holds promise for further development. Since the technique relies on a scalar measurement in a liquid flow field, it is currently restricted to very fine scale measurements because of the large Schmidt number involved with dye flow markers in liquids. Thus, as demonstrated so far the method is over

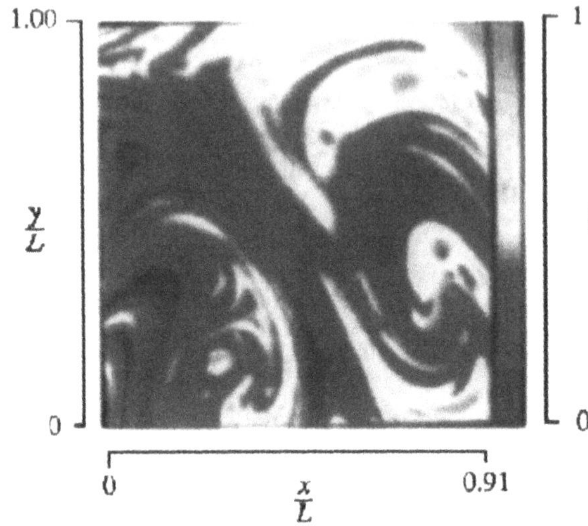

FIGURE 8-14a *The measured scalar field, ζ(x,t), for Sc >>1 as obtained by Dahm et al. 1991. Resolution is* 256^2 *for one plane with 256 colors for the concentration field (Dahm et al. 1992). (Reprinted by permission of American Institute of Physics.)*

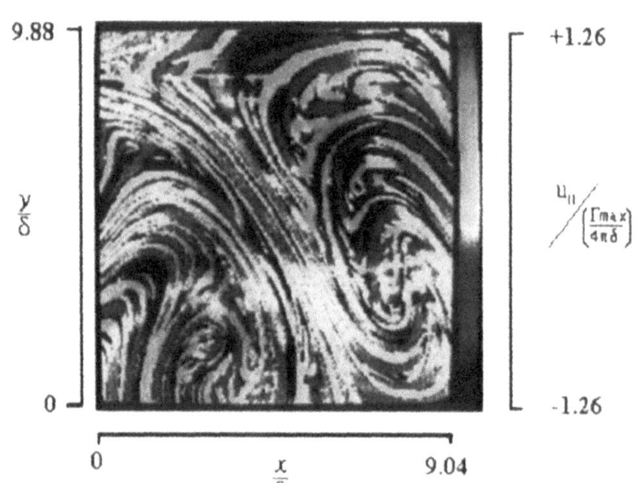

FIGURE 8-14b *The velocity component field* $u_{\parallel}$ *(x,t) along the local scalar gradient vector direction* $\hat{e}_{\nabla\zeta}$ (**x**,t) *obtained vis the scalar transport equation (Dahm et al. 1992). (Reprinted by permission of American Institute of Physics.)*

resolved since the scales associated with the high Schmidt number scalar transport is 45 times that needed for the velocity field. To use the method as suggested for large scale motions would require even more massive data handling, which is at the limit now with present technology.

It is interesting to speculate what might be done by combining this scalar technique with the previously mentioned larger scale PTV approach. The recognition for the need of such large scale measurements, has led the authors in the final article (Dahm *et al.* 1996), to suggest a possible extension to large scale motions. Thus, they suggest a plausible means to obtain the full-field measurements by intentionally using under-resolved scalar field data. The means suggested, although not as yet tested, would remove the need for fully-resolved measurements that have been presented so far.

Merkel *et al.* (1994, 1995) have recently modified the LIF technique to use a scanning mirror and flow tomography to investigate turbulent mixing. The LIF technique dates back to the earlier work of Dimotakis and his coworkers (Koochesfahani and Dimotakis, 1986; Dahm and Dimotakis, 1990). The present effort is based on a proposal by Maas (1993) that suggests an optimal fitting technique to match the scalar field evolution. In the more recent work, fifty consecutive planes were recorded by use of a high-speed photodiode camera, which gave a 2-ms exposure time for each plane. A small volume (15x15x3 mm^3) within a turbulent jet, near the intermittent boundary, was investigated. This work, also at a large Schmidt number, is restricted in its view to a relatively small scale. This, in turn, requires that the resolution for Δx, Δy and Δz must be smaller than the scalar diffusivity to allow meaningful measurement in all three space directions. Since the spatial resolution is set by the optical detector, with 512x512 pixels being typical, the LIF technique is still restricted to very fine scale measurements. It is expected that this limitation can be eliminated by the use of newer and higher resolution cameras and faster and much larger data storage systems. It also may be possible to apply the recent suggestions of Dahm *et al.* (1996) to the correlation technique; however, this suggestion has not been evaluated. Figure 8-15 shows an example of the scalar and velocity fields obtained by the correlation approach. In contrast to the work of Dahm and his coworkers, this work does not invert the scalar transport equations, but rather uses a correlation technique to obtain the presented results.

Long and his coworkers (Frank *et al.*, 1994, 1996) have studied LIF and scattering techniques for combustion problems. There can be disadvantages for any of the techniques that range from very low signal to noise ratio for Raman scattering to LIF limitations associated with the chemistry of the tagged molecules. Figure 8-16 provides two examples. The first is a pseudo-color example of the concentration field of a 2-D Freon jet mixing into air. The second is an example for a non-reacting jet that shows the 2-D fields of the simultaneous

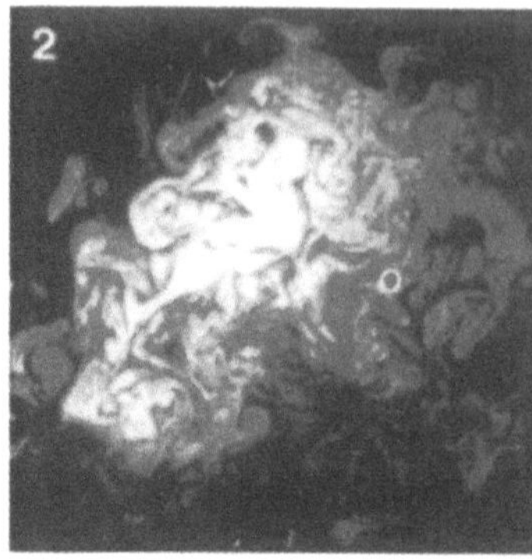

FIGURE 8-15a *Cross section of jet at $x/d_0 = 200$ from the orifice. The location of measurement is marked by the circle. (Merkel et al. 1994). (Reprinted by permission of Merkel et al.)*

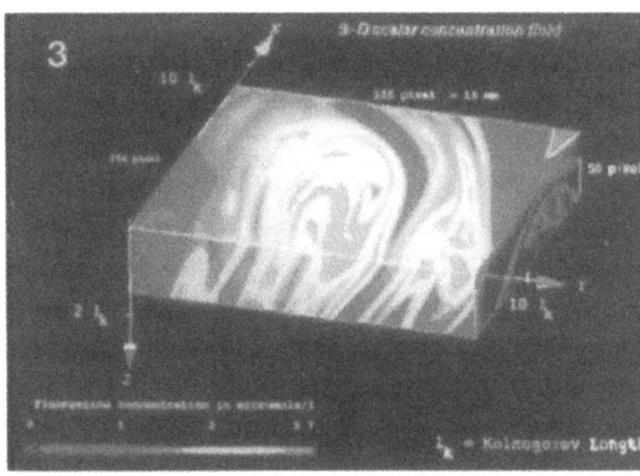

FIGURE 8-15b *3-D scalar concentration field within the marked region of* **a**). *(Merkel et al. 1994). (Reprinted by permission of Merkel et al.)*

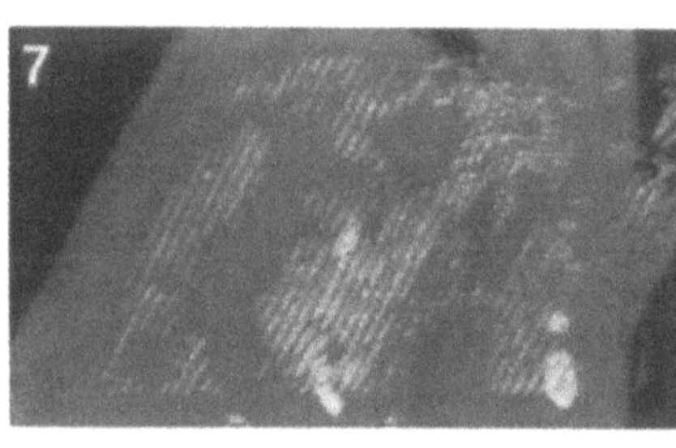

FIGURE 8-15c *The velocities of the centroids of the volumes at the middle plane of* **b**). *(Merkel et al. 1994). (Reprinted by permission of Merkel et al.)*

scalar and velocity maps (Long, 1993; Yip *et al.*, 1987). The scalar field uses LIF and the velocity field uses PIV for the evaluation. This is one of the few multi-dimensional measurements for both the scalar and velocity fields. There have been no attempts to extend this result to 3-D by using scanning techniques, but clearly, this should be possible.

Montemagno and Gray (1995) have used a variation on the LIF technique to make volume measurements in a porous media by matching the index of refraction of media to the fluid and tagging the fluid with laser dye chemicals. The complexity here is to bring together the many subareas that had to be studied into the final picture.

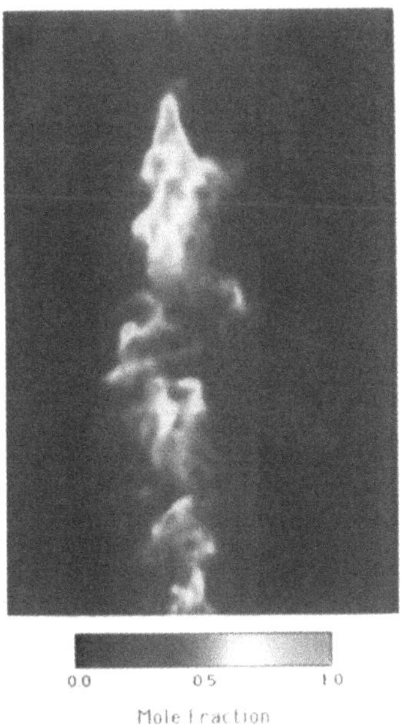

FIGURE 8-16a *Concentration measurements for a jet of Freon in air (Long 1993). (Reprinted by permission of Academic Press)*

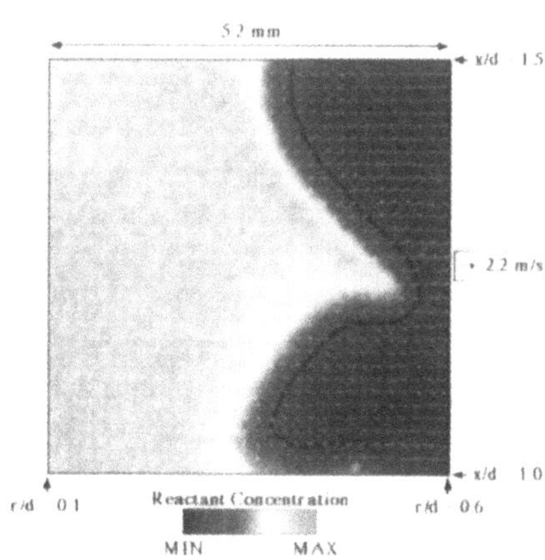

FIGURE 8-16b *Simultaneous scalar and velocity map for a non-reacting nozzle flow using biacetyl fluorescence and PIV (Frank et al. 1996). (Reprinted by permission of Frank et al.)*

8.3.5 Interferometry, Holographic, and Tomographic Techniques for Scalar Measurements

Before laser techniques became available, classic interferometry was one of the major techniques used in the investigation of high speed flows, combustion and heat transfer phenomena. Basically, in these studies either density, temperature or pressure changes, and as a consequence, the refractive index distribution also changes. Hence, when two light waves emitting simultaneously from the same light source, one penetrating the test section and the other passing outside the test section, interference fringes are created because of the optical path difference between the two light waves. However, this happens only if the optical path differences are exactly within the wavelength of the light at the plane of interference. A consequence of this is that classical interferometry is expensive and the adjustment time both extensive and tedious. However, by this technique, the scalar fields can be extracted from the fringe patterns. With the further development of lasers and holography, a new approach, holographic interferometry, has become important. For this new interferometry technique, both a double exposure and a real-time method have been presented. A new review book on "Optical Measurements, Techniques and Applications" has been edited by Mayinger (1994) and is an excellent starting place for this entire subarea. There are reviews on holography, holographic interferometry, light scattering, light emission and tomography (Hesselink, 1988; Mewes, 1991). In the limited space available here, we will not try to compete with the book. A number of interferometric techniques parallel PIV (but for scalar fields) and are 2-D in nature. However, a great deal of effort has been made to extend the methods to 3-D by using holographic and tomographic methods. The combination of holographic interferograms with tomographic methods involve scanning or a complex set of beams combined with stereoscopic viewing to get the basic data. For example, Söllor *et al.* (1994) have used holographic interferograms together with tomographic reconstruction techniques to investigate unsteady flow fields. In their experiments, the full-view reconstruction uses data from two holograms and the analysis uses 20 views of these spread over 180^0. Ruff and Zhang (1993) have studied applications of combining holographic interferometer with tomography to investigating flow and heat transfer phenomena. For this purpose, three holographic planes with different orientation (at 30, 90 and 150^0) were used to record double-exposure interference patterns, simultaneously. The tomographic data were then acquired by examining 30 views over a full angle of 140^0. The reconstruction operator was designed to use all three holographs in the reconstruction so as to provide a high degree of accuracy in the final result. By this method, 3-D temperature fields were obtained and one reconstructed field is shown in Figure 8-17. Figure 8-18 shows some results for the temperature during the growth and condensation of a steam bubble and for a viscous mixing system as a function of time. Many more examples can be found in the recent review book cited above.

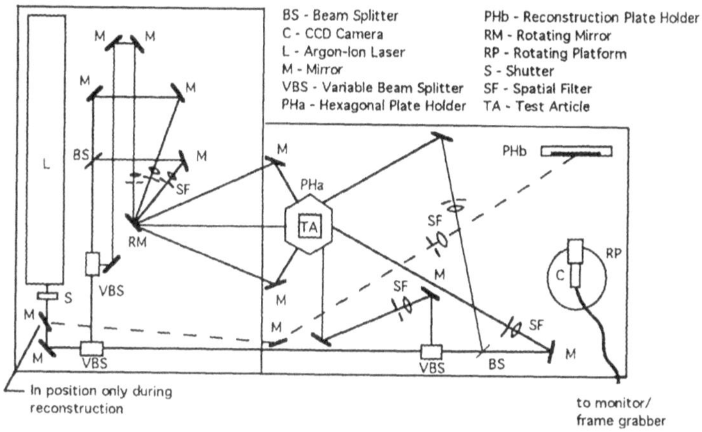

FIGURE 8-17a *Multidirectional holographic interferometry recording and reconstruction system (Ruff and Zhang 1993). (Reprinted by permission of Int. Soc. of Optical Eng.)*

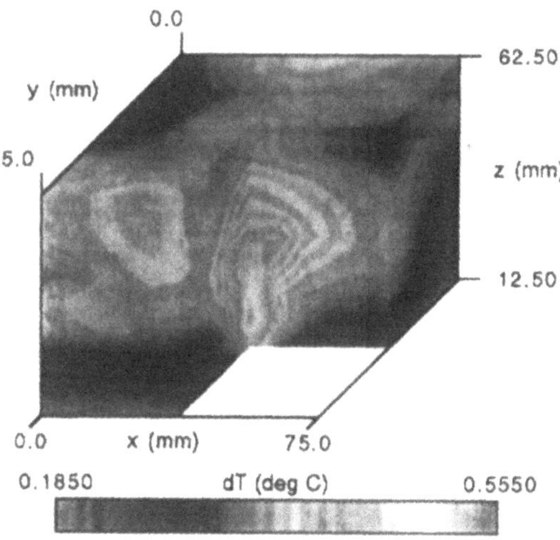

FIGURE 8-17b *Reconstructed 3-D temperature measurements on the outside and a cut-away portion of the natural convection test section (Ruff and Zhang 1993). (Reprinted by permission of Int. Soc. of Optical Eng.)*

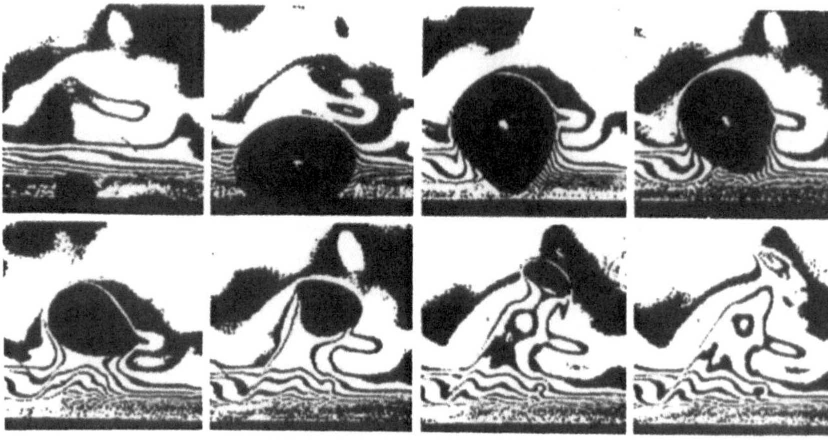

FIGURE 8-18a *Temperature field for the growth and condensation of a steam bubble obtained by the real-time method (Mayinger 1994). (Reprinted by permission of Springer-Verlag.)*

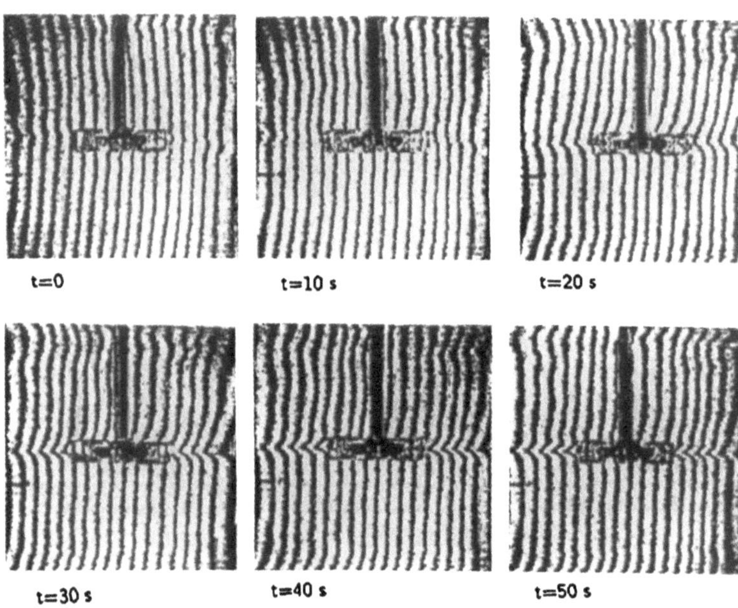

FIGURE 8-18b *Interferometric images for dissipation in a mixing vessel (Ostendorf 1987, see Mayinger 1994). (Reprinted by permission of Springer-Verlag.)*

8.3.6 Nuclear Magnetic Resonance (NMR)

The last of the pure 3-D techniques to be discussed here, but clearly not the last technique that will be introduced in the quest for 3-D imaging, is nuclear magnetic resonance (NMR). There are several excellent reviews on NMR and flow measurements. A recent review by Caprihan and Fulushima (1990) is a good starting place. The references cited are extensive and include the half dozen previous reviews on the subject. In the recent review by Gladden (1994), readers can find more details about the principles of NMR and its applications, especially in chemical engineering. From the point of view of application, NMR techniques can be classified into three categories: (I) spectroscopy experiments which explore the chemical and physical environment of a particular atomic or molecular species, (ii) diffusion measurement and (iii) imaging, in which spatially resolved maps and objects are obtained (see Gladden, 1994). For flow measurements, we focus attention on NMR imaging (sometimes called NMRI). The NMRI approach is a true non-invasive, three-dimensional method. Most people know of NMR because of its use in medical measurements. The basic facilities can be expensive, but their existence in medical research areas is making them more available today. Several strategies exist for measurement of flow fields by NMRI. Three of main approaches are steady-state, time-of-flight and phase-shift techniques. The principles of these techniques and their application are described fully elsewhere (Caprihan *et al.*, 1986; Altobelli *et al.*, 1992; Gladden, 1994). In brief, in time-of-flight, the nuclei within a slice are exited by a selective radio frequency (RF) pulse and magnetically "tagged". These tagged spins are detected at a time Δt later, from which 3-D velocity distribution of these spins can be obtained. It is clear that when the materials are correctly selected, both the velocity and concentration can be measured (Majors, 1989; Altobelli *et al.*, 1992; Nakagawa *et al.*, 1993; Turnet *et al.*, 1995), simultaneously. However, ferromagnetic constructions and electrical conductors must be avoided. In addition, NMRI techniques applied to liquids are more efficient and more common than to gas flows or solids. The small signal-to-noise ratio, among other factors, in gases or solids results in poor quality of measurements.

The NMRI technique uses an induction coil surrounding the sample to image nuclear spin density that results from the nuclear spin system rearrangement. An initial magnetic pulsed field orients the nuclear spin system and then it relaxes back toward a random state. Because of the relaxation time and that tomographic reconstruction is needed to extract the 3-D details, there are time limitations (Altobelli *et al.*, 1992) (currently of the order of 10-ms, at best). Consequently, the technique has been used mostly for steady or quasi-steady laminar flows because of the rather low data acquisition rate. However, modifications to allow turbulent and unsteady flows to be investigated have been reported and new approaches are being considered. One research group currently employing NMRI techniques is that of Powell (Li *et al.*, 1994, 1995) and coworkers at University of California at Davis in the study of turbulent pipe flows. A significant advantage of applying NMRI to measure turbulent boundary layers is that the data near the wall, $1 < y^+ < 10$, can be obtained, which is more difficult by other conventional techniques.

Figure 8-19 provides several examples of recent results. Included are measurements by Gleeson and Woessner (1991) on 3-D NMR imaging of pore connectivity in limestone and the recent results from Li *et al.* (1994, 1995) on the measurements of low Reynolds number transition and turbulent flows. As an imaging technique, NMRI has some advantages over other scattering techniques such as optical, ultrasonic or X-ray. The major advantage of NMRI is that it does not suffer from shadows and other opacity effects because the typical fluids usually do not affect the output nuclear spin density signal. This can allow the NMRI technique to measure an optical opaque sample or a flow field in an optical opaque mold. The other advantage of NMRI is that it does not need a model for interpretation of the raw data in contrast to other methods. A straight approach to observe the particle velocity profile, which is somewhat similar to PIV, is to

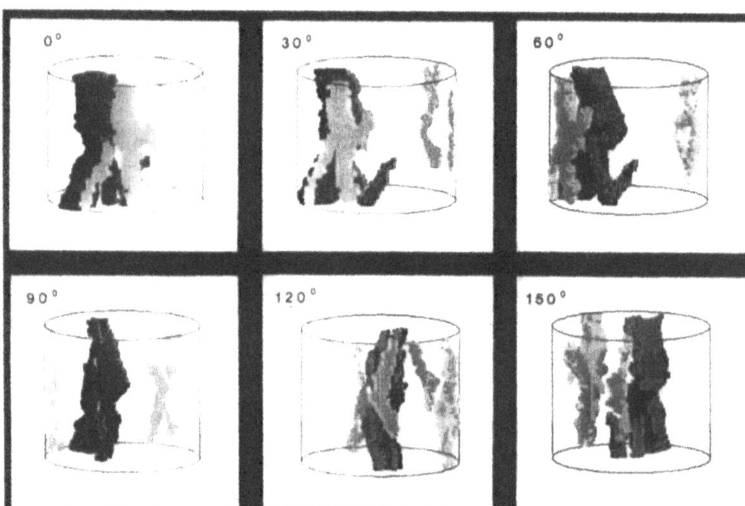

FIGURE 8-19a *3-D reconstruction from merging 2-D flow-weighted images of water saturated limestone with the water flowing at 7.5 ml/min. Rendering shows the flow viewed at the noted angles. The high flow values are dark (Gleeson and Woessner 1991). (Reprinted by permission of Pergamon Press.)*

use particles containing a nuclear species not found in the suspending fluid. However, these added particles must possess all the characteristics needed for PTV and in addition must have the correct NMRI properties.

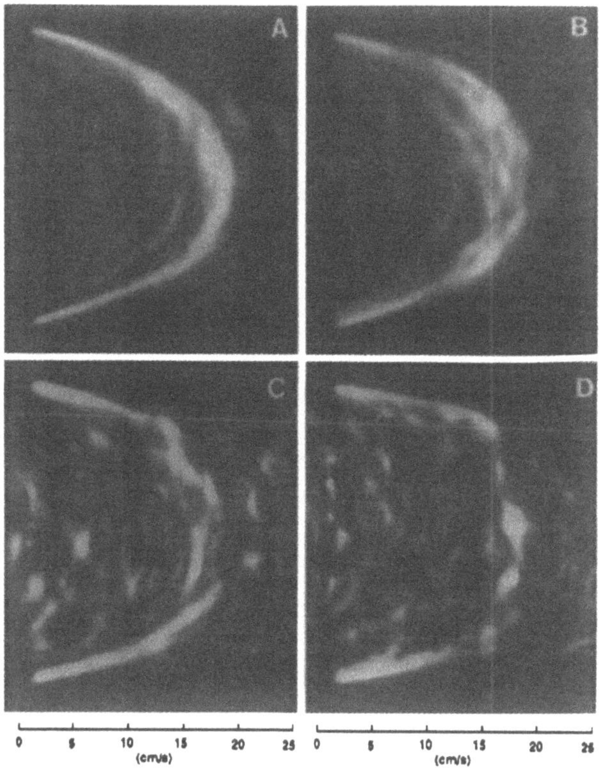

FIGURE 8-19b *Joint spatial-velocity distribution images of water at 1970, 2090, 2310 and 3080 Reynolds numbers. The vertical axis is the position across the pipe and the horizontal axis represents the axial velocity (Li et al. 1994) (Reprinted by permission of Elsevier Science Ltd.)*

8.4 ACKNOWLEDGMENTS

The authors would like to acknowledge their colleagues - Professors Yann Guezennec and Jeffery Chalmers and Dr. R.V. Venkat. In addition, thanks must be extended to the outstanding students in Chemical and Mechanical Engineering at The Ohio State University and the entire crew from the W.W. Clyde Chair (Univ. of Utah) Course on Full-Field Measurements in Turbulent Shear Flows that was given in the Fall of 1994. Thanks must also be extended to the many colleagues who kindly provided reprints and comments on their full-field efforts.

8.5 REFERENCES

Adrian, R.J., "Particle-imaging techniques for experimental fluid mechanics," Ann. Rev. Fluid Mech., 23, 261-304 (1991).

Adrian, R.J., ed. Selected Paper on Laser Doppler Velocimetry, SPIE Vol. MS 78, SPIE Optical Engineering Press, 617 pp (1993).

Adrian, R.J., "Bibliography of particle velocimetry using imaging methods: 1917-1995," TAM Rep. No. 817, UILU-ENG-96-6004, Theor. and Appl. Mech., Univ. Il. (1996), also will be available as TSI Model 600090 PIV Bibliography Vers. 1.0.

Altobelli, S.A., Caprihan, C., Fukushima, E., and Majors, P.D., "NMR Imaging Studies of Velocity and Concentration Distributions in Flows," in Magnetic Resonance Microscopy, Methods and Applications in Materials Science, Agriculture and Biomedicine (Bluemich, B., and Kuhn, W., eds.), 1992.

Barnhart, D.H., Adrian, R.J., and Papen, G.C., "Phase-conjugated holographic system for high-resolution particle image velocimetry," Applied Optics, Vol. 33, No. 30, 7159-7170 (1994).

Brücker, C., and Althaus, W., "Study of vortex breakdown by particle tracking velocimetry (PTV) Part 1: Bubble-type vortex breakdown," Exps. in Fluids, 13, 339-349 (1992).

Caprihan, A., Davis, J.G., Altobelli, S.A., and Fukushima, E., "A New Method for Flow Velocity Measurement: Frequency Encoded NMR," Magnetic Resonance in Medicine, 3, 352-362 (1986).

Caprihan, A., and Fukushima, E., "Flow Measurements by NMR," Physics Reports (Rev. Section of Physics Letters), 198, 195-235 (1990).

Chu, C.-C., and Liao, Y.-Y., "A quantitative study of the flow around an impulsively started circular cylinder," Exps. in Fluids, 13, 137-146 (1992).

Dahm, W.J.A., and Dimotakis, P.E., "Mixing at large Schmidt numbers in the self-similar far field of turbulent jets," J. Fluid Mechs., 217, 299-330 (1990).

Dahm, W.J.A., Southerland, K.B., and Buch, K.A., "Direct, high-resolution, four-dimensional measurements of the fine scale structure of Sc>>1 molecular mixing in turbulent flows," Phys. Fluids, A3, 1115 (1991).

Dahm, W.J.A., Su, L.K., and Southerland, K.B., "A scalar imaging velocimetry technique for fully resolved four-dimensional vector velocity field measurements in turbulent flows," Phys. Fluids, A4, 2191-2206 (1992).

Dahm, W.J.A., Su, L.K., and Tacina, K.M., "Four-Dimensional Measurements of Vector Fields in Turbulent Flows," AIAA 96-1987, 27th AIAA Fluid Dynamics Conference, New Orleans, LA, June 17-20, 1996.

Falco, R.E., and Chu, C.-C., "Measurement of two-dimensional fluid dynamic quantities using a photo chromic grid tracing technique," SPIE, 814, 706 (1987).

Frank, J.H., Lyons, K.M., Marran, D.F., Long, M.B., Starner, S.H., and Bilger, R.W., "Mixture Fraction Imaging in Turbulent Non-premixed Hydrocarbon Flames," Proceedings, manuscript by private communication (1994).

Frank, J.H., Lyons, K.M., and Long, M.B., "Simultaneous Scalar/Velocity Field Measurements in Turbulent Gas-Phase Flows," Combustion and Flame (in press) (Oct. 1996).

Gad-el-Hak, M., ed. Advances in Fluid Mechanics Measurements. New York: Springer-Verlag. 606 pp. 1989.

Gladden, L.F., "Nuclear magnetic resonance in chemical engineering: principles and applications," Chemical Engineering Science, Vol.49, No.20, 3339-3408 (1994).

Gleeson, J.W., and Woessner, D.E., "Three-dimensional and flow-weighted NMR imaging of pore connectivity in a limestone," Magnetic Resonance Imaging, Vol. 9, 879-884 (1991).

Grant, I., ed. Selected Paper on Particle Image Velocimetry, SPIE Vol. MS 99, SPIE Optical Engineering Press, 712 pp (1994).

Guezennec, Y.G., Brodkey, R.S., Trigue, N.T., and Kent, J.C., "Algorithms for Fully Automated Three-Dimensional Particle Image Velocimetry," Exps. in Fluids, 17, 209-219 (1994).

Guezennec, Y.G. and Kiritsis, N., "Statistical Investigation of Errors in Particle Image Velocimetry," Exps. in Fluids, 10, 138-146 (1990).

Guezennec, Y.G., Zhao, Y., and Gieseke, T.J., "High-speed 3-D scanning particle image velocimetry (3-D SPIV) technique," 7th Int. Symp. on Applications of Laser Techniques to Fluid Mechanics, July 11-14, Lisbon, Portugal (1994).

Hassan, Y.A., "Measurements of two-phase flows with digital image velocimetry," in Exp. & Comp. Asps. of Validation of Multiphase Flow CFD Codes, 180, 37-46 (1994).

Hesselink, L., "Digital image processing in flow visualization," Ann. Rev. Fluid Mech., 20, 4231-485 (1988).

Hill, R.B., and Klewicki, J.C., "Data reduction methods for flow tagging velocity measurements," Exps. in Fluids, 20, 142-152 (1996).

Hussain, F., Meng, H., Liu, D., Zimin, V., Simmons, S., and Zhou, C., "Recent Innovations in Holographic Particle Velocimetry," Proc. 7th ONR Propulsion Meeting, (Roy, G., and Givi, P., eds.), 233-249 (1994).

Kasagi, N., and Matsunaga, A., "Three-dimensional particle-tracking velocimetry measurement of turbulence statistics and energy budget in a backward-facing step flow," Int. J. Heat and Fluid Flow, 16, 477-485 (1995).

Koochesfahani, M.M., and Dimotakis, P.E., "Mixing and chemical reactions in a turbulent liquid mixing layer," J. Fluid Mechs., 170, 83-112 (1986).

Li, T.-Q., Seymore, J.D., Powell, R.L., McCarthy, K.L., Odberg, L., and McCarthy, M.J., "Turbulent pipe flow studied by time-averaged NMR imaging: Measurements of velocity profile and turbulent intensity," Magnetic Resonance Imaging, 12, 923-934 (1994).

Li, T.-Q., Odberg, L., Powell, R.L., and McCarthy, M.J., "Quantitative Measurements of Flow Acceleration by Means of Nuclear Magnetic Resonance Imaging," J. Magnetic Resonance, 109B, 213-217 (1995).

Long, M.B., "Multi-Dimensional Imaging in Combusting Flows by Lorenz-Mie, Rayleigh and Raman Scattering," Instrumentation for Flows with Combustion (Taylor, A.M.P.K., ed.), Academic Press, 468-508 (1993).

Maas, H.G., "Determination of velocity fields in flow tomography sequences by 3-D least squares matching," Proc. 2nd Conf. on Optical 3D Measurement Techniques, Zürich (1993).

Maas, H.G., Gruen, A., and Papantoniou D., "Particle tracking Velocimetry in Three-dimensional flows: Part I Photogrammetric determination of particle coordinates," Exps. in Fluids, 15, 133-146 (1993).

Majors, P.D., Givler, R.C., and Fukushima, E., "Velocity and Concentration Measurements in Multiphase Flows by NMR," J. of Magnetic Resonance, 85, 235-243 (1989).

Makik, N.A., Dracos, Th., and Papantoniou, D., "Particle tracking Velocimetry in Three-dimensional flows: Part II Particle Tracking," Exps. in Fluids, 15, 279-294 (1993).

Mayinger, F., (ed.) Optical Measurements: Techniques and Applications, Springer-Verlag. (1994).

Meinhart, C.D., Prasad, A.K., and Adrian, R.J., "A parallel digital processor system for particle image velocimetry," Meas. Sci. Technol., 4, 619-626 (1993).

Meng, H., and Hussain, F., "Holographic particle velocimetry: a 3D measurement technique for vortex interactions, coherent structures and turbulence," Fluid Dynamics Research, 8, 33-52 (1991).

Meng, H., and Hussain, F., "Instantaneous flow field in an unstable vortex ring measured by holographic particle velocimetry," Physics of Fluids, 7 (1), 9 (1995).

Merkel, G.J., Dracos, T., Rys, P., and Rys, F.S., "Turbulent Mixing investigated by Laser Induced Fluorescence," Proc. 5th Europ. Turb. Conf. (1994).

Merkel, G.J., Rys, P., Rys, F.S., and Dracos, T., "Concentration and velocity field measurements in turbulent flows by Laser Induced Fluorescence Tomography," Proc. EU-ROMEC Workshop on Imaging Techniques and Analysis in Fluid Dynamics, Rome (1995).

Mewes, D., "Measurement of Temperature Fields by Holographic Tomography," Exp. Thermal and Fluid Sci., 4, 171-181 (1991).

Miles, R.B., and Nosenchuck, D.M., "Three-Dimensional Quantitative Flow Diagnostics," in Advances in Fluid Mechanics Measurements (Gad-el-Hak., M., ed.), Lecture Notes in Engineering, Springer-Verlag, Berlin, 1989.

Montemagno, C.D., and Gray, W.G., "Photoluminescent volumetric imaging: A technique for the exploration of multiphase flow and transport in porous media," Geophysical Research Letters, 22, 425-428 (1995).

Nakagawa, M., Altobelli, S.A., Caprihan, C., Fukushima, E., and Jeong, E.-K., "Non-invasive measurements of granular flows by magnetic resonance imaging," Exps. in Fluids, 16, 54-60 (1993).

Ostendorf, W., "Einsatz der optischen Tomographie zum Messen von Temperaturfeldern in Rührgefäßen," Dissertation Universitität Hanover, 1987

Popovich, A.T., and Hummel, R.L., "A new method for non-disturbing turbulent flow measurements very close to a wall," Chem. Engr. Sci., 22, 21-25 (1967)

Prasad, A.K., and Adrian, R.J., "Stereoscopic particle image Velocimetry applied to liquid flows," Exps. in Fluids, 15, 49-60 (1993).

Racca, R.G., and Dewey, J.M., "A method for automatic particle tracking in a three-dimensional flow field," Exps. in Fluids, 6, 25-32 (1988).

Raffel, M., Gharib, M., Ronneberger, O., and Kompenhans, J., "Feasibility study of three-dimensional PIV by correlating images of particles within parallel light sheet planes," Exps. in Fluids, 19, 69-77 (1995).

Reese, J., Chen, R.C., and Fan, L.-S., "Three-dimensional particle image velocimetry for use in three-phase fluidization systems," Exps. in Fluids, 19, 367-378 (1995).

Ruff, G.A., and Zhang, Y., "Interferometric tomography in a three-dimensional differentially heated enclosure," in Optical Diagnostics in Fluid and Thermal Flow (Cha, S.S., and Trolinger, J.D.,eds.), SPIE Proc., 2005, 602-610 (1993).

Seeley, L.E., Hummel, R.L., and Smith, J.W., "Experimental velocity profiles in laminar flow around spheres at intermediate Reynolds numbers," J. Fluid Mechs., 68, 591-608 (1975).

Simmons, S., Meng, H., Hussain, F., and Liu, D., "Advances in holographic particle velocimetry," in Optical Diagnostics in Fluid and Thermal Flow (Cha, S.S., and Trolinger, J.D.,eds.), SPIE Proc., 2005, 1001-119 (1993).

Slama, C., ed. Manual of Photogrammetry, Fourth Edition, American Society of Photogrammetry 1980.

Söllor, C., Wenskus, R., Middendorf, P., Meier, G.E.A., and Obermeier, F., "Interferometric tomography for flow visualization of density fields in supersonic jets and convective flow," Applied Optics, Vol. 33, No. 14, 2921-2932 (1994).

Turnet, M.A., Cheung, M.K., McCarthy, M.J., and Powell, R.L., "Magnetic resonance imaging study of sedimenting suspensions of noncolloidal spheres," Physics Fluids, 7, 904-911 (1995).

Venkat, R.V., "Study of hydrodynamics due to turbulent mixing in animal cell microcarrier bioreactors," Dissertation of The Ohio State University, 1995

Weinstein, L.M., and Beeler, G.B., "Flow Measurements in a Water Tunnel Using a Holocinematographic Velocimetry," AGARD-CP-413, 16 (1987).

Yip, M., Lam, J.K., Winter, M., and Long, M.B., "Time-Resolved Three-Dimensional Concentration Measurements in a Gas Jet," Science, 235, 1209-1211 (1987).

9

Radioactive Tracer Techniques

Jian Gang Sun and Michael Ming Chen

9.1 INTRODUCTION

The motions of solids play central role in determining various unique characteristics of fluidization systems. Among these characteristics are the high heat and mass transfer rate, high solids mixing rate, and high erosion rate of bed internals. The motion of individual particles is important to the understanding of the mechanisms of solids dynamics and its formulations. Despite its importance, however, experimental techniques for measuring particle motion in fluidization systems without disturbing the flow field are limited. Among those, the radioactive-tracer technique has been shown to be capable of providing detailed information on local instantaneous particle motion and on the distribution of mean and statistical parameters.

The radioactive-tracer method was first used to study the mixing of catalysts in commercial fluidized beds in two steps. To obtain maximum sensitivity, gamma emitters were selected as the radioactive source. In these experiments, the tracers were made from catalyst particles tagged with a gamma-emitting radioisotope. After the tracers were released into the bed, their subsequent mixing with the bed particles was detected by sensors. In early studies (Singer et al., 1957; Overcashier et al., 1959), samples were withdrawn from various locations in the bed at specified time intervals and their intensity was measured by a sodium iodide (NaI) scintillation detector. In studies made somewhat later (May, 1959; Hull and Rosenberg, 1960), several scintillation detectors were mounted at various locations around the bed to monitor the variation in local radiation, indicating the state of local solids mixing and feed velocity in the riser. The radioactive-tracer method was also developed to determine the motion of an individual tracer particle. Kondukov et al. (1964) used six scintillation detectors around the bed as three pairs along the three Cartesian coordinates. The tracer particle was made of clear plastic into which was inserted a small piece of radioactive Co^{60} metal imbedded in it, with its size and weight matched with bed particles. With a proper calibration process, the tracer position was determined from readings obtained by these detectors in the x, y, and z directions. Velzen et al. (1974) later applied a similar tracer method to study solids motion in a sprouted bed. They used a single scintillation detector fixed at the top of the bed to determine the axial motion of the tracer in their small-diameter bed. These studies, however, provided only limited qualitative information about the particle motion because adequate instrumentation and efficient data processing schemes were not available.

The radioactive-tracer technique was perfected for fluidized-bed application through the development of a computer-aided particle-tracking facility (CAPTF) (Lin, Chen, and Chao, 1985; Moslemian, 1987; Sun, 1989). Considerable effort was expended to develop the efficient photon-counting instrumentation and automated data reduction and processing schemes. In Section 9.2, the principle of radiation detection and a theoretical model of the CAPTF is presented. The instrumentation of the CAPTF and the data reduction schemes are described in Section 9.3. Sample results obtained by the CAPTF are presented in Sections 9.4 and 9.5, and a conclusion is presented in Section 9.6.

9.2. PRINCIPLES OF RADIATION DETECTION

The radiation detection process whereby the CAPTF detects radiation in a fluidized bed is schematically illustrated in Fig. 9.1. The radioactive tracer particle emits gamma photons at a certain average rate in all directions. These photons pass through the surrounding solid particles and the wall of the bed. Some of them reach the scintillation detector, which consists of a NaI crystal coupled with a photomultiplier. The interaction of the photon with the crystal produces fluorescent spikes that are picked up and amplified by the photomultiplier and converted into electrical pulses that are further amplified and counted by associated electronics.

The count rate of the detector signal represents the number of photons received by the scintillation detector. The number of photons received is, in turn, related to the position of the tracer and the detector. A theoretical determination of the relationship between the count rate and the tracer/detector position would not only provide a better understanding for the operation of the CAPTF, but also have practical importance in optimizing the system. This relationship, which accounts for various physical and geometrical influencing factors, has been established by Sun (1985) and is briefly described below.

9.2.1 Factors that Affect Radiation Measurement

Many factors affect a gamma radiation measurement (Knoll, 1979; Tsoulfanidis, 1983). The most important factors relevant to the CAPTF are the characteristics of the radioactive source, the interaction of gamma rays with matter, the position of the source relative to that of the detector, the efficiency of the scintillation detector, and the dead-time behavior of the whole measurement system. These factors are separately discussed in the following subsections.

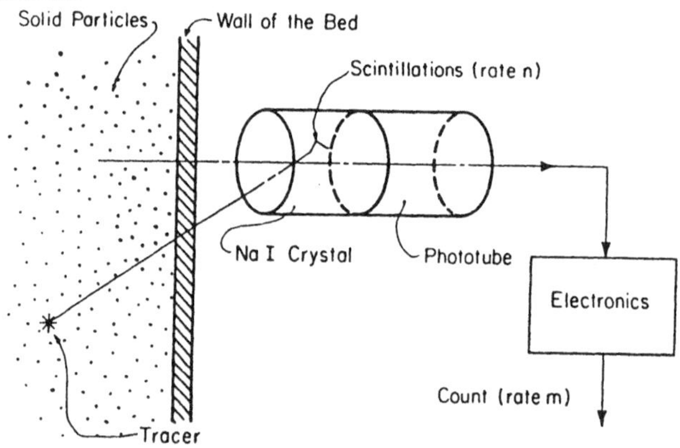

FIGURE 9.1 *Process of radiation detection in a fluidized bed*

9.2.1.1 Radioactive Source

A radioactive source may affect a measurement by its geometrical configuration and physical properties. In the CAPTF, the radioactive tracer

Table 9.1 Decay data for nuclide Sc^{46} and Na^{24}

Isotope	$t_{1/2}$	E(MeV)	Y(%)
Sc46	83.7 days	0.889	99.98
		1.121	99.99
Na24	15 h	1.37	100
		2.75	100

Y: Gamma ray photon yield per disintegration

particles are dynamically identical to the bed particles. Generally, the particle diameter is on the order of 1 mm, which is very small when compared with the attenuation length of gamma rays in the 0.1-10 MeV range. Therefore, the tracer particles can be treated as isotropic point radioactive sources.

The radioactive isotopes used in the tracer particles are made of scandium-46 (Sc^{46}) and sodium-24 (Na^{24}), their physical properties are listed in Table 9.1.

The decay of the activity of a radioactive material is described by the equation

$$S = S_0 e^{-\frac{t}{t_m}}, \tag{9.1}$$

where S_0 and S are activities at time zero and t, respectively; and t_m is the mean decay time, which is 120.8 days for Sc^{46} and 21.6 h for Na^{24}. The activity unit is Curie (Ci), which is equivalent to $3.70*10^{10}$ disintegrations per second. Table 9.1 shows that, for Sc^{46}, each disintegration yields two photons with energy levels of 0.889 and 1.121 MeV, respectively, and for Na^{24}, each disintegration yields two photons with energy levels of 1.37 and 2.75 MeV, respectively.

9.2.1.2 Interaction of Gamma Rays with Matter

Gamma rays are electromagnetic radiation. The term photon is used when gamma rays are treated as particles with associated energies. The interaction of gamma rays with bed particles and a fluidization column can be treated separately as attenuation and scattering effects.

Attenuation. Many interactions can occur between photons and matter. However, the photoelectric effect, the Compton effect, and the pair production effect are the three major interactions involved in the gamma radiation. The Compton effect is predominant when the gamma ray energy is in the range of 0.5-5 MeV, and the photoelectric and pair production effects are important only in lower and higher energy ranges, respectively.

The total probability of an interaction can be represented by the total linear attenuation coefficient m, which is a function of the gamma ray energy. In the literature, m is usually expressed as a product of the material density r and a total mass attenuation coefficient a, i.e., m = ra. For a parallel beam of monoenergetic gamma rays passing through a material of thickness r, the intensity of the exit beam I can be expressed as

$$I = I_0 e^{-\alpha \rho r}, \tag{9.2}$$

where I_0 is the initial gamma ray intensity.

For the fluidized bed system illustrated in Fig. 9.1, the materials that cause gamma ray attenuation are the bed particles and the fluidization column wall. Thus, Eq. 9.2 can be expressed as

$$I = I_0 e^{-(\alpha_p \rho_p r_p + \alpha_w \rho_w r_w)}, \tag{9.3}$$

where a_p and a_w, r_p and r_w, and r_p and r_w are the total mass attenuation coefficients, material densities, and gamma ray penetrating distances for the bed particles and the fluidization column wall, respectively.

Scattering. The entire beam of all of the gamma rays that reach the detector consists of two components: an unscattered beam and a scattered beam. Commonly, it is convenient to express the entire beam in terms of a buildup function B, i.e.,

$$I = BI_u , \tag{9.4}$$

where I is the total beam and I_u is the unscattered beam given by Eq. 9.3.

Compton scattering, the predominant component that contributes to the buildup function B for photon energy levels in the range of 0.5-5 MeV, can be considered a collision between a photon and a free electron in the medium (Tsoulfanidis, 1983; Tait, 1980; Segre, 1953). After the collision, the direction of motion of the photon is changed and associated with a change of photon energy. The energy E of a photon scattered through an angle q with the incident direction is described by the equation

$$E = \frac{E_0}{1 + (1 - \cos\theta) E_0 / (mc^2)}, \tag{9.5}$$

where E_0 is the incident energy of the photon and mc^2 is a constant equal to 0.511 MeV. Detailed analysis (Segre, 1953; 1964) showed that, for an incident photon with an energy at ˜1 MeV, the distribution of the scattered photons is within a small range of the change in angle q. This implies that the energy change of the photons due to Compton scattering is also small (about the order of q^2) and many scattered photons will build up to the main beam. Therefore, the buildup function B may still reach an appreciable value.

For the case of an infinite plate shield placed between a source and a detector, Berger's formula gives the buildup function B as (Tsoulfanidis, 1983)

$$B = 1 + a\mu r e^{b\mu r}, \qquad (9.6)$$

where m is the linear attenuation coefficient, r is the distance of gamma ray penetration, and a and b are two parameters that are dependent on both material and gamma ray energy. Equation 9.6 may be used for the fluidization bed system with a simple modeling of distance r. The parameters a and b can be determined by fitting experimental data.

9.2.1.3 Geometrical Configuration of the Detection System

The geometry effect of the detection system concerns the size and shape of the radioactive source and the detector, and the distance between them. For the fluidized-bed configuration, these factors can be accounted for completely by the solid angle W of the detector with respect to the point source. The solid angle represents the fraction of photons emitted from the source that reaches the detector.

9.2.1.4 Efficiency of the Detectors

When a photon enters a detector, it may or may not produce a signal or it may produce a signal lower than the discriminator threshold and, therefore, it is not counted. This effect is accounted for by the detector efficiency h, defined as the ratio of the number of photons recorded to the number of photons that impinge upon the detector per unit time. Statistically, the probability that a photon has at least one interaction in the detector NaI crystal is $1 - e^{-\mu r}$, where m is the linear attenuation coefficient of NaI and r is the distance that the photon travels in the crystal. For an isotropic radioactive point source, the detector efficiency can be expressed as (Tsoulfanidis, 1983)

$$\eta = \frac{1}{\Omega}\int_{\Omega}(1 - e^{-\mu r})d\Omega, \qquad (9.7)$$

where Ω is the solid angle. When the spherical coordinate that originates at the point source is used, $d\Omega$ is expressed as $\sin\theta d\theta d\phi$. Here r becomes a function of θ and ϕ. The solution of Eq. 9.7 indicated that η varies from 0.3 to 0.5 for a NaI crystal that is 2 in. long and 2 in. in diameter, depending on the distance between the source and the detector in the system (Sun, 1985). For the detectors that are used in the CAPTF, which contains a NaI crystal that is 2 in. long and 2 in. in diameter, η is very weakly dependent on the angle θ, which has been experimentally demonstrated by Lin (1981).

9.2.1.5 Dead-Time Effect

An electric pulse signal in the radiation detection circuit is characterized by a short rise time followed by a long decay time. When several successive interactions occur too closely in a short period of time, the detection system my not be able to distinguish them and some counts will be lost. The minimum time needed for a system to distinguish two successive events and record them as two counts is called the dead time of the counting system. Dead-time losses may be particularly important in cases with high counting rates. The dead time may arise from a detector or from the associated electronics. In the CAPTF, the dead time of the system is determined by the detectors because of the long decay time of the interactions of the photon with the NaI crystals within the detectors.

Two models of dead-time behavior have been commonly used: the paralysable model and the nonparalysable model (Knoll, 1979). Experimental data suggested that the paralysable model is suitable to describe the current detection system (Sun, 1985). For this model, the statistical relationship of the recorded count rate m to the true scintillation rate n is expressed as

$$m = ne^{-\tau n}, \tag{9.8}$$

where t is the dead time of the system. From Eq. 9.8 we note that there is a maximum observable rate for the paralysable model, above which the detector will be "saturated." This behavior restricts the maximum sensitivity and accuracy of the radiation detection system.

9.2.2 Relationship between Tracer Position and Detector Count Rate

The various effects that influence the detector count rate were described in the previous sections. If all of the correlation equations are considered, the detector count rate can be expressed as a function of tracer position. In the following section, this formulation is presented and its prediction is compared with measured data.

9.2.2.1 Formulation

If the various effects described above are considered, the scintillation rate n can be expressed as

$$n = ABS\Omega\eta e^{-(\alpha_p\rho_p r_p + \alpha_w\rho_w r_w)}, \tag{10.9}$$

where A is a constant, B is the buildup factor, S is the activity of the tracer particle, Ω is the solid angle, η is the detector efficiency, and the last term is due to the absorption of gamma rays by matter. Substitution of the expressions for the detector efficiency h in Eq. 9.9 gives

$$n = ABS_0 e^{-t/t_m} e^{-(\alpha_p\rho_p r_p + \alpha_w\rho_w r_w)} \int_\Omega (1 - e^{-\mu r})\sin\theta d\theta d\phi \ . \tag{9.10}$$

The recorded count rate m is affected by the dead-time behavior of the detection system and is determined by the paralysable model in Eq. 9.8.

Equations 9.10 and 9.8 are the basic formulations that relate the position of the tracer particle to the count rate of the detection system. The parameters in these equations have been determined for the CAPTF (Sun, 1985).

9.2.2.2 Comparison of Theoretical Predictions with Experimental Data

Equations 9.8 and 9.10 were used to predict the recorded count rates for given positions of the tracer particle. The experiments were performed with tracer particles of differing radioactive activities. The predicted results and the experimental data, shown in Figs. 9.2 and 9.3, are in good agreement. Because the experimental data in Figs. 9.2 and 9.3 were actual calibration data, the good agreement between the theoretical predictions and the data indicate that calibration curves may be generated from Eqs. 9.8 and 9.10. These equations have also been used to study solids mixing in a fluidized bed with the CAPTF (Moslemian, 1987).

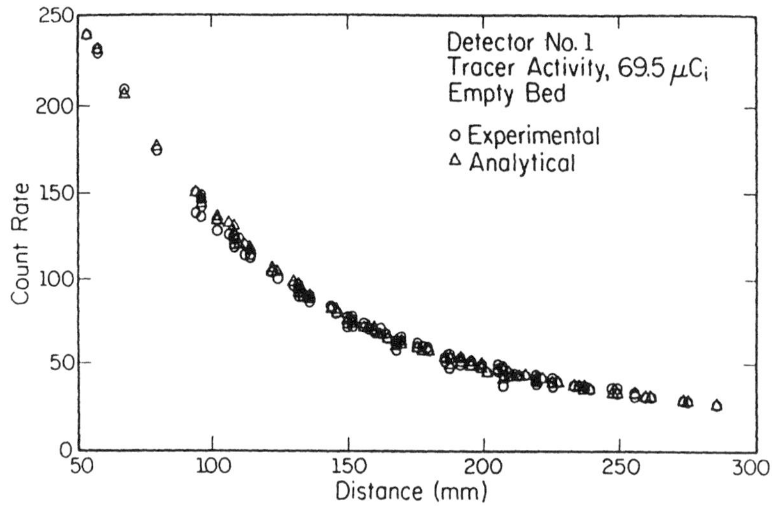

FIGURE 9.2 *Comparison of experimental calibration data with analytical predictions in an empty fluidized bed.*

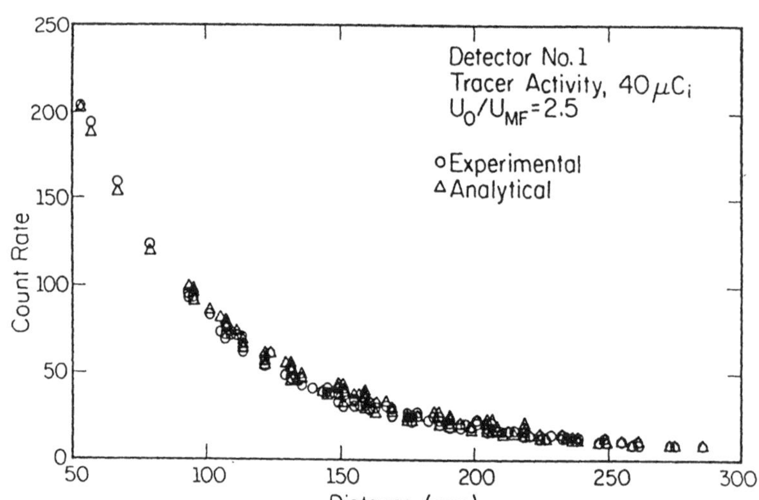

FIGURE 9.3 *Comparison of experimental calibration data with analytical predictions in a fluidized bed at* $u_o/u_{mf} = 2.5$

9.3. THE COMPUTER-AIDED PARTICLE-TRACKING FACILITY

9.3.1 Principles of Operation

The CAPTF can be used in two modes of operation. In the single-particle tracking mode, a radioactive particle, made of Sc^{46} and dynamically identical to the bed particles under study, is introduced into the fluidized bed. As the tracer particle moves with other particles, its gamma radiation is continuously monitored by an array of 16 strategically arranged scintillation detectors that surround the bed. The count rate of each detector is automatically converted by an on-line computer to the distance between the tracer and the detector according to a previously established calibration. The computer then proceeds to calculate the instantaneous position of the tracer from the 16 distances, taking full advantage of the redundancy provided by the large array of detectors. Time differentiation of the position data yields the local instantaneous velocities. After a test run of many hours, a large number of such instantaneous velocity measurements are available for each "location" in the bed, identified by a numbered small-sampling volume. The ensemble average of all velocities for each sampling volume then yields the mean particle velocity for the location. By subtracting the mean from the instantaneous velocity, the fluctuating components of the velocity can also be obtained. From these, the statistical quantities of the solids fluctuating motions are readily computed. Counting the number of occurrences in each sampling volume enables us to determine the distribution of occurrence probability for the entire bed.

The CAPTF can also be operated in the swarm-particle tracking mode. In this mode, the CAPTF employs a small amount (usually 10 g) of radioactive particles as tracers. The tracer particles are simply the bed particles (soda-lime glass beads), except that they had been activated in a nuclear reactor to convert the sodium in the glass to its radioactive isotope Na^{24}. After introduction of the tracers into the bed, their subsequent migration and dispersion were monitored by the 16 scintillation detectors. Initially, the detector signals (count rates) show transients; then they settle down to statistically stationary values that represent the uniformly mixed condition. The transient portion of the detector signals is related to the mixing rate, and the time variation of the signals in the statistically stationary state provides information on the fluctuating frequency of the bulk solids motion. Therefore, this technique is useful for the study of solids mixing and fluctuations in fluidized beds.

The CAPTF was developed by Lin, Chen, and Chao (1981, 1985), and improved later by Liljegren (1983) and Moslemian (1987). The advantage of this technique is that the flow field is not disturbed by the facility and, therefore, the measurement gives the actual movement of particles inside the bed.

9.3.2 Hardware Implementation

9.3.2.1 Radioactive Tracer Particle

For single-particle tracking, the radioactive tracer particle was made from a miniature scandium ingot with a specific gravity of 2.89 g/cm^3 which is only slightly higher than that of the glass particles in the bed (2.5 g/cm^3). The scandium particle was coated with a layer of polyurethane so its size and mass matched that of the glass spheres. The coating also serves to prevent abrasive loss of radioactive material in the erosive fluidized environment. The coated tracer was irradiated in a nuclear reactor to obtain the Sc^{46} isotope, which has a half-life of 84 days. Figure 10.4 shows a typical energy spectrum for Sc^{46}. The two distinct peaks at 0.89 and 1.12 MeV are due to the primary emission of Sc^{46}. The tracer particle was reirradiated to activities in the range of 400-600 mCi. This relatively high-intensity source is needed

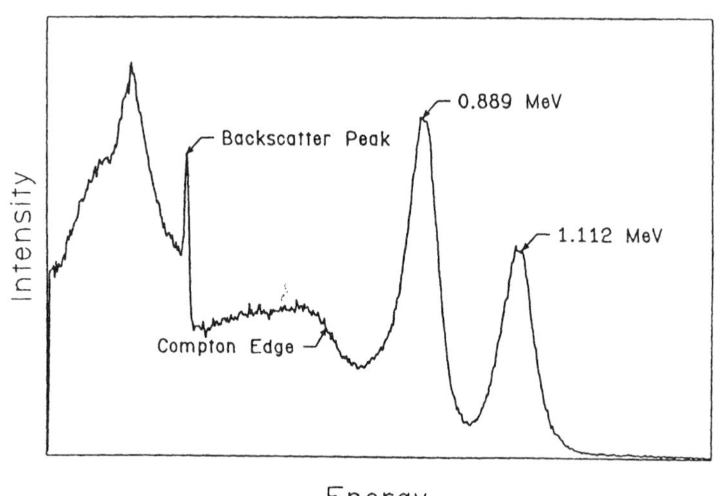

Figure 9.4 *Typical spectrum of Sc^{46}*

to improve data accuracy because of the statistical nature of the radiation count rate measurement.

For swarm-particle tracking, the tracer particles are simply the soda-lime glass beads in the bed. The glass beads contain ~10% of sodium by weight, which can be converted to its radioactive isotope in a nuclear reactor. The Na^{24} isotope emits gamma radiation at 1.37 and 2.75 MeV, and has a half-life of 15 h. The irradiated activity of the 10-g glass particles was ~400 mCi.

9.3.2.2 Scintillation Detector Array

Sixteen Bicron Model 2M2/2 scintillation detectors, composed of 2-in. (51-mm)-long, 2-in. (51-mm)-diameter NaI(T1) crystals with integral 10-stage photomultiplier tubes (PMT), were used to continuously monitor the gamma ray emission from the tracer. They are strategically arranged around the perimeter of the bed, as illustrated in Fig. 9.5. The rise time of the current pulses generated at the anode of the PMT is ≈80 ns; the decay time is ≈430 ns. Therefore, the total time required for processing of each pulse is ≈0.5 ms, which corresponds to a maximum count rate of ≈2 MHz.

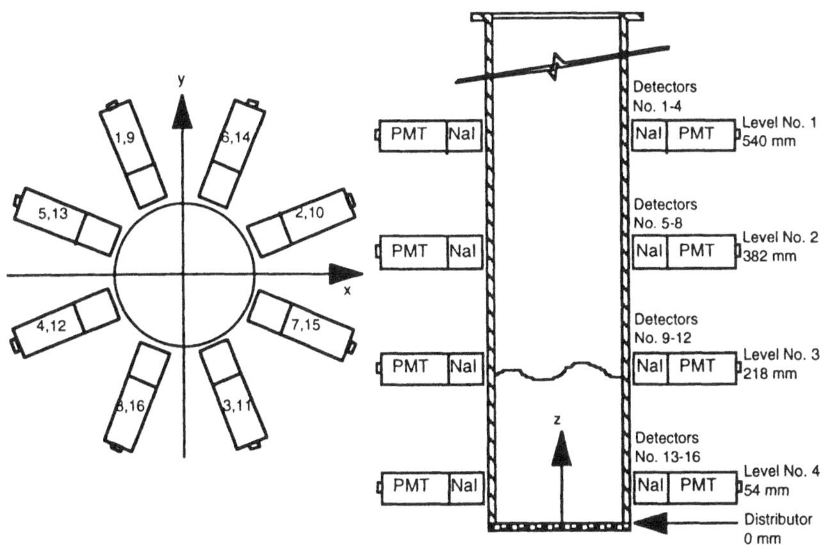

FIGURE 9.5 *Arrangement of dectectors around a cylindrical fluidized bed*

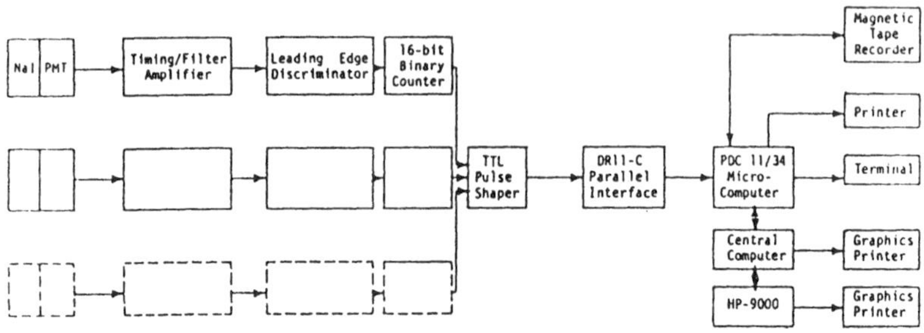

FIGURE 9.6 *Schematic diagram of the signal-processing instruments and data acquisition system*

9.3.2.3 Data Acquisition Electronics

The data acquisition system utilizes a direct photon-counting scheme, as shown schematically in Fig. 9.6. The pulse signals from the scintillation detector are further amplified by high-speed timing/filter amplifiers. The amplified signals have a noisy background originated mainly from secondary emissions due to the interaction of gamma rays with bed materials and from gamma rays that had only partially deposited their energy with the NaI(T1) crystal. Because most of the secondary emissions consist of gamma rays of fairly low energy, their contributions can be effectively removed by employing a leading-edge discriminator. Referring to Fig. 9.4 for the energy spectrum of Sc^{46}, the discriminator threshold may be set at the base of the Compton edge. Pulses of greater magnitude than the threshold energy are presumably from the gamma rays which come directly from the tracer to the detector where they deposit all of their energy in the interaction with the crystal. Consequently, these pulses are converted into logic pulses by the discriminator units, and are then counted by the 16-bit binary digital pulse counters. Outputs of the counters are fed into an on-line computer through a transistor-transistor logic (TTL) pulse shaper. The use of the TTL device with the digital counters permits simultaneous measurement of the output of all of the detectors without time delay. The DR11-C parallel interface is a general-purpose module suitable for interfacing logic signals and a minicomputer.

9.3.2.4 Fluidized Bed System

The fluidized bed used in this study was constructed from a 190-mm (7.5-in.)-i.d. plexiglass tube. The air distributor was made of sintered plastic plate with nominal pore spacing of 90 μ.

9.3.3 Software Implementation

9.3.3.1 Data Acquisition and Reduction Method

The data acquisition, reduction, and storage were controlled by an on-line computer. The sampling rate was determined by an input variable in the data acquisition software, and was usually set at 30 ms. The numbers of counts (referred to as count rates) of the 16 detectors recorded during the sampling duration were converted into distances between the tracer and the detectors by using previously established calibration. The computer then proceeded to calculate the instantaneous location of the tracer particle. Because the tracer position was usually calculated in real time, only the coordinates of the instantaneous tracer position needed to be stored. The total duration of the experiment depended on the specified data accuracy, but it usually took at least 5 h to ensure significant sampling of the entire bed volume.

9.3.3.2 Calibration Curves

A monotonic relationship between intensity (i.e., count rate) and distance between the tracer and each detector was established by calibration. The density dependence of gamma ray attenuation through the bed made it necessary to calibrate in situ because of the inhomogeneity of the bubbling fluidized bed. The procedure involved positioning the tracer in a large number of distributed locations within the bed and then measuring the count rates of all detectors at each tracer location.

Of particular concern was whether the density-distance relationship would vary with the angle from the axis of the cylindrical detector. As it turned out, an empirical center of the crystal near its geometrical center could be found such that the angular dependence was virtually eliminated (Lin, 1981). Thus, a single calibration curve that relates intensity to distance can be established for each detector.

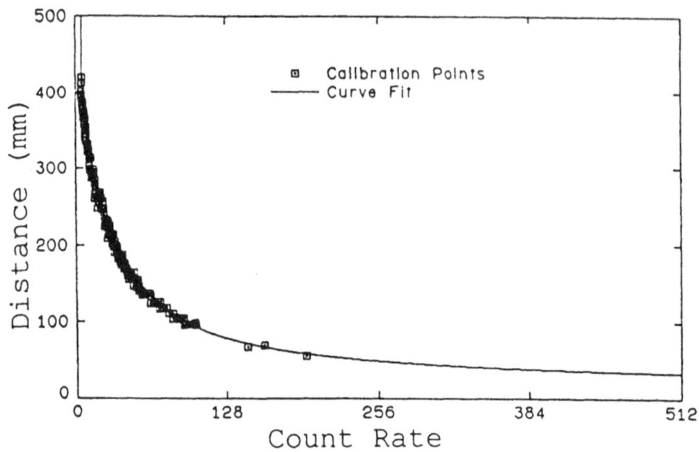

FIGURE 9.7 *Typical calibration data for 500- μm glass particles at u_o = 54.8 cm/s*

It is desirable to express the calibration data in functional form with a curve fit for real-time processing. Polynomial fits of various orders by the least-square method may be used in various regions of the data to represent the intensity-to-distance relationship with the following form:

$$r = \sum_{n=0}^{N} a_n (1/I)^n . \tag{9.11}$$

Here, r is the distance from the tracer to the empirical center of the crystal, I is the intensity of gamma rays (or count rate), a_n's are the coefficients of the curve fit, and N is the number of the polynomial fit and was usually selected between 3 and 7. Figure 9.7 shows a typical set of calibration data and the polynomial-curve fit.

9.3.3.3 Computation of Instantaneous Position of the Tracer

In principle, only three detectors are needed to determine the tracer position. The availability of measured distances from 16 detectors resulted in data redundancy for location determination. To take advantage of this planned redundancy, a weighted least-square method based on an linearization scheme was used to determine the optimum tracer position.

If we denote the position of the tracer by (x, y, z), the position of the NaI crystal of ith detector by (x_i, y_i, z_i), and the measured distance between the tracer and the ith detector by r_i, then, an error function F can be defined by

$$F(x,y,z)=\sum_{i=1}^{16}\frac{1}{\sigma_i}\left[(x-x_i)^2+(y-y_i)^2+(z-z_i)^2-r_i^2\right]^2, \tag{9.12}$$

where σ_i is a weighting factor and, for simplicity, is taken to be a function of r_i only. The function F represents the measurement error. By differentiating Eq. 9.12 with respect to *x*, *y*, and *z*, and setting the resulting expressions to zero, we can determine the optimum position of the tracer by solving the resulting set of three equations. However, these equations are nonlinear with respect to *x*, *y*, and *z* and require iteration for their solution. To overcome this drawback, Lin (1981) developed a linear regression scheme in which a new independent variable u ($= x^2 + y^2 + z^2$) is defined, and the position of the tracer is then determined by differentiating F(x, y, z, u) with respect to *x*, *y*, *z*, and *u* and setting the resulting expressions to zero. Thus,

$$2\sum_{i=1}^{16}\frac{x_i^2}{\sigma_i}x+2\sum_{i=1}^{16}\frac{x_iy_i}{\sigma_i}y+2\sum_{i=1}^{16}\frac{x_iz_i}{\sigma_i}z-\sum_{i=1}^{16}\frac{x_i}{\sigma_i}u=\sum_{i=1}^{16}\frac{x_id_i}{\sigma_i}, \tag{9.13a}$$

$$2\sum_{i=1}^{16}\frac{y_ix_i}{\sigma_i}x+2\sum_{i=1}^{16}\frac{y_i^2}{\sigma_i}y+2\sum_{i=1}^{16}\frac{y_iz_i}{\sigma_i}z-\sum_{i=1}^{16}\frac{y_i}{\sigma_i}u=\sum_{i=1}^{16}\frac{y_id_i}{\sigma_i}, \tag{9.13b}$$

$$2\sum_{i=1}^{16}\frac{z_ix_i}{\sigma_i}x+2\sum_{i=1}^{16}\frac{z_iy_i}{\sigma_i}y+2\sum_{i=1}^{16}\frac{z_i^2}{\sigma_i}z-\sum_{i=1}^{16}\frac{z_i}{\sigma_i}u=\sum_{i=1}^{16}\frac{z_id_i}{\sigma_i}, \tag{9.13c}$$

and

$$2\sum_{i=1}^{16}\frac{x_i}{\sigma_i}x+2\sum_{i=1}^{16}\frac{y_i}{\sigma_i}y+2\sum_{i=1}^{16}\frac{z_i}{\sigma_i}z-\sum_{i=1}^{16}\frac{1}{\sigma_i}u=\sum_{i=1}^{16}\frac{d_i}{\sigma_i}, \tag{9.13d}$$

where $d_i = x_i^2 + y_i^2 + z_i^2 - r_i^2$. It was found that the solution of Eqs. 9.13a-d is always very close to the original set of nonlinear equations for various types of errors (Sun, 1985).

9.3.3.4 Computation of Instantaneous Velocity of the Tracer

After obtaining the tracer position data, the instantaneous tracer velocity $\underline{v}$ at time t is obtained by simply dividing the distance between two consecutive tracer positions by the sampling duration δt as follows:

$$\underline{v}(\underline{a},t) = \frac{\underline{s}(\underline{a},t+\delta t)-\underline{s}(\underline{a},t)}{\delta t}\,, \tag{9.14}$$

where $\underline{s}$ denotes the position of the tracer, which is initially at $\underline{a}$. The tracer velocity in each sampling duration is taken to be a constant. The present technique is incapable of resolving velocity variations of time scales smaller than the sampling duration.

9.3.3.5 Computation of Mean Velocity and Density Distributions of Solids

The tracer position and velocity data obtained as described above represent a Lagrangian description of the motion of a single particle in the bed. However, it is usually desirable to present the data in Eulerian form. To this end, the Lagrangian data are used to evaluate local means of dynamic variables as functions of position in the fluidized bed. This is accomplished by dividing the cylindrical bed into imaginary sampling compartments in a cylindrical coordinate system, with the origin at the bottom center of the bed, as shown in Fig. 9.8. For this fluidized bed system, 10 radial, 16 circumferential, and 50 axial subdivisions were chosen, to give a total of 8000 compartments. By running the experiments for sufficiently long times, the tracer particle typically appears many times in each sampling compartment. Ensemble averages of the Lagrangian quantities of the tracer when it appears in a sampling compartment give the values of the corresponding Eulerian quantities for that compartment. The resulting data are then averaged circumferentially because of the near axisymmetry of the data.

The mean density and velocity of the solids may be evaluated on the basis of statistical probability. Let us denote V_c as the volume of a compartment; n as the local particle number density, which is unknown; and N as the total number of particles in the bed. The size d_p and the mass m_p of the particles in the bed are assumed to be uniform. Then, nV_c/N is the probability of finding the tracer particle in the compartment at a particular time. The value of nV_c/N is also the fractional time during which the tracer is found in the compartment. Therefore, if the total duration of an experiment is Δt, we have

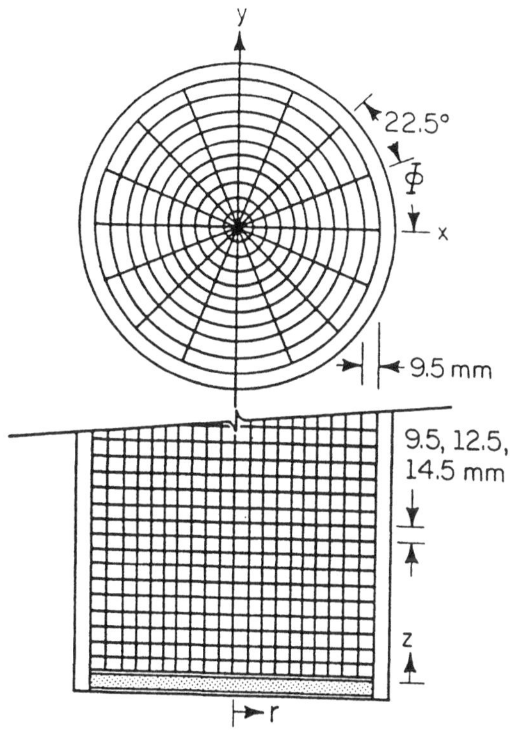

FIGURE 9.8 *Configuration of data reduction compartments in cylindrical coordinates*

$$\frac{nV_c}{N} = \frac{\sum_k \delta t_k}{\Delta t}, \tag{9.15}$$

where δt_k is the duration of the kth residency of the tracer in the compartment. If we multiply the numerator and denominator of the left-hand-side term in Eq. 9.15 by the particle mass m_p, noticing that $Nm_p = M$ is the total mass of the bed, the mean density ρ (= nm_p) of the solids is then determined from

$$\rho = \frac{M}{V_c \Delta t} \sum_k \delta t_k . \tag{9.16}$$

Correspondingly, the local mean solids velocity can be computed from

$$\bar{\underline{v}} = \frac{\sum_k \delta t_k \underline{v}_k}{\sum_k \delta t_k} = \frac{M}{\rho V_c \Delta t} \sum_k \delta t_k \underline{v}_k . \tag{9.17}$$

Equations 9.16 and 9.17 can be viewed as a form of conditional time average. However, for a general compartment in cylindrical coordinates, the computation of the residence duration for a particle in straight-line motion is quite complex and time consuming. Instead, we count only the number of occurrences of the tracer in the compartments. One occurrence is assumed to be associated with one sampling duration δt, and is assigned at the center location of the tracer trajectory in the period δt. Let us denote N_0 as the total number of occurrences of the tracer in the bed ($=\Delta t/\delta t$), and n_0 as the number of occurrences of the tracer in the compartment V_c. The mean solids density and velocity may then be computed from the conditional ensemble averages

$$\rho = \frac{M n_0}{V_c N_0} \tag{9.18}$$

and

$$\bar{\underline{v}} = \frac{M}{\rho V_c N_0} \sum_k \underline{v}_k , \tag{9.19}$$

where the summation is performed only when the tracer occurs in the compartment.

With the tracer occurrences, we simplified the continuous tracer trajectory into discrete-point tracer trajectory. Therefore, the probability of finding the tracer in a compartment, which is nV_c/N, as discussed above, should also be equal to the fractional occurrence of the tracer in the compartment n_0/N_0. This shows that, when $\Delta t \to \infty$, which is equivalent to $N_0 \to \infty$, Eqs. 9.18 and 9.19 are identical to Eqs. 9.16 and 9.17, respectively. But when Δt is finite, Eqs. 9.18 and 9.19 would be less accurate because they are derived from simplified discrete-point motion of the tracer particle.

9.3.3.6 Estimation and Measurement of Data Accuracy

Two intrinsic random events are associated with tracer position measurement in the CAPTF. One is the random emission of gamma photons from the source; the other is the random attenuation of the photon path due to fluctuation of solids density in the fluidized bed. The effect of these random events on the accuracy of the measured tracer position may be determined from the analytical expressions described in Section 10.1. For practical consideration, however, an approximate estimation of the accuracy is derived below.

Consider a gamma source at a distance r from the detector. By assuming that the scintillation crystal is sufficiently small, an approximate relationship between the source activity S and the rate of scintillation n at the crystal is

$$n = \frac{d_c^2}{16r^2}\eta S. \tag{9.20}$$

Here, the first factor at the right-hand-side, which represents the fraction of gamma photons intercepted by the scintillating crystal, is equal to the solid angle extended by the detector crystal to the source, with d_c denoting the crystal diameter. The efficiency of the crystal η depends on both the material and the size of the crystal.

In the radioactive-particle tracking methodology, the distance is inferred by measuring the scintillation rate n. Thus, the approximate relationship between the accuracy of the distance measurement and the scintillation rate from Eq. 9.20 is

$$\frac{|\delta r|}{r} = \frac{|\delta n|}{2n}\,. \tag{9.21}$$

Because of the finite length of the output pulses from the scintillation detectors (on the order of 1 μs), the maximum practical count rate n_{max} is $\approx 10^5$ counts/s for the gamma photon energies in the vicinity of 1 MeV.

The duration of the counting period δt depends on application, and is usually in the range of 10-30 ms. For a stochastic process, the accuracy with which one can determine the average value is on the order of $1/N^{1/2}$. Hence, the maximum accuracy for count rate is $\approx(\delta t\ n_{max})^{-1/2}$, i.e.,

$$\frac{|\delta n|}{n} \approx (\delta t\, n_{\max})^{-1/2} . \tag{9.22}$$

This maximum accuracy is attained only when the tracer is at a minimum distance from the detector. For particle-tracking measurements, the system must be able to measure accurately in a range of distances. Let the closest working distance be denoted r_o where the maximum count rate $n_{\max}$ is obtained. Then, the count rate at a mean working distance r_m is

$$n_m = n_{\max} \frac{r_o^2}{r_m^2} . \tag{9.23}$$

When the above results are combined, the expected accuracy for distance for a mean working distance r_m is

$$|\delta r| \approx \frac{r_m^2}{2r_o} \frac{1}{(\delta t\, n_{\max})^{1/2}} . \tag{9.24}$$

According to the above estimation, for r_o on the order of 100 mm, r_m in the range from r_o to $\approx 2r_o$, δt of 30 ms, and $n_{\max}$ of 10^5 counts/s, the mean error for distance measurement is in the range of 1-4 mm.

The measurement accuracy was experimentally determined by positioning the tracer in known locations inside the bed (Moslemian, 1987). The apparent tracer positions were calculated from the linear regression formulation Eq. 9.13, based on the measured detector count rates and the predetermined calibrations. Sufficient data were taken at each location to allow for statistical determination of mean and standard deviations of the tracer position and velocity. In general, the axial errors were often greater than the radial errors because of the longer axial distance of the bed that the detectors had to monitor. For an empty bed, the mean axial error in determining the tracer position was ≈4.7 mm and the mean radial error was 3.9 mm. The corresponding standard deviations were 1.6 mm in the axial direction and 1.2 mm in the radial direction. These deviations were due to the statistical nature of the radiation detection and are the minimum deviations obtainable for the tracer position. The measured mean axial and radial velocities were approximately zero (<1 cm/s) at all locations inside the bed. However, the standard deviations of the velocities were 7.6 cm/s in the axial direction and 5.3 cm/s in the radial direction. By comparing these measurements with estimated values given above, it is seen that the measurements were not taken under optimum conditions and additional improvements in accuracy could be achieved.

9.4 SOLIDS DYNAMICS IN FLUIDIZED BEDS

The single-particle tracking mode of the CAPTF was used to study solids dynamics in fluidized beds. Most of the following data were obtained from a 19-cm (7.5-in.)-i.d. cylindrical fluidized bed. Some data were also obtained from a two-dimensional (2-D) bed with a cross-sectional area of 40 x 3.8 cm^2. The bed particles were soda-lime glass spheres with diameters that ranged from 425 to 600 μm with a mean of 500 μm, and diameters from 600 to 850 μm with a mean of 705 μm. They have a specific gravity of 2.50 g/cm^3. These glass spheres are Class B particles according to Geldart's classification (1973). They are characterized by the formation of bubbles at or near the minimum fluidization velocity u_{mf}, which was determined experimentally by the usual pressure drop method. It was found that u_{mf} = 21.9 and 30.2 cm/s for the 500- and 705-mm particles, respectively.

9.4.1 Mean Velocity and Density Distribution of Solids

Figure 9.9 shows a typical result of the circumferentially averaged solids circulation pattern, vector plot of solids velocity, and density distribution field for the 500-mm particles at $u_0/u_{mf} = 2$. The averaged circulation pattern in Fig. 9.9a exhibits two counter-rotating vortices: particles in the lower vortex descended in the center and ascended near the wall (AWDC) and those in the upper vortex ascended in the center and descended near the wall (ACDW). In the velocity vector plot Fig. 9.9b, the magnitudes of the velocity vectors were normalized by the magnitude of the maximum velocity. The starting points of the vectors denote the center of each sampling compartment, and the lengths of the vectors are proportional to the magnitudes of the velocities. It is apparent that the solids velocities are usually higher near the centerline of the bed than near the wall. The solids density was evaluated from the repeated appearance of the radioactive tracer particle in each sampling compartment, Eq. 9.18. From the density contour plot in Fig. 9.9c, the density is uniform at a given height only in an upper portion of the bed.

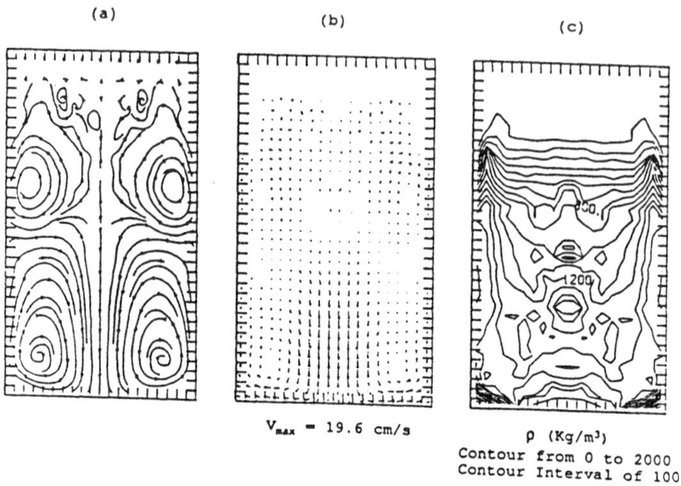

FIGRUE 9.9 *Solids mean dynamic behavior (circumferentially averaged) in a cylindrical fluidized bed for 500-μm glass particles at u_o/u_{mf} = 2 (a) recirculation pattern, (b) mean velocity vector field, and (c) density distribution*

The circumferentially averaged solids circulation patterns for the 500-μm particles at u_0/u_{mf} = 1.5, 2, and 4 are plotted in Fig. 9.10, which shows that the mean dynamic behavior of the solids depends strongly on the air flow rate. The lower vortex is predominant at low gas flow rate (Fig. 9.10a); its size diminishes as gas velocity increases. At very high gas velocities it was shown that the lower vortex will completely disappear (Moslemian, 1987). The average recirculation patterns for larger particles (700 μm and 2 mm) were not significantly different from those for the 500-μm particles. However, large differences were observed in the absolute magnitude of the solids velocities. Those variations were mostly due to the higher superficial gas velocities required for the larger particles. As expected, the density of the solids decreased with increasing fluidization velocity.

The foregoing observations can be interpreted in terms of the bubble behavior in fluidized beds. Werther and Molerus (1973) reported that very close to the distributor region, intensified bubble activity exists in an annular region near the wall. As bubbles detach and rise, they tend to move toward the center. If the bed is sufficiently deep, they will eventually merge at the center. Because the solids are carried upward in the wake of the bubbles, they basically move along the bubble tracks. Therefore, the solids would

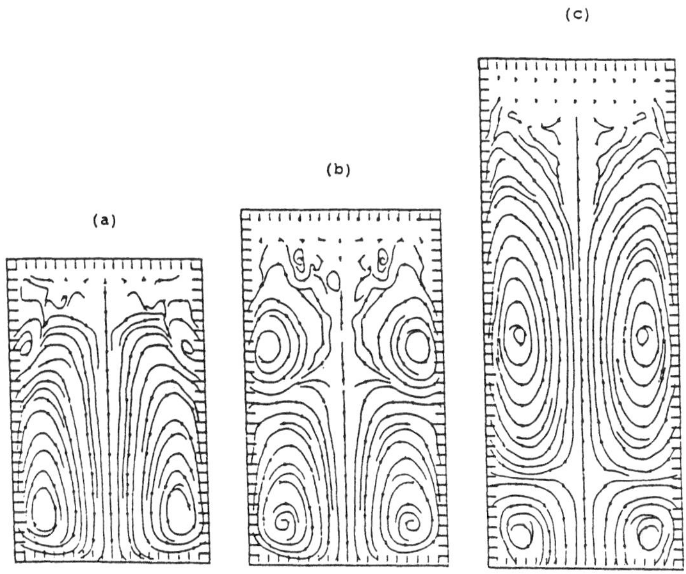

FIGURE 9.10 *Effect of superficial velocity on solids circulation patterns in a cylindrical fluidized bed for 500-mm glass particles at (a) u_o/u_{mf} = 1.5, (b) u_o/u_{mf} =2, (c) u_o/u_{mf} = 4*

ascend near the wall and descend at the center (AWDC) in the lower vortex. The elevation that separates the two vertical vortices marks the approximate location of complete bubble coalescence. Above this elevation, solids ascend at the center and descend near the wall (ACDW).

9.4.2 Solids Flow in Presence of Bed Internals

A very complex solids flow pattern will result when solid obstacles exist in the fluidized bed. The solids recirculation pattern in a cylindrical bed with a single sphere was presented by Lin, Chen, and Chao (1985), and in a 2-D bed with a single and multiple cylinders by Ai (1991). It was demonstrated that large obstacles would not only affect the local solids velocity, but also the global solids circulation patterns.

A comprehensive study of the effects of internal rod bundles on bed hydrodynamics was compiled (Chen, Chao, and Liljegren, 1983). It was found that, qualitatively, the flow pattern of the solids in the bed was not significantly affected by the presence of distributed tube banks.

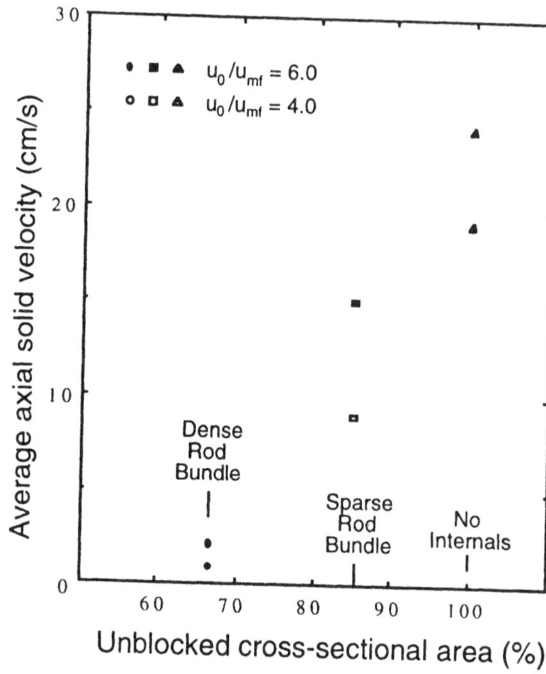

FIGURE 9.11 *Effect of internal rod bundles on the magnitude of solids velocity*

Quantitatively, however, the magnitudes of the solids circulation velocity was significantly reduced. A sample result is shown in Fig. 9.11.

9.4.3 Conservation of Mass for the Solids

With the availability of the ensemble-averaged solids velocities and densities, the consistency of the data can be assessed by determining if the solids mass flow through any closed imaginary surfaces in the bed is conserved. Ideally, the net mass flow should vanish. To carry out such a continuity check, mass flow rates were calculated at a number of enclosed imaginary surfaces. The imbalance of the solids flow across a surface was determined from $2\left|\dot{M}_{in} - \dot{M}_{out}\right| / (\dot{M}_{in} + \dot{M}_{out})$, where $\dot{M}_{in}$ and $\dot{M}_{out}$ are the average incoming and outgoing flow rates through the surface, respectively. The value was found to be generally <10% (Lin et al., 1985; Ai, 1991).

9.4.4 Lagrangian Autocorrelations of Fluctuating Velocities

The chaotic motion of the solids in gas fluidized beds necessitates the measurement of the fluctuating and mean velocities of the solids for thorough understanding of their dynamic behavior. The statistical information of the fluctuating velocity may be obtained from the Lagrangian autocorrelations.

The Lagrangian autocorrelation coefficient $R_{aa}(\underline{x},t)$ at a given position $\underline{x}$ is defined by Tennekes and Lumley (1972) as

$$R_{\alpha\alpha}(\underline{x},\tau)=\frac{<v'_{\alpha}(\underline{a},t)v'_{\alpha}(\underline{a},t+\tau)>}{<v'^{2}_{\alpha}(\underline{a},t)>}, \tag{9.25}$$

where $v'_{\alpha}(\underline{a},t)$ is the fluctuating velocity in either axial ($\alpha = z$) or radial ($\alpha = r$) direction, $\underline{a}$ is the initial position of the tracer, and $<>$ denotes the ensemble average.

Figures 9.12 and 9.13, respectively, show some sample results of the axial and radial autocorrelation coefficients R_{zz} and R_{rr} at six axial locations at approximately midradius of the 19-cm-i.d. fluidized bed for 500-μm glass beads at u_0 = 54.8 cm/s (Moslemian, 1987). Zero crossing of the time axis provides a measure of the correlation time of the random motion of the solids. The Lagrangian correlation between the axial motions in Fig. 9.12 strongly depended on the location within the bed. The correlation time was shortest near the distributor region (Lin et al., 1985). It increased to a maximum at the elevation that separates the two vertical vortices where the particles exhibited longest memory. The correlation time generally decreased at higher values of z. The overshoots and decays of the axial correlations indicate the existence of harmonic sloshing motion in this direction. It was estimated that the maximum frequency in the axial direction ranged from 1.3 to 5.1 Hz. On the other hand, the radial correlation times indicated in Fig. 9.13 were smaller than those in the axial direction and they were insensitive to both axial and radial locations (Lin, et al., 1985; Moslemian, 1987). The experimental results of Moslemian (1987) also indicated that the Lagrangian autocorrelation coefficients were generally independent of changes in fluidization velocity and particle size.

The Lagrangian velocity autocorrelations can be used to evaluate several important quantities that characterize the fluctuating motion of solids,

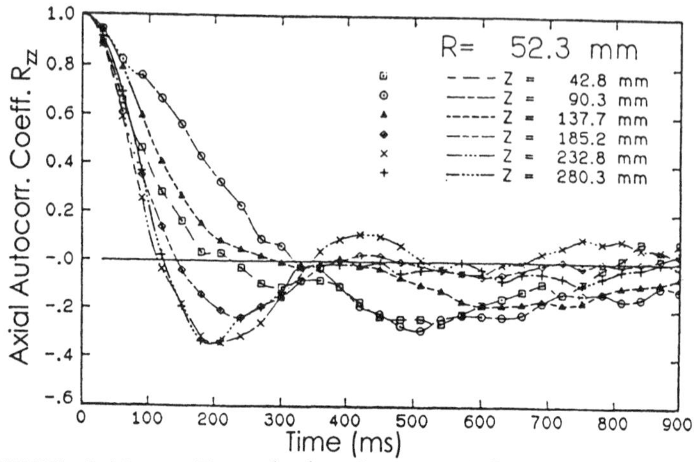

FIGURE 9.12 *Typical distribution of Lagrangian autocorrelation coefficients in axial direction for 500-μm glass particles at u_o = 54.8 cm/s*

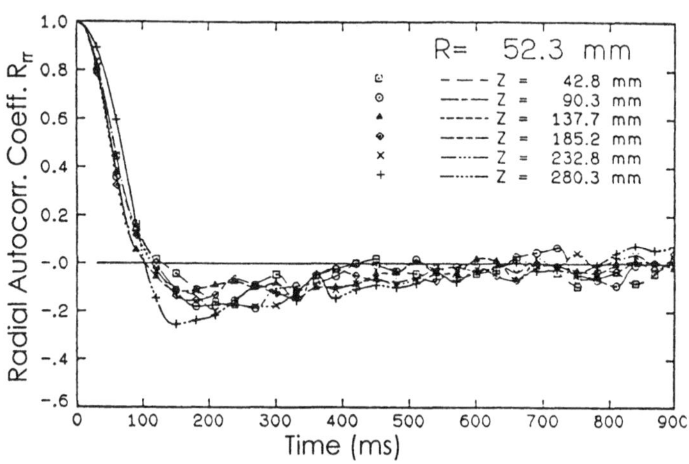

FIGURE 9.13 *Typical distribution of Lagrangian autocorelation coefficients in radial direction for 500-mm glass particles at u_o = 54.8 cm/s*

such as the root-mean-square (RMS) velocities, the Lagrangian integral time scales, and the dispersion coefficients. Those values, similar to the Lagrangian autocorrelations, were generally independent of changes in fluidization velocity and particle size, although some axial values may be affected by the gas velocity.

9.4.5 Turbulent Reynolds Stresses

Figure 9.14 shows a sample result of the turbulent Reynolds shear

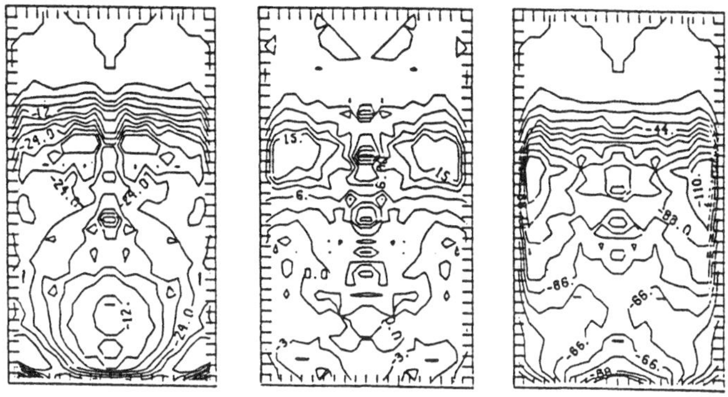

FIGURE 9.14 *Solids turbulent Reynolds stresses in a fluidized bed for 500-μm glass particles at* $u_o/u_{mf} = 2$

stress $-\rho\langle v'_r v'_z\rangle$, radial normal stress $-\rho\langle v'^2_r\rangle$, and axial normal stress $-\rho\langle v'^2_z\rangle$ in a 19-cm-i.d. fluidized bed for 500-μm glass particles at $u_o/u_{mf} = 2$. The shear stress $-\rho\langle v'_r v'_z\rangle$ was negative in the lower vortex in the regions near the distributor plate and the wall (see mean circulation pattern in Fig. 9.9), but its value was small in the lower vortex. The normal stresses $-\rho\langle v'^2_r\rangle$ and $-\rho\langle v'^2_z\rangle$ were larger near the centerline than near the wall in the lower vortex. All Reynolds stresses reached their maximum at elevations near the center of the upper vortex. The maximum of the shear stress $-\rho\langle v'_r v'_z\rangle$ was approximately at the center of the upper vortex, whereas the radial normal stress $-\rho\langle v'^2_r\rangle$ reached its maximum close to the centerline, and the axial normal stress $-\rho\langle v'^2_z\rangle$ had its maximum near the wall. If the values of $-\rho\langle v'_r v'_z\rangle$, $-\rho\langle v'^2_r\rangle$, and $-\rho\langle v'^2_z\rangle$ are compared, it is apparent that the variations in $-\rho\langle v'_r v'_z\rangle$ and $-\rho\langle v'^2_r\rangle$ were of the same order of magnitude. Both were smaller by a factor of four than the values of $-\rho\langle v'^2_z\rangle$. This difference was attributed to the high fluctuations in the axial velocities. Experiments with other particle sizes and fluidization velocities showed that the influence of particle size on the state of the turbulent Reynolds stresses in a gas fluidized bed is less important than fluidization velocity.

It may be of interest to note that summation of the normal components of the velocity correlations $\langle v'_i v'_j\rangle$ represents the solids kinetic energy, or "granular temperature," which is a primary parameter in the kinetic theory of granular particles.

9.4.6 Mass and Momentum Conservation in Fluidized Beds

The general time-averaged conservation laws for a gas/solid system have been derived for a control volume in terms of continuum gas velocity and mass and momentum fluxes of discrete particles (Sun, 1989). Taking advantage of the small gas/solid density ratio, the general results were simplified so that only the sums of the contributions of the discrete particles were needed. Because the fluxes can be extracted from particle-tracking measurements, the equations were used to evaluate the mass and momentum balances in a fluidized bed. As expected, the results indicated that body force and pressure drop are the two dominant balancing forces. The momentum flux terms, including the granular translational and collisional stresses, are the next high-order terms and they are approximately one order of magnitude smaller than the pressure and the body force. The interface interaction terms have the smallest value.

9.4.7 Mass Flux and Solids Mean Density

Based on the method for deriving the general time-averaged conservation laws, another independent formulation to evaluate the solids density can be obtained from the solids mass flux. The mean mass flux m_i in the ith direction is expressed (Sun, 1989) as

$$\overline{m}_i = \frac{M}{\delta A_i \Delta t} \sum sign(v_i), \qquad (9.26)$$

where M is the total mass of bed particles, v_i is the velocity of the tracer particle in the ith direction, and the summation accounts for only the crossings of the tracer particle through the sampling area δA_i for the duration of the experiment Δt. The solids density is then determined from

$$\rho = \frac{M}{\delta A_i \Delta t \, \overline{v}_i} \sum sign(v_i), \qquad (9.27)$$

where $\overline{v}_i$ is the mean velocity obtained from the volume-averaging method of Eq. 9.19.

At each location, two solids densities can be calculated from Eq. 9.27

by using the radial and axial mass fluxes and velocity distributions. It was shown (Sun, 1989) that, in regions where solids flow predominantly in one direction, the solids densities are very close to each other and to those evaluated from the occurrence method defined in Eq. 9.18; whereas in regions where solids flow in different directions in adjacent compartments, the densities might become unrealistically high or low. The reason was the inconsistency in the calculation of the mass flux and the mean velocity, because the flux was evaluated from surface averaging whereas the velocity was evaluated from volume averaging.

9.4.8 Momentum Fluxes and Particulate Stresses

With the same area-averaging method and notations used in Eqs. 9.26 and 9.27, the time-averaged momentum flux $\bar{p}_{ij}$ of the particulate phase can be expressed (Sun, 1989) as

$$\bar{p}_{ij} = \frac{M}{\delta A_i \Delta t} \sum sign(v_i) v_j = \rho \bar{v}_i \bar{v}_j + \frac{M}{\delta A_i \Delta t} \sum sign(v_i') v_j' \ , \qquad (9.28)$$

where ρ is the mean solids density, v_i and v_j are the velocity components in the i and j directions, respectively, and they can be expressed as a summation of mean and fluctuation components, i.e., $v_i = \bar{v}_i + v_i'$ and $v_j = \bar{v}_j + v_j'$. The last term in the above equation gives rise to the mean particulate stress

$$\tau_{ij} = -\frac{M}{\delta A_i \Delta t} \sum sign(v_i') v_j' \ . \qquad (9.29)$$

This particulate stress represents the kinetic components of granular momentum transfer, and includes both the "viscous" contribution due to the small-scale random motion of individual particles as well as the "macroscopic turbulence" contributions due to collective random motions such as eddies and bubbles (Sun, Chen, and Chao, 1990). The complete granular stress should consist of this particulate stress component and a collisional stress component.

Figure 9.15 shows the circumferentially averaged distributions of measured particulate stresses τ_{rr}, τ_{rz}, and τ_{zz} in the 19-cm-i.d. fluidized bed that contains 500-μm glass spheres at $u_0/u_{mf} = 2$. The two shear stresses τ_{rz} and τ_{zr} were found to be essentially identical and, therefore, the particulate

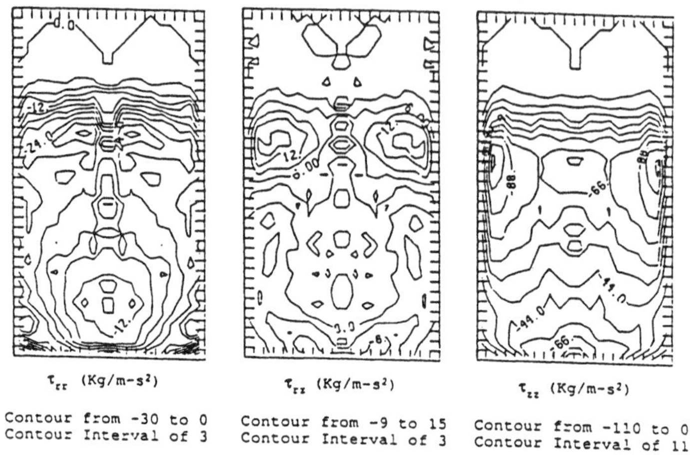

FIGURE 9.15 Solids particulate stresses in a fluidized bed for 500-μm glass particles at $u_o/u_{mf} = 2$

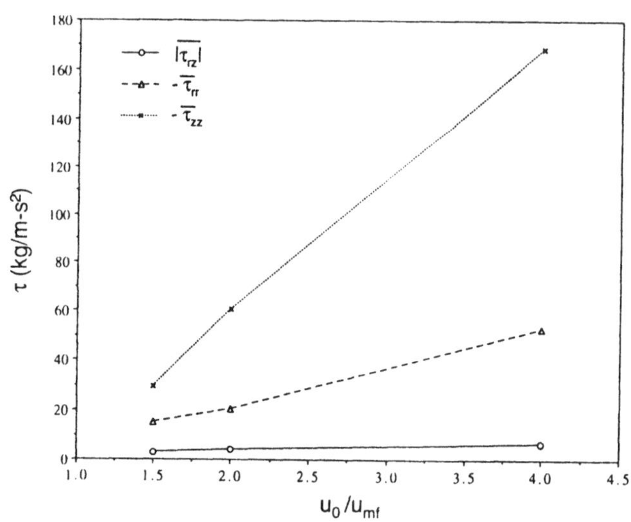

FIGURE 9. 16 *Variation of density-weighted, volume-averaged particulate stresses with* u_o/u_{mf}

stress tensor is symmetrical. These particulate stresses can be compared with the corresponding turbulent Reynolds stresses $-\rho\langle v_r'^2\rangle$, $-\rho\langle v_r' v_z'\rangle$, and $-\rho\langle v_z'^2\rangle$ shown in Fig. 9.14, where both the distribution patterns and the magnitudes of τ_{ij} and $-\rho\langle v_i'v_j'\rangle$ are similar. This result suggests that the density fluctuation ρ' and the velocity fluctuations $v_i'v_j'$ are not well correlated in fluidized beds.

To develop some insight about the manner that particulate stresses change with fluidization velocity, the magnitudes of the measured stresses were weighted with the local solids bulk density and averaged over the entire bed. Figure 9.16, which shows the variation of the weighted average stresses $|\overline{\tau_{rz}}|$, $-\overline{\tau_{rr}}$, and $-\overline{\tau_{zz}}$ with the velocity ratio u_0/u_{mf}, reveals that the magnitude of these stresses increases approximately linearly with gas velocity, suggesting that the gas momentum is the driving force for generating the solids fluctuations.

9.4.9 Particle Velocity Distributions

The complete velocity distribution function $f(\underline{v}, \underline{x})$ is usually defined in six spaces (v_x, v_y, v_z, x, y, z). However, for convenience in evaluating the particle speed distribution in a specific direction in the velocity space, the vector velocity $\underline{v}$ is represented by its magnitude, the speed v, and an angle ωin the vector direction of $\underline{v}$. The velocity distribution function is, therefore, denoted by $f^*(v, \omega, \underline{x})$, such that $f^*(v, \omega, \underline{x})d\omega dv$ represents the fraction of particles located at $\underline{x}$ with velocity vectors in an element of solid angle dw centered about the vector $\underline{v}$ and with speed between v-$dv/2$ and v+$dv/2$, and

$$\int_0^\infty dv \int_0^{4\pi} f^*(v, \omega, \underline{x})\, d\omega = 1. \tag{9.30}$$

The function $f^*(v, \omega, \underline{x})$ is the normalized function and is assumed to be time independent. The normalized speed distribution function $f(v, \underline{x})$ is then introduced such that $f(v, \underline{x})dv$ is the fraction of particles in the speed range dv, centered at speed v, moving in all possible directions at location $\underline{x}$. Clearly,

$$f(v, \underline{x}) = \int_0^{4\pi} f^*(v, \omega, \underline{x})\, d\omega\,, \tag{9.31}$$

and

$$\int_0^\infty f(v, \underline{x})\, dv = 1\,. \tag{9.32}$$

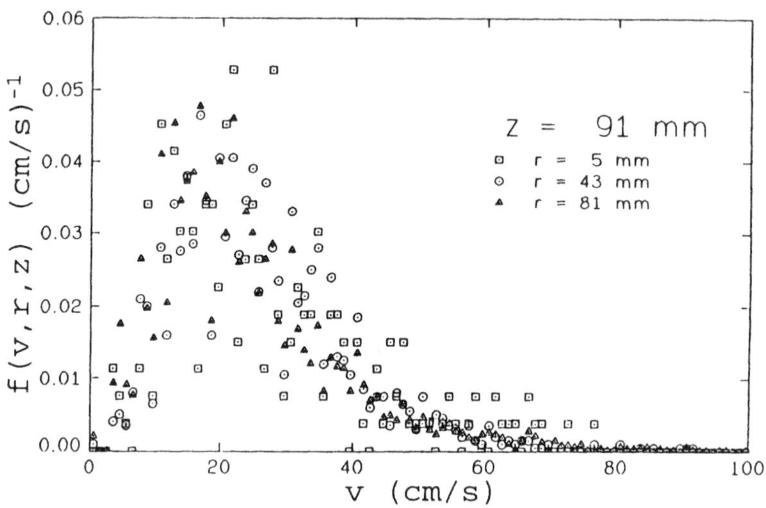

FIGURE 9.17 *Solids mean speed distributions in a cylindrical fluidized bed for 500-μm glass particles at* $u_o/u_{mf} = 2$

Although the complete velocity distribution function in the six spaces can be evaluated from experimental data obtained by the CAPTF, only the particle speed distributions are presented for illustration here. Figure 9.17 shows the speed distribution functions at axial position $z = 91$ mm and three radial positions $r = 5$, 43, and 81 mm in the 19-cm-i.d. fluidized bed for 500-μm glass particles at $u_o/u_{mf} = 2$. These results, however, were obtained in the presence of macroscopic disturbances such as bubbles. When measurements are taken under steady-flow conditions, the true velocity distribution function will be obtained and its shape may be determined.

The velocity distribution functions of particles that hit various regions of a single rod and rod bundles in the 2-D bed were measured with the CAPTF (Ai, 1991). The data were important in modeling tube erosion in fluidized beds because erosion of immersed tubes is a highly localized phenomenon and the erosion rate is dependent upon the impact velocity and impact frequency of the particles. The sampling interval was 5 ms and the test run was typically 40 h in these studies. Both directional and speed distribution functions of particles that hit the various local regions of the single round rod have been evaluated. The cylindrical surface is equally divided into 16 segments. Figure 9.18 shows the results of the directional function of particle for Surfaces 6 and 12. The location of the surface is

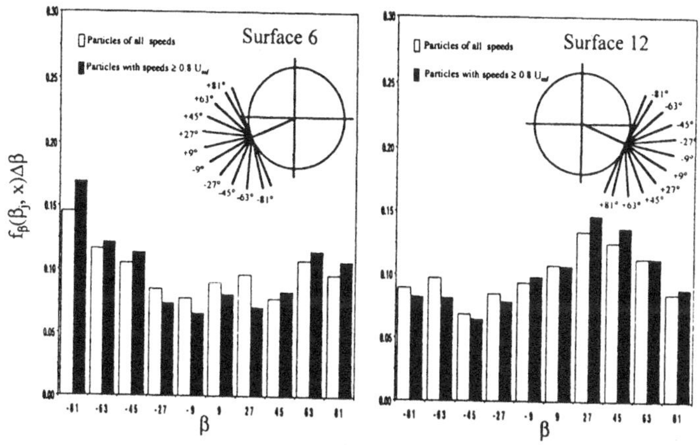

FIGURE 9.18 *Directional distribution functions of particles that hit two surfaces of a 12.7-cm-diameter cylinder*

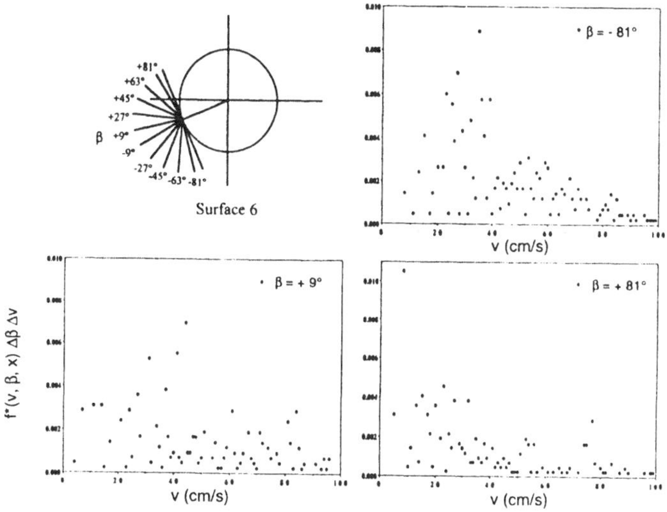

FIGURE 9.19 *Speed distribution functions of particles that hit Surface 6 of a 12.7-cm-diameter cylinder along various directions*

illustrated in the figure. Two directional distribution functions are shown at each surface. The empty bars indicate the fractions of particles that hit the surface region in the 10 wedge directions (with $\Delta\beta = 18°$) at all speed ranges, and the filled bars indicate the fractions of particles at speeds greater than a

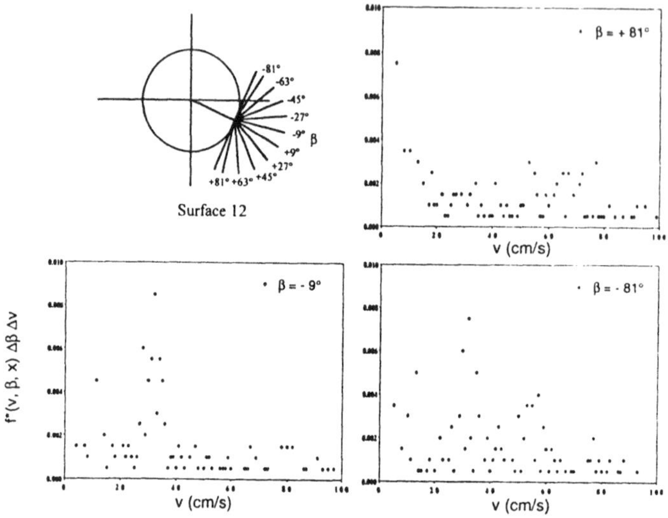

FIGURE 9.20 *Speed distribution functions of particles that hit Surface 12 of a 12.7-cm-diameter cylinder along various directions*

critical speed (= $0.8u_{mf}$). The particle speed distributions are shown in Figs. 9.19 and 9.20 for a few directions on Surfaces 6 and 12, respectively. It is apparent from Figs. 9.18-9.20 that there is a lack of symmetry between both the directional and speed distribution functions of particles that hit the pair of symmetrically located Surfaces 6 and 12. It is known, however, that bubble behavior is sensitive to distributor design; the lack of symmetry could be the result of a small nonuniformity in the air distribution system.

9.5 SOLIDS MIXING AND FLUCTUATION IN FLUIDIZED BEDS

The swarm-particle tracking mode was used to investigate solids mixing and fluctuations in fluidized beds. After introduction of tracers into the bed, their subsequent migration and dispersion were monitored by the 16 scintillation detectors that surround the bed as illustrated in Fig. 9.5. After the initial transients, the detector signals (count rates) settled down to statistically stationary values that represent the uniformly mixed condition. The transient portion of the detector signals is related to the mixing of the tracer particles in the bed, and the time variations of the signals in the statistically stationary state provide information on the fluctuating frequency of the motion of the bulk solids.

Solids mixing in the fluidized bed has been studied both

experimentally and numerically, with good agreement for certain ranges of operating conditions. Solids mixing can be viewed as the consequence of two processes: (1) convective mixing due to the large-scale circulation pattern of the solids and, (2) diffusive mixing due to the small-scale random motion of the solids. Both sets of data are readily available through the use of the CAPTF. The numerical study was based on a finite-difference computation of a convection/diffusion equation, with the solids diffusivity computed from the integral time scale of the velocity autocorrelation function according to Taylor's dispersion formula.

Examination of the characteristics of solids global fluctuation in gas-fluidized beds revealed that sloshing is a dominant mechanism in bubbling fluidized beds. Two modes of sloshing are present, namely, the axisymmetric mode and the antisymmetric mode. A standing-surface wave model was then developed to predict the global fluctuation frequency of the solids sloshing and the model predictions were in good agreement with experimental data.

9.5.1 Solids Mixing

In the experimental measurement of the solids mixing process, the radioactive particles were released from the top center of the cylindrical bed column into the bed free surface under each operating condition. The detector outputs were sampled at 50-ms intervals. A sample result of the average number of counts at each detector level was plotted versus time in Fig. 9.21 for the 500-μm glass particles at u_0 = 54.8 cm/s (Moslemian, 1987). In the experiment, the sampling of the detectors was initiated slightly sooner than the release of the radioactive particles to ensure the recording of the events near zero time. As the particles were released, the outputs of the detectors positioned above the free surface (Levels 1 and 2 shown in Fig. 9.5) reached their maximum within a small fraction of second because of the passage of the falling tracer particles. The steady-state averaged count rate did not overlap for the four levels because of the variation in density distribution seen by detectors at different levels. From the curves in Fig. 9.21, three types of information could be obtained. The asymptotic steady-state values of the count rates were measures of the mean density distributions of the radioactive particles in the bed. The time required to reach these asymptotic values was an indication of the mixing rate of the bed particles. The shapes of the curves yielded information on the manner in

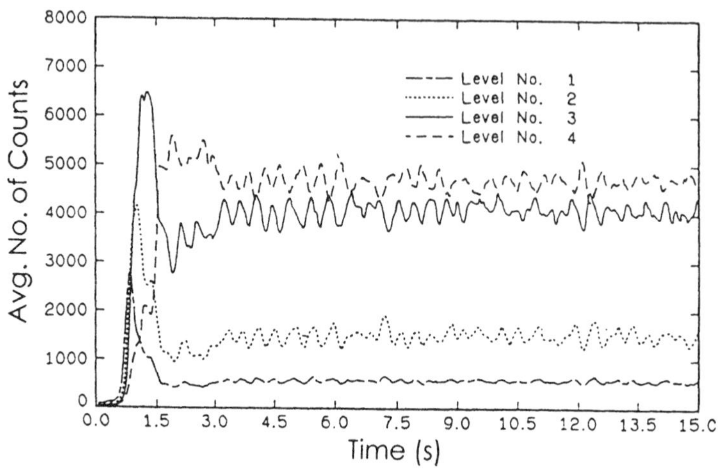

FIGURE 9.21 *Mixing of 10 g of 500-µm radioactive (Na^{24}) glass particles in a bed of 500-µm glass particles at u_o = 54.8 cm/s*

which the mixing was accomplished.

In Fig. 9.21, each experimental curve showed an overshoot before settling into its asymptotic value. This overshoot was a consequence of the large-scale solids recirculation in contrast to smaller scale diffusion. The overshoot was characteristic at lower velocities and it disappeared at greater velocities. The mixing behavior is similar for particles of differing size.

A convection/diffusion equation was used to numerically simulate the mixing process (Moslemian, 1987). The convective contributions were modeled through the mean solids velocity distributions from the single-particle tracking measurements. The diffusive terms were evaluated by computing the dispersion rates in the radial and axial directions with the Lagrangian velocity autocorrelations in the respective directions. The radial solids dispersion coefficients were ≈10 cm²/s, and the axial dispersion coefficients were an order of magnitude higher than those in the radial direction. Averaged coefficients were used in the simulations because their variations in the bed were not large. To compare the predictions of the numerical simulations with the experimental data from the swarm-particle tracking measurements, the predicted distributions of the mass of the radioactive particles were converted to compatible detector count rates through the theoretical relationship between the position of the tracers and the outputs of individual detectors, Eqs. 9.8 and 9.10. A sample of the numerically simulated count rates, shown in Fig. 9.22 along with the

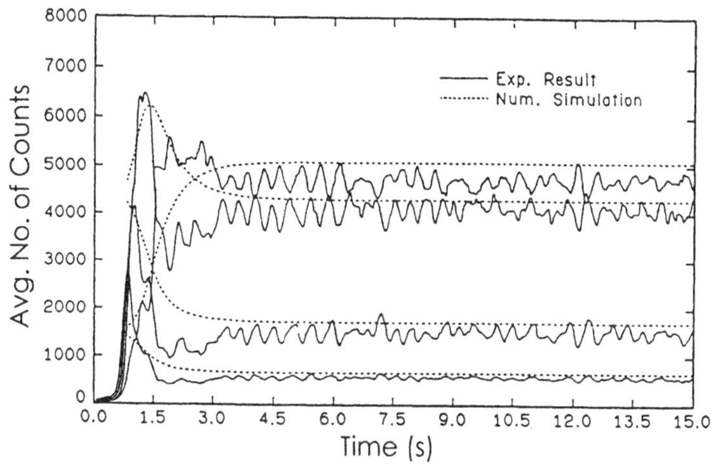

FIGURE 9.22 *Numerical and experimental mixing results for 500-μm glass particles at u_o = 54.8 cm/s*

experimental result, shows that the model predicted the shape of the curves reasonably well. The apparent overestimation of the steady-state detector count rates was probably due to the overestimation of activity of the tracer particles and overestimation of detector efficiency in this study.

9.5.2 Solids Global Fluctuation

The solids global fluctuations are evidence from the time variations of the detector signals in the statistically stationary state, shown in Fig. 9.21, which represents the uniformly mixed condition. To determine the mode (sloshing versus slugging) and the frequency of the global fluctuating motion of particles, it is necessary to examine the signals of individual detectors because they represent the local density variation near the detectors (Sun, Chen, and Chao, 1994). As shown in Fig. 9.5, the detectors were arranged in four levels, located at 540, 382, 218, and 54 mm above the distributor plate. At each level, there were four detectors, 90° apart in a horizontal plane. They were also staggered vertically. In this investigation, the detectors at Level 3, Nos. 9-12, were most relevant because they were approximately at the same level as the free surface of the bubbling bed, where the fluctuations were the strongest. Experiments were performed for the 500- and 705-μm glass particles at various fluidization velocities. The height of the static bed was set at 190 mm in all of the experiments. The detector signals were recorded at 30-ms interval.

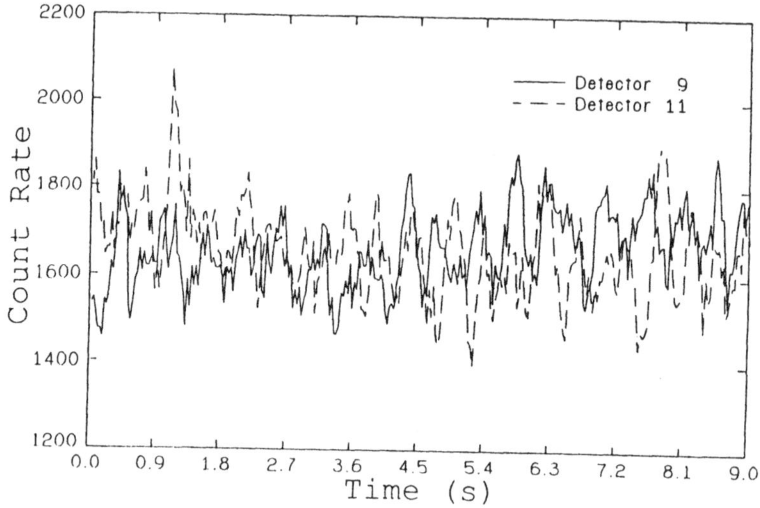

FIGURE 9.23 *Variation of count rates from two diametrically opposite detectors for 500-μm glass particles at* $u_0/u_{mf} = 2$

From visual observation of the fluidized bed in operation, at least two distinct modes of fluctuation have been noticed, namely, an axisymmetric mode and an antisymmetric mode of the solids sloshing motion. The existence of the two modes can be identified by examining the signals of the diametrically opposite detector pairs, 9/11 and 10/12 at Level 3. The signals from detectors 9 and 11, reproduced in Fig. 9.23 for the 500-μm particles at $u_0/u_{mf} = 2$, reveal that most of the time the fluctuations are in phase, indicating the presence of axisymmetric oscillations. Occasionally, however, out-of-phase fluctuations are also found (between time periods 4.5-5.4 s and 7.2-8.1 s in Fig. 9.23), indicating the presence of antisymmetric sloshing motion. In passing, we note that the out-of-phase fluctuations are revealed in the figure only when the vertical plane along which the antisymmetric sloshing occurred was parallel to the axis of Detectors No. 9 and 11. When the antisymmetric sloshing was perpendicular to the detector axis, the out-of-phase fluctuations were not revealed. The latter, however, can be seen from the signals of detector pair 10/12. As it turned out, for the operating conditions used in this investigation, the antisymmetric sloshing was dominant.

To extract the mode, frequency, and other information from the detector signals, their cross-correlations, autocorrelations, and power spectral density functions were examined. For discrete detector signals of zero mean, acquired at a sampling time interval δt, $x_n = x(n\delta t)$ and $y_n = y(n\delta t)$, $n = 1, 2, \ldots, N$, N being the length of one set of signals, the cross-correlation function C_{xy} can be evaluated (Bendat and Piersol, 1986) from

$$C_{xy}(j\delta t) = \frac{1}{N-j}\sum_{n=1}^{N-j} x_n y_{n+j}, \qquad \mathrm{j} = 0,\ 1,\ ¼,\ \mathrm{m}, \tag{9.33}$$

where j is the lag number and m is the maximum lag number ($m < N$). The autocorrelation function C_{xx} can also be evaluated from Eq. 9.33 by simply replacing y with x. The power spectral density function is determined directly from the Fourier transform of the signal. Using a Hamming window function to reduce power leakage to side lobes in the frequency domain, we find that the discrete Fourier transform of the signal x_n is

$$X(f_k) = \delta t\sqrt{\frac{8}{3}}\sum_{n=1}^{N-1} x_n\left[1-\cos^2\left(\frac{\pi n}{N}\right)\right]\exp\left(-\frac{i2\pi kn}{N}\right), \quad \mathrm{k} = 1, 2, \ldots, \mathrm{N}\text{-}1, \tag{9.34}$$

where $f_k = k/(N\,\delta t)$, $i = \sqrt{-1}$, and the power spectral density function for an average of n_d sets of signals of x_n is

$$S_{xx}(f_k) = \frac{1}{n_d N\delta t}\sum_{j=1}^{n_d}\left|X_j(f_k)\right|^2. \tag{9.35}$$

Figures 9.24-9.26 show, respectively, the cross-correlations, autocorrelations, and power spectrum of the detector signals for the 500-µm glass particles at $u_0/u_{mf} = 2$. In Fig. 9.24, the curves for the opposite detector pairs, 9/11 and 10/12, are of particular interest. They exhibit nearly zero values of correlation at zero time lag, indicating significant antisymmetric sloshing motion. The autocorrelations, shown in Fig. 9.25, reveal the existence of both near-periodic and random fluctuations. The power spectrum of the detector signals in Fig. 9.26 shows the dominant fluctuations of the solids motion in the bed. The dominant frequencies for the 500- and 705-mm glass particles at $u_0/u_{mf} = 1.5, 2, 3$, and 4 are listed in Table 9.2. The

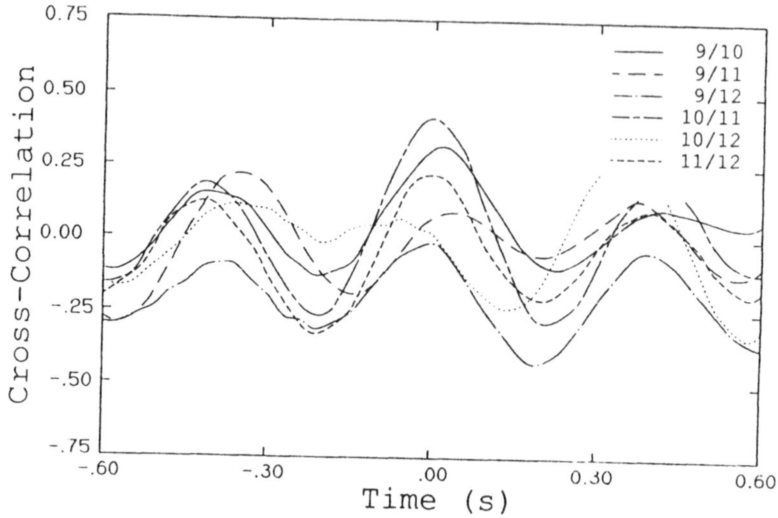

FIGURE 9.24 *Cross-correlations of detector signals for 500 μm glass particles at $u_o/u_{mf} = 2$*

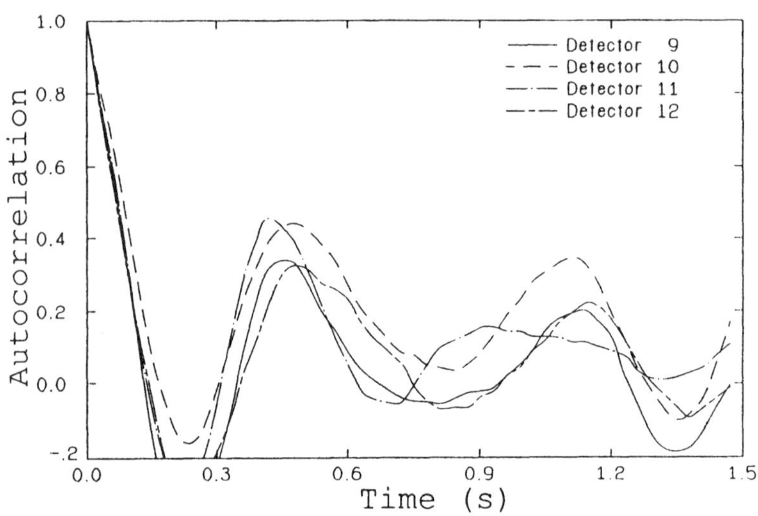

FIGURE 9.25 *Autocorrelations of detector signals for 500-μm glass particles at $u_o/u_{mf} = 2$*

most striking feature of the data is that the dominant frequency of the solids fluctuations is essentially independent of particle size and the fluidization velocity.

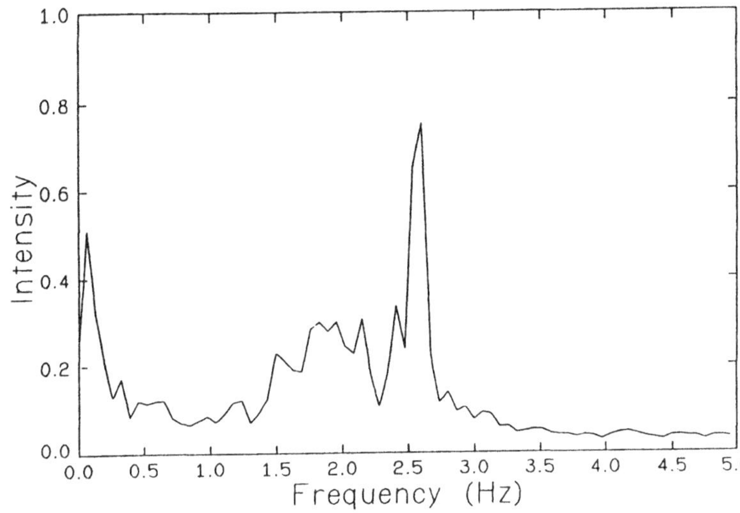

FIGURE 9.26 *Averaged power spectrum of detector signals for 500-μm glass particles at $u_o/u_{mf} = 2$*

Table 10.2 Dominant fluctuation frequencies (Hz) of 500- and 705-μm glass particles in a 19-cm-i.d. cylindrical fluidized bed with static bed height of 19 cm

d_p		u_o/u_{mf}		
μm	1.5	2	3	4
500	2.15	2.61	2.48	2.38
705	2.20	2.28	2.28	2.48

A standing surface wave model was developed to predict the global fluctuation frequency of solids sloshing in beds of intermediate and shallow depth. The axisymmetric and the antisymmetric modes of sloshing in cylindrical beds are the full- and half-wave modes of the standing surface waves. The model predictions for the sloshing frequency were found to be in good agreement with experimental data of this study and others in the literature, as shown in Fig. 9.27. More importantly, it was found that, although the excitation for bed fluctuations originates from bubbles, the fluctuation frequency is controlled by surface waves.

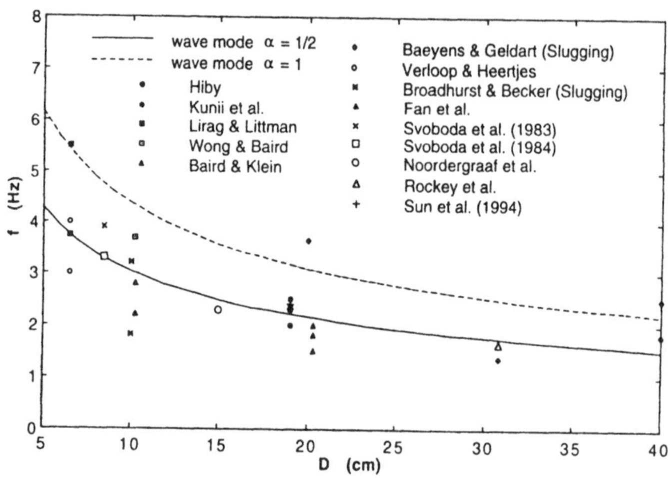

FIGURE 9.27 *Predicted frequency and experimental data for cylindrical beds of intermediate depth (References are listed in Sun et al., 1994)*

9.6 CONCLUSION

Radioactive tracer techniques have long been used to study particle motion in solids fluidization systems. The advantage of this technique is that the flow field is not disturbed by the measurement facility and, therefore, the measurement of the motion of the tracers represents the actual movement of particles in the system. The tracer particles are usually made of gamma-emitting radioisotopes, and their gamma radiation is measured directly by scintillation detectors. Factors that affect gamma radiation measurement were identified as the characteristics of the radiation source, interactions of gamma rays with matter, the tracer's position relative to the detector, detector efficiency, and dead time of the measurement system.

A computer-aided particle-tracking facility (CAPTF) has been developed to measure the motion of radioactive tracers in fluidized beds. This achievement was the first successful attempt to use the radioactive tracer technique to obtain detailed quantitative information on solids dynamic data in fluidized beds. The CAPTF makes use of one or more radioactive tracer particles that are dynamically identical to the bed particles under study. In the single-particle tracking mode, the gamma radiation from the radioactive tracer is continuously monitored by 16 scintillation detectors to provide information on the tracer's instantaneous location. Time differentiation of the position data yields the local instantaneous velocities. After a test run of many hours, a large number of such instantaneous velocity measurements are available for each "location" in the bed, identified by a numbered small sampling volume. The ensemble average of all of the velocities for each sampling volume then yields the mean particle velocity for the location. Counting the number of occurrences in each sampling volume enables the determination of mean solids density distribution. By subtracting the mean from the instantaneous velocity, the fluctuating components of the velocity can also be obtained. From these, statistical quantities such as the RMS velocities, Lagrangian autocorrelations, and turbulent Reynolds stresses are readily computed. The single-particle tracking data also provide the particle velocity distribution function, the mass and momentum fluxes, and the particulate stresses. In the swarm-particle tracking mode of the CAPTF, a collection of radioactive tracer particles was introduced in selected locations in the bed. Their subsequent dispersion and migration were monitored by the 16 scintillation detectors. The detector signals yield fundamental information on solids mixing and fluctuation in fluidized beds. The combination of the velocity data from the single-particle tracking studies and the solids mixing and fluctuation data from the swarm-particle tracking studies has yielded a heretofore unachieved complete description of solid-particle behavior in fluidized beds. Such data should be most valuable in helping to make a significant step toward achieving a rational formulation of the governing conservation equations and the scaling laws that are derivable from them.

NOTATION

B	buildup function
C	correlation function
d_c	scintillation crystal diameter
d_p	particle diameter
E	photon energy
f	velocity distribution function; frequency
I	gamma ray intensity
m	recorded count rate

M	total solids mass in fluidized bed
$\dot{m}$	mass flux
m_p	particle mass
mc^2	constant (= 511 KeV)
n	scintillation rate; local particle number density
N	total number of particles in fluidized bed
n_0	number of occurrence of tracer in a compartment
N_0	number of occurrence of tracer in whole bed
$\dot{p}$	momentum flux
r	distance; radius coordinate
R	correlation coefficient
s	tracer position
S	radiation activity; power spectrum function
t	time
$t_{1/2}$	half decay time
t_{m}	mean decay time
u_0	superficial gas velocity
u_{mf}	minimum fluidization gas velocity
v	particle velocity component; particle speed
$\underline{v}$	particle velocity
v'	particle fluctuating velocity
$\bar{v}$	particle mean velocity
V_{c}	compartment volume
$x, y, z,$	Cartesian coordinates
X	Fourier transform of a discrete signal
$x_i, y_i, z_i,$	position of ith detector
α	mass attenuation coefficient
δt	detector sampling period
Δt	experiment duration
η	detector efficiency
θ, ϕ	angles
μ	linear attenuation coefficient
ρ	density
σ	weighting factor
τ	dead time; stress
Ω	solids angle

REFERENCES

Ai, Y.-H., 1991, Solids Velocity and Pressure Fluctuation Measurements in Air Fluidized Beds, M. S. Thesis, Univ. of Illinois, Urbana-Champaign.

Bendat, J. S. and Piersol, A. C., 1986, Random Data: Analysis and Measurement Procedure, 2nd Ed., Wiley, New York.

Chen, M. M., Chao, B. T., and Liljegren, J. C., 1983, The Effects of Bed Internals on the Solids Velocity Distribution in Gas Fluidized Beds, paper presented at the IVth International Conference on Fluidization, Kashikojima, Japan, May 29-June 3, 1983.

Geldart, D., 1973, Types of Gas Fluidization, Powder Technol., Vol. 7, pp. 285-292.

Hull, R. L. and Rosenberg, A. E. von, 1960, Radiochemical Tracing of Fluid Catalyst Flow, Ind. Eng. Chem., Vol. 52, pp. 989-992.

Knoll, G. F., 1979, Radiation Detection and Measurement, Wiley, New York.

Kondukov, N. B., Kornilaev, A. N., Skachko, I. M., Akhromenkov, A. A., and Kruglov, A. S., 1964, An Investigation of the Parameters of Moving Particles in a Fluidized Bed by a Radioisotopic Method, Int. Chem. Eng., Vol. 4, pp. 43-47.

Liljegren, J. C., 1984, Effects of Immersed Rod Bundles on Gross Solids Circulation in a Gas Fluidized Bed, M. S. Thesis, Univ. of Illinois, Urbana-Champaign.

Lin, J. S., 1981, Particle-Tracking Studies for Solids Motion in a Gas Fluidized Bed, Ph. D. Thesis, Univ. of Illinois, Urbana-Champaign.

Lin, J. S., Chen, M. M., and Chao, B. T., 1985, A Novel Radioactive Particle Tracking Facility for Measurement of Solids Motion in Gas Fluidized Beds, AIChE. J., Vol. 31, pp. 465-473.

May, W. G., 1959, Fluidized-Bed Reactor Studies, Chem. Eng. Prog., Vol. 55, pp. 49-56.

Moslemian, D., 1987, Study of Solids Motion, Mixing, and Heat Transfer in Gas-Fluidized Beds, Ph. D. Thesis, Univ. of Illinois, Urbana-Champaign.

Overcashier, R. H., Todd, D. B., and Olney, R. B., 1959, Some Effects of Baffles on a Fluidized System, AIChE. J., Vol. 5, pp. 54-60.

Segre, E., 1953, Experimental Nuclear Physics, Vol. 1, Wiley, New York.

Segre, E., 1964, Nuclei and Particles, W. A. Benjamin, New York.

Singer, E., Todd, D. B., and Guinn, V. P., 1957, Catalyst Mixing Patterns in Commercial Catalytic Cracking Units, Ind. Eng. Chem., Vol. 49, pp. 11-19.

Sun, J. G., 1985, Data Processing Problems for Radioactive Particle Tracking Measurement, M. S. Thesis, Univ. of Illinois, Urbana-Champaign.

Sun, J. G., 1989, Analysis of Solids Dynamics and Heat Transfer in Fluidized Beds, Ph. D. Thesis, Univ. of Illinois, Urbana-Champaign.

Sun, J. G., Chen, M. M., and Chao, B. T., 1990, Radioactive Particle Tracking Measurement of the Mean Particulate Stress in a Fluidized Bed, paper presented at AIChE. Winter Annual Meeting, Chicago, Nov. 11-15, 1990.

Sun, J. G., Chen, M. M., and Chao, B. T., 1994, Modeling of Solids Global Fluctuations in Bubbling Fluidized Beds by Standing Surface Waves, Int. J. Multiphase Flow, Vol. 20, pp. 315-338.

Tait, W. H, 1980, Radiation Detection, Butterworths, London.

Tennekes, H. and Lumley, J. L., 1972, A First Course in Turbulence, p. 224, MIT Press, Cambridge, MA.

Tsoulfanidis, N., 1983, Measurement and Detection of Radiation, Mc-Graw Hill, New York.

Velzen, D. van, Flamm, H. J., Langenkamp, H., and Casile, A., 1974, Motion of Solids in Sprouted Beds, Can. J. Chem. Eng., Vol. 52, pp. 156-161.

Werther, J. and Molerus, O., 1973, The Local Structure of Gas Fluidized Beds II: The Spatial Distribution of Bubbles, Int. J. Multiphase Flow, Vol. 1, pp. 123-138.

Index

E

F

G

H

I

K

L

N

O

P

Kolibri 360 GmbH
Ferdinand-Jühlke-Straße 7
99095 Erfurt, Deutschland
produktsicherheit@kolibri360.de